Lecture Notes in Management and Industrial Engineering

Editor
Adolfo López-Paredes
Valladolid, Spain

This bookseries provides a means for the dissemination of current theoretical and applied research in the areas of Industrial Engineering & Engineering Management. The latest methodological and computational advances that both researchers and practitioners can widely apply to solve new and classical problems in industries and organizations constitute a growing source of publications written for and by our readership. The aim of this bookseries is to facilitate the dissemination of current research in the following topics:

- Strategy and Enterpreneurship
- Operations Research, Modelling and Simulation
- Logistics, Production and Information Systems
- Quality Management
- Product Management
- Sustainability and Ecoefficiency
- Industrial Marketing and Consumer Behavior
- Knowledge and Project Management
- Risk Management
- Service Systems
- Healthcare Management
- Human Factors and Ergonomics
- Emergencies and Disaster Management
- Service Systems
- Education

More information about this series at http://www.springer.com/series/11786

Cesáreo Hernández • Adolfo López-Paredes
José M. Pérez-Ríos
Editors

Managing Complexity

Challenges for Industrial Engineering and Operations Management

Springer

Editors
Cesáreo Hernández
INSISOC
University of Valladolid
Valladolid
Spain

José M. Pérez-Ríos
INSISOC
University of Valladolid
Valladolid
Spain

Adolfo López-Paredes
INSISOC
University of Valladolid
Valladolid
Spain

ISSN 2198-0772 ISSN 2198-0780 (electronic)
ISBN 978-3-319-04704-1 ISBN 978-3-319-04705-8 (eBook)
DOI 10.1007/978-3-319-04705-8
Springer Cham Heidelberg New York Dordrecht London

Library of Congress Control Number: 2014939519

Printed on acid-free paper

Springer is part of Springer Science+Business Media (www.springer.com)

Preface

This volume collects a selection of the best papers presented at the joint 7th International Conference on Industrial Engineering and Industrial Management, CIO 2013, and the XIX International Conference on Industrial Engineering and Operations Management, ICIEOM 2013. They are a good sample of the state of the art in the field of Industrial and Operations Management.

The field of Industrial Engineering (IE) is defined by the Institute of Industrial Engineers (IIE) [http://www.iienet2.org/]: "*Industrial Engineering is concerned with the design, improvement, and installation of integrated systems of men, materials, equipment and energy. It draws upon specialized knowledge and skill in the mathematical, physical and social sciences together with the principles and methods of engineering analysis and design to specify, predict, and evaluate the results to be obtained from such systems*".

A more recent definition of IE is proposed by a research group involved in the IESE project (Industrial Engineering Standards in Europe) [http://www.iestandards.eu/]: "*The branch of engineering that engages in the study of how to describe, evaluate, design, modify, control and improve the performance of complex systems, viewed over time and within their relative context.*"

These two definitions show the evolution of the field, both in methods and scope, following the evolution of Management for sustained wealth generation. The focus of Management has evolved from product design and manufacturing towards developing distinctive capabilities of the firm to be shared with other firms, all along the project management. Managing for competition has been replaced by collaborative management.

This change of focus and the support of new communications technologies has broaden the very scope of the field: not just from products to services but to the analysis, design and control to improve the performance of any physical landscape populated by social agents. This new focus demands new skills and methods to manage complexity. Thus the title of the volume: Managing Complexity.

The papers presented in this book, address methods, questions and applications to the Business Strategy, Modelling and Simulation in Operations Management, Logistics and Production, Service Systems, Innovation and Knowledge, and Project Management.

The contributions have been arranged in four chapters:

- Management
- Production
- Logistics and Supply Chain
- Methods and Applications.

We want to express our gratitude to all the contributors and reviewers.

October 2013
Valladolid, Spain

Cesáreo Hernández
Adolfo López-Paredes
José M. Pérez-Ríos

Contents

Contributors

M. Jesús Álvarez TECNUN, Universidad de Navarra, San Sebastian, Spain

Ángel F. Agudo-Peregrina Departamento de Ingeniería de Organización, Administración de Empresas y Estadística, Universidad Politécnica de Madrid Av, Madrid, Spain

Maria del Mar Eva Alemany Díaz Centro de Investigación de Gestión e Ingeniería de la Producción (CIGIP), Universitat Politècnica de València, Valencia, Spain

Rocío Alfaro Cátedra PROTHIUS, Dpto. de Organización de Empresa, Universitat Politècnica de Catalunya, Barcelona, Spain

Carlos Andrés Romano Research Group in Reengineering, Operations Management, Group Work and Logistics Excellence, Industriales, Universitat Politècnica de València Cno, Valencia, Spain

Beatriz Andres Research Centre on Production management and Engineering (CIGIP), Universitat Politècnica de València (UPV), Alcoy, Spain

José Alberto Arauzo Grupo INSISOC. EII, Universidad de Valladolid, Valladolid, Spain

Sabina Asensio Cuesta Departamento de Proyectos de Ingeniería, Escuela Técnica Superior de Ingeniería de Diseño, Universitat Politècnica de València Cno, Valencia, Spain

Peter Bajor Grupo GIO. ETSII, Universidad Politécnica de Cartagena, Cartagena, Spain

Cristina Batalla Cátedra PROTHIUS, Dpto. de Organización de Empresa, Universitat Politècnica de Catalunya, Barcelona, Spain

Joaquín Bautista Dpto. Organización de Empresas, ETSEIB, UPC Catalunya, Barcelona, Spain

Tamara Borreguero-Sanchidrián Grupo IOL, Dpto. Ingeniería de Organización, Administración de Empresas y Estadística. ETSII, UPM Madrid, Madrid, Spain

Andrés Boza Centro de Investigación de Gestión e Ingeniería de Producción (CIGIP), Universitat Politècnica de València, Valencia, Spain

Alberto Cano Cátedra PROTHIUS, Dpto. de Organización de Empresa, Universitat Politècnica de Catalunya, Barcelona, Spain

Jesús Antonio Carrillo-Castrillo Dpto. de Organización Industrial y Gestión de Empresas II. ETSI, Universidad de Sevilla, Camino de los Descubrimientos, Seville, Spain

Julián Chaparro-Peláez Departamento de Ingeniería de Organización, Administración de Empresas y Estadística, Universidad Politécnica de Madrid Av, Madrid, Spain

Manuel Chica Serrano European Centre for Soft Computing, Mieres, Asturias, Spain

Ernesto Cilleruelo Foresight Technology and Management (FTM) Group, Department of Industrial Engineering, University of the Basque Country UPV/EHU, Bilbao, Spain

Ramon Companys Pascual Dpto. de Organización de Empresas, Escola Tècnica Superior d'Enginyers Industrials de Barcelona, Universitat Politècnica de Catalunya, Barcelona, Spain

Óscar Cordón García European Centre for Soft Computing, Mieres, Asturias, Spain

Ana María Coves Department of Management, Institute of Industrial and Control Engineering, Universitat Politècnica de Catalunya, Barcelona, Spain

Llanos Cuenca Centro de Investigación en Gestión e Ingeniería de Producción (CIGIP), Universitat Politècnica de València, Valencia, Spain

José Ignacio Díez Department of Civil Engineering, Higher Polytechnic School, University of Burgos, Burgos, Spain

Sergio Damas Arroyo European Centre for Soft Computing, Mieres, Asturias, Spain

María Victoria de la Fuente Aragón Grupo GIO. ETSII, Universidad Politécnica de Cartagena, Cartagena, Spain

Rocio de la Torre Department of Management/ETSEIB, Universitat Politècnica de Catalunya, Barcelona, Spain

Ricardo del Olmo Department of Civil Engineering, Higher Polytechnic School, University of Burgos, Burgos, Spain

Uwe Dombrowski Institute for Advanced Industrial Management, Technische Universität Braunschweig, Braunschweig, Germany

Nekane Errasti Departamento de Mecánica y Producción Industrial, Escuela Politécnica Superior, Mondragon Unibertsitatea, Mondragon, Spain

Isabel Fernández Quesada Grupo de Ingeniería de Organización, Dpto. de Administración de Empresas, Escuela Politéc, de Ingeniería de Gijón, Universidad de Oviedo, Gijón, Spain

Pedro Fernández-Cardador Universidad Politécnica de Madrid, Madrid, Spain

V. S. Fuertes-Miquel Dpto. de Ingeniería Hidráulica y Medio Ambiente, Universitat Politècnica de València, Valencia, Spain

Pedro Gómez Gasquet Centro de Investigación de Gestión e Ingeniería de la Producción (CIGIP), Universitat Politècnica de València Cno, Valencia, Spain

José Manuel Galán Departamento de Ingeniería Civil.EPS, Área de Organización de Empresas, Burgos, Spain

Jaione Ganzarain Departamento de Mecánica y Producción Industrial, Escuela Politécnica Superior, Mondragon Unibertsitatea, Mondragon, Spain

Jesús García Arca Grupo de Ingeniería de Organización (GIO), Escola de Enxeñería Industrial, Vigo, Spain

Nazario García Fernández Grupo de Ingeniería de Organización, Dpto. de Administración de Empresas, Escuela Politéc, de Ingeniería de Gijón, Universidad de Oviedo, Gijón, Spain

Fausto Pedro García Márquez INGENIUM GROUP, ETSI Industriales, Universidad de Castilla-La Mancha, Ciudad Real, Spain

Isidro Peña García-Pardo Dpto. de Administración de Empresas, Facultad de Derecho y Ciencias Sociales, Universidad de Castilla-La Mancha, Ciudad Real, Spain

Álvaro García-Sánchez Grupo IOL, Dpto. Ingeniería de Organización, Administración de Empresas y Estadística. ETSII, UPM Madrid, Madrid, Spain

Jose P. Garcia-Sabater Grupo ROGLE. Dpto. de Organización de Empresas, Universidad Politécnica de Valencia, Valencia, Spain

Gaizka Garechana Foresight Technology and Management (FTM) Group, Department of Industrial Engineering, University of the Basque Country UPV/EHU, Bilbao, Spain

Javier Gavilanes-Trapote Foresight Technology and Management (FTM) Group, Department of Industrial Engineering, University of the Basque Country UPV/EHU, Vitoria, Spain

Angel Manuel Gento Municio Dpto. de Organización de Empresas y CIM, Escuela de Ingenierías Industriales, Universidad de Valladolid, Valladolid, Spain

José Luis González Álvarez Unidad de Protección Civil y Emergencias, Delegación del Gobierno en Castilla y León, Valladolid, Spain

A. Trinidad González-Portela Garrido Grupo de Ingeniería de Organización (GIO), Escola de Enxeñería Industrial, Vigo, Spain

Rafael Grosso de la Vega Dpto. de Organización Industrial y Gestión de Empresas II. ETSI, Universidad de Sevilla, Camino de los Descubrimientos, Seville, Spain

José Guadix Martín Dpto. de Organización Industrial y Gestión de Empresas II. ETSI, Universidad de Sevilla, Camino de los Descubrimientos, Seville, Spain

Sara Hatami Departamento de Estadística e Investigación Operativa, Facultad de Ciencias Matemáticas, Universitat de València, València, Spain

Eva-Maria Hellmich Institute for Advanced Industrial Management, Technische Universität Braunschweig, Braunschweig, Germany

Víctor Hermano INSISOC Research Group, Department of Business Management and Marketing, Business School of Universidad de Valladolid, Valladolid, Spain

Ángel Hernández-García Departamento de Ingeniería de Organización, Administración de Empresas y Estadística, Universidad Politécnica de Madrid, Madrid, Spain

Juan Ignacio Igartua Departamento de Mecánica y Producción Industrial, Escuela Politécnica Superior, Mondragon Unibertsitatea, Mondragon, Spain

Santiago Iglesias-Pradas Departamento de Ingeniería de Organización, Administración de Empresas y Estadística, Universidad Politécnica de Madrid, Madrid, Spain

Carmen Jaca Department of Industrial Management, School of Industrial Engineers, TECNUN, Universidad de Navarra, San Sebastian, Spain

Francisco-Cruz Lario Esteban Centro de Investigación en Gestión e Ingeniería de Producción, Universitat Politècnica de València, Valencia, Spain

Jaso Larruscain Foresight Technology and Management (FTM) Group, Department of Industrial Engineering, University of the Basque Country UPV/EHU, Vitoria, Spain

Juan José Lavios Grupo INSISOC. EPS, Universidad de Burgos, Burgos, Spain

Amaia Lusa Department of Management/ IOC/ETSEIB, Universitat Politècnica de Catalunya, Barcelona, Spain

Miguel Ángel Manzanedo Department of Civil Engineering, Higher Polytechnic School, University of Burgos, Burgos, Spain

Natalia Martín-Cruz INSISOC Research Group, Department of Business Management and Marketing, Business School of Universidad de Valladolid, Valladolid, Spain

Manuel Mateo Department of Management/ETSEIB, Universitat Politècnica de Catalunya, Barcelona, Spain

Ricardo Mateo School of Economics and Business Administration, University of Navarra, Pamplona, Spain

Ana María Mejías Sacaluga Grupo OSIG, Dpto. de Organización de Empresas y Marketing, Escuela de Ingeniería Industrial, Universidad de Vigo, Vigo, Spain

Tim Mielke Institute for Advanced Industrial Management, Technische Universität Braunschweig, Braunschweig, Germany

Joan Ignasi Moliné Department of Management, Institute of Industrial and Control Engineering, Universitat Politècnica de Catalunya, Barcelona, Spain

Manuel Monterrey Meana Grupo de Ingeniería de Organización, Dpto. de Administración de Empresas, Escuela Politéc, de Ingeniería de Gijón, Universidad de Oviedo, Gijón, Spain

Javier Muñoz Sanz Escuela de Ingenierías Industriales, Universidad de Valladolid, Valladolid, Spain

Giovanni Mummolo Department of Mechanics, Mathematics and Management, Politecnico di Bari, European Academy for Industrial Management, Bari, Italy

Mohammad Namakshenas Department of Industrial Engineering, College of Engineering, Shahed University, Tehran, Iran

Luis Onieva Dpto. de Organización Industrial y Gestión de Empresas II. ETSI, Universidad de Sevilla, Camino de los Descubrimientos, Seville, Spain

Miguel Ortega-Mier Grupo IOL, Dpto. Ingeniería de Organización, Administración de Empresas y Estadística. ETSII, UPM Madrid, Madrid, Spain

Angel Ortiz Centro de Investigación de Gestión e Ingeniería de Producción (CIGIP), Universitat Politècnica de València, Valencia, Spain

José Pérez Ríos Grupo INSISOC, Dpto. de Organización de Empresas y CIM, Escuela Técnica Superior de Ingeniería Informática, Universidad de Valladolid, Valladolid, Spain

Luis Paipa-Galeano Universidad de La Sabana, Chía, Cundinamarca, Colombia

Juan E. Pardo Froján Grupo OSIG, Dpto. de Organización de Empresas y Marketing, Escuela de Ingeniería Industrial, Universidad de Vigo, Vigo, Spain

Javier Parreño Escuela Técnica Superior de Ingenieros Industriales, Universidad Politécnica de Madrid, Madrid, España

José Antonio Pascual Ruano Dpto. de Organización de Empresas y CIM, Escuela de Ingenierías Industriales, Universidad de Valladolid, Valladolid, Spain

Félix J. Pascual-Miguel Departamento de Ingeniería de Organización, Administración de Empresas y Estadística, Universidad Politécnica de Madrid Av, Madrid, Spain

David Pérez Perales Centro de Investigación en Gestión e Ingeniería de Producción, Universitat Politècnica de València, Valencia, Spain

Raúl Pino Díez Grupo de Ingeniería de Organización, Dpto. de Administración de Empresas, Escuela Politéc, de Ingeniería de Gijón, Universidad de Oviedo, Gijón, Spain

Raúl Poler Centro de Investigación de Gestión e Ingeniería de Producción (CIGIP), Universitat Politècnica de València, Valencia, Spain

Borja Ponte Blanco Grupo de Ingeniería de Organización, Dpto. de Administración de Empresas, Escuela Politéc, de Ingeniería de Gijón, Universidad de Oviedo, Gijón, Spain

David J. Poza INSISOC. E.I. Industriales, Universidad de Valladolid, Valladolid, Spain

J. Carlos Prado Prado Grupo de Ingeniería de Organización (GIO), Escola de Enxeñería Industrial, Vigo, Spain

Rosa Río-Belver Foresight Technology and Management (FTM) Group, Department of Industrial Engineering, University of the Basque Country UPV/ EHU, Vitoria, Spain

Mario Ramírez Ferrero Grupo INSISOC, Universidad de Burgos, Burgos, Spain

Alfonso Redondo Castán Dpto. de Organización de Empresas y CIM, Escuela de Ingenierías Industriales, Universidad de Valladolid, Valladolid, Spain

Imma Ribas Vila Dpto. de Organización de Empresas, Escola Tècnica Superior d'Enginyers Industrials de Barcelona, Universitat Politècnica de Catalunya, Barcelona, Spain

Ana Isabel Rodriguez-Escudero Facultad de Ciencias Económicas y Empresariales, Dpto. de Organización de Empresas y CIM, Universidad de Valladolid, Valladolid, Spain

Lorenzo Ros-McDonnell Grupo GIO. ETSII, Universidad Politécnica de Cartagena, Cartagena, Spain

Félix Roux Escuela Técnica Superior de Ingenieros Industriales, Universidad Politécnica de Madrid, Madrid, España

Felipe Ruiz Escuela Técnica Superior de Ingenieros Industriales, Universidad Politécnica de Madrid, Madrid, España

Rubén Ruiz Grupo de Sistemas de Optimización Aplicada, Instituto Tecnológico de Informática, Universitat Politècnica de València, Ciudad Politécnica de la Innovación, Edifico 8G, València, Spain

Cristina Ruiz Martín Grupo INSISOC, Dpto. de Organización de Empresas y CIM, Escuela de Ingenierías Industriales, Universidad de Valladolid, Valladolid, Spain

Lourdes Sáiz Department of Civil Engineering, Higher Polytechnic School, University of Burgos, Burgos, Spain

Pablo Sánchez Mayoral Grupo INSISOC, Dpto. de Organización de Empresas y CIM, Escuela Técnica Superior de Ingeniería Informática, Universidad de Valladolid, Valladolid, Spain

Rosa Mª Sánchez-Saiz Departamento de Ingeniería Civil.EPS, Área de Organización de Empresas, Burgos, Spain

Rashed Sahraeian Department of Industrial Engineering, College of Engineering, Shahed University, Tehran, Iran

Cristina Santandreu-Mascarell IGIC. Dpto. de Organización de Empresas, Universidad Politécnica de Valencia, Valencia, Spain

Javier Santos Department of Industrial Management, School of Industrial Engineers, TECNUN, Universidad de Navarra, San Sebastian, Spain

José Ignacio Santos Departamento de Ingeniería Civil.EPS, Área de Organización de Empresas, Burgos, Spain

Tom Taylor Dashdot & President of Association for Project Management, London, UK

Maria Valero-Herrero Dpto. de Organización de Empresas, Universidad Politécnica de Valencia, Valencia, Spain

Iván Velasco Jiménez Grupo INSISOC, Dpto. de Organización de Empresas y CIM, Escuela Técnica Superior de Ingeniería Informática, Universidad de Valladolid, Valladolid, Spain

Pilar I. Vidal-Carreras Grupo ROGLE. Dpto. de Organización de Empresas, Universidad Politécnica de Valencia, Valencia, Spain

Elisabeth Viles Department of Industrial Management, School of Industrial Engineers, TECNUN, Universidad de Navarra, San Sebastian, Spain

Marta Zárraga Rodríguez ISSA, Universidad de Navarra, San Sebastián, Spain

Part I
Management

Looking at the Future of Industrial Engineering in Europe

Giovanni Mummolo

Abstract The paper aims at identifying competence required to Industrial Engineers to face with current and future challenges of the EU. In order to properly address the topic, a reference framework is proposed. The framework provides a context analysis of major industrial and societal challenges which affect grand challenges as identified by the European '2020' strategy. Knowledge exploitation of Key Enabling Technologies and Sustainable Urban Development are considered major industrial and societal challenges which, in turn, are significant opportunity of growth for industry. Major gaps between industry needs and competence provided by Industrial Engineering (IE) curricula are identified. A multi-university collaborative network could be the appropriate answer to bridge educational gaps and pursue a unifying view of IE in the EU, respecting educational and industrial traditions of different Countries of the European Higher Educational Area (EHEA).

Keywords EU industrial and societal challenges · IE competence gap analysis · Future of industrial engineering in the EU

1 Context Analysis

In order to address the role of IE in Europe a context analysis of industrial and societal challenges in the EU is proposed.

Industrial challenges relate to EU industrial system capability to gain a role of global competitor in knowledge-based manufacturing and services; societal challenges, finalized to the 'well-being' of EU Citizens, represent at the same time an ethic commitment and opportunities of growth for industry. Industrial and societal challenges are mutual dependent: economic growth provides resources to face societal challenges which, in turn, represent a growing market for industry.

Industrial and societal challenges provide significant impacts on the grand challenges of the EU 2020 strategy:

G. Mummolo (✉)
Department of Mechanics, Mathematics and Management, Politecnico di Bari, European Academy for Industrial Management, viale Japigia, 182, 70126, Bari, Italy
e-mail: mummolo@poliba.it

C. Hernández et al. (eds.), *Managing Complexity*, Lecture Notes in Management and Industrial Engineering, DOI 10.1007/978-3-319-04705-8_1,

1. Economic Growth and Jobs' creation;
2. Energy and Climate Change;
3. 'Well-being' and Social Welfare.

Industrial and Societal challenges require knowledge and skills referable to IE. Challenges will be briefly discussed with reference to the framework in Fig. 1. "Technological Issues" and "non-Technological Issues" are identified as major industrial challenges while 'Healthily Aging Society' and "Sustainable Urban Development" are considered as major societal challenges.

The proposed framework is finalized to point out IE competence required to face with the EU grand challenges.

1.1 Industrial Challenges

1.1.1 Technological Issues

Key Enabling Technologies (KETs)

A strategic plan to boost EU economies should include conception and production of KETs—related goods and services.

A definition of KETs-based product is [3]:

> (a) an enabling product for the development of goods and services enhancing their overall commercial and social value;
> (b) induced by constituent parts that are based on nanotechnology, micro/nano electronics, industrial biotechnology, advanced materials and/or photonics;
> (c) produced by (but not limited to) advanced manufacturing technologies.

KETs for Manufacturing and Services

The 'Horizon 2020' framework program pays attention to the development and application of KETs to face with the grand challenges of the EU.

The KETs global current market volume was estimated around 800 billion € in 2006/2008 and is expected to grow to about 1.3 trillion USD by 2015 [7] (see Table 1). A specific KET is of interest for different final products; furthermore, different KETs are often integrated in the same more complex product. Feeding KETs based products in more a complex product generates a value chain capable to support EU industry growth and competitiveness. Major market share of specific KETs are estimated for Photonics and Micro-Nano Electronics. The annual growth rate of photovoltaic industry is estimated over 40% by 2015. Advanced Materials and Manufacturing Systems are considered the major drivers to sustain the applications of specific KETs. Low cost 3D-printers technology enable low volume productions, in this way overcoming the limits of scale economies (The Economist, April, 27, 2013).

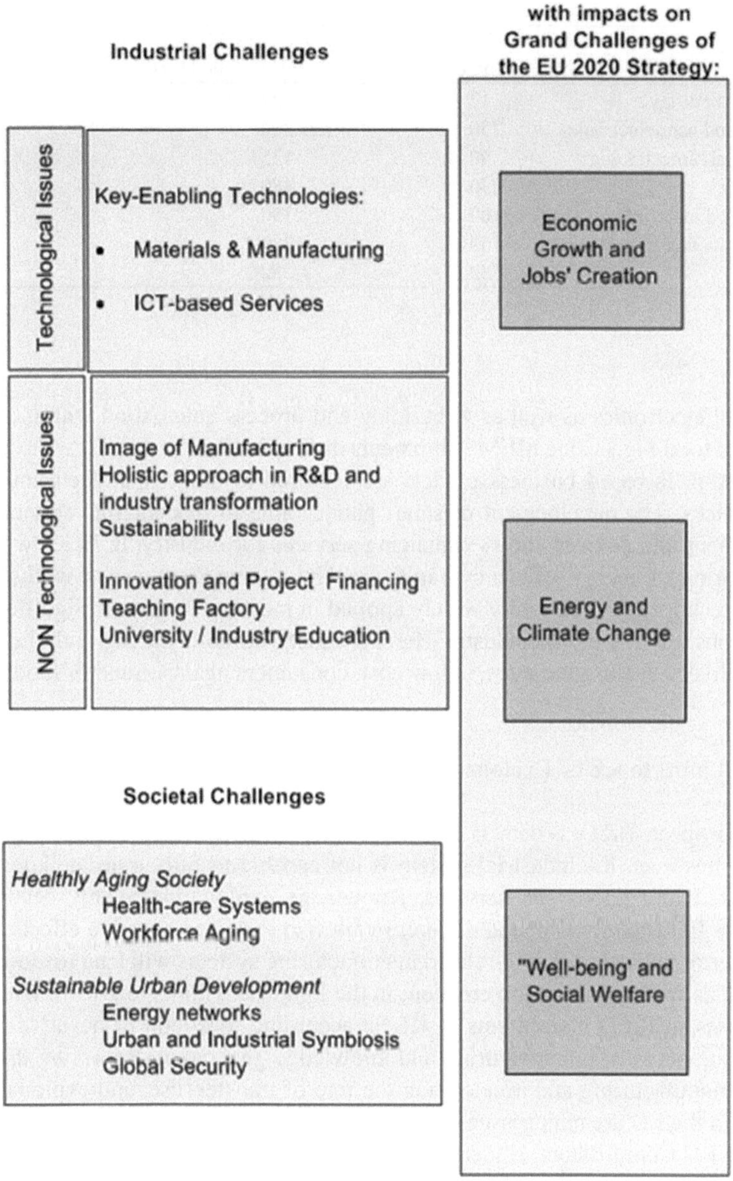

Fig. 1 Framework of industrial and societal challenges in the EU

Green chemistry is boosting industrial biotechnologies in new (e.g. bio-fuels production) as well as in classical production sectors (e.g. automotive).

MEMS (micro—electromechanical systems) and NEMS (at nano level) technologies are pervading the value chain of aerospace, automotive, biomedical, telecommunication, and electronic industry. The high-value MEMS market is due to

Table 1 Global market potentials for KETS [7]

	Current market size (bn USD)	Expected size (2012/15 bn USD)	Expected annual growth rate (%)
Nanotechnology	12	27	16
Micro and nanoelectronics	250	300	13
Industrial biotechnology	90	125	6
Photonics	230	480	8
Advanced materials	100	150	6
Adv. manufacturing systems	150	200	5
Total	*832*	*1,282*	

medical electronics as well as to building and process automation with more than 80 % of total high-value MEMS shipments during 2012.

New ICTs-based business models are required to make manufacturing closer to services. The development of smart phones able to link internet services with equipment and devices allows enhancing services for industry. ICTs allow remote monitoring of energy efficiency, faults prevention, and diagnosis of workstations. RFID technology is currently widely applied in industrial logistics. Significant applications are in the food industry. Here, product traceability along with the supply and delivery chains guarantee, at low cost, consumers against fraud of foodstuff.

Major Limits to KETs' Exploitation

The European R&D system is a global leader in generating base knowledge for KETs; however, the industrial system is not capable to fully translate knowledge generated into goods and services. Knowledge exploitation mainly occurs outside the EU (mainly China and Korea) with two expected negative effects. In the short term, the lack of KETs-related manufacturing systems will tend to slow down growth as well as inhibit job creation; in the long term, limits to growth will cause decreases in R&D investments in KETs, according to a loop of negative mutual influence between manufacturing and knowledge generation. Reversing the trend of de-manufacturing and accelerating the rate of transfer, use, and exploitation of KETs in the EU are major issues to stimulate growth and jobs' creation.

The EU Commission [3] identified three main causes limiting industrial deployment of KETs in the EU:

i. Lacking of a common industrial policy at the member state level: integration of different KETs within the same product or service would require a synergic integration of the overall production value chain.
ii. Capital intensive production along with the insufficient access to source of risk capital for SMEs represent a major limit for KETs exploitation since industrial research and experimental development often require investments of one order

of magnitude higher than the ones required for base research. Venture capital in the EU fell from 22 to 3 billion € from 2000 to 2010. Public-private partnerships along with coordination of public resources are required to tackle the current financial crisis of risk capital.

iii. Shortage of skill and competences in the labor market capable of handling the multi-disciplinary nature of KETs. Estimates indicate that by 2015, in the EU, up to 700.000 ICT skilled technicians will be needed and 400.000 new job positions in the KETs-area of nanotechnology are expected. The gap between KETs related job position demand and skills available tends to increase because of rapid industry growth and retirement of skilled workers. The social dimension is considered a crucial factor for attracting human capital.

1.1.2 Non-Technological Issues

Research and innovation require skilled competence in the very next future. Interrelated higher education and industry innovation is becoming a priority for manufacturing and service sectors.

With reference to manufacturing, a survey analysis of academia and industry actions in some EU Countries is shown in [16]. Here, Non-Technological Issues (NTIs) are investigated to assess their impacts on 'Growth and Jobs', 'Energy and Climate Change', and 'Social Welfare'.

NTIs considered in the consultation were:

1. Image of manufacturing (mainly for young people);
2. Establish a European Research and Innovation Area for Manufacturing;
3. An holistic approach in R&D programmes and industry transformation;
4. New infrastructures (ICTs and transport infrastructures);
5. Innovation & Project Financing;
6. Teaching Factory;
7. Sustainability issues and European Research Area dimension.

Results of consultation, sorted by Western and Eastern countries, shows how all NTIs are expected to impact on growth and jobs creation, green technologies and social welfare. According to the investigation, the (European) 'Teaching Factory' theme is considered in the EU eastern Countries the most important NTI and the second score for 'growth and jobs' creation for the EU western Countries; here 'Innovation and Project Financing' scored first. The need of a dual (University—Industry) education is pointed out.

1.2 Societal Challenges

Transition to low-carbon and to a resource efficient economy plays a central role in the Europe 2020 strategy. A worldwide consciousness on environmental issues stimulates the demand of eco-technologies finalized to transform renewable energy, increase energy efficiency, recycle and re-use materials.

However, the recent economic crisis and the current stagnation of the economy requires coupling environmental challenges with EU growth to recovery industry competitiveness with major world competitors.

The recent European Commission Memo [4] recognizes societal challenges in the EU as economic opportunities of growth given an increasing demand of 'well being' and quality of life expressed worldwide. A major contribution by the EU R&D system and industry is expected to tackle the following societal challenges:

- Healthily Aging Society and Workforce Aging
- Sustainable Urban Development

1.2.1 Healthily Aging Society and Work Force Aging

Ageing of European population impacts health and welfare systems as well as labor market. The Old-age Dependency Ratio (ODR), i.e. the ratio of people aged 65 years or older over people aged 15–64 years is projected to increase from 25.4%, as per 2008 evaluation, to 53.5% forecasted for 2060: i.e., for every person aged 65 years or older only two persons will be in the working age (instead of four people as per the current situation). By 2060, mean life expectancy will reach 84.5 years (75.7 years, 2012) for men and 89.0 years (82.1 years, 2012) for women [17]. In the view of population aging, welfare system requires to be conceived and planned properly in advance.

Health Long-Care Systems

Intelligent products allow elderly and disabled people an active and healthy lifestyle. Wearable robot [14] and intelligent vehicles are applications of smart technologies effectively tackling physical disabilities of elderly and disabled people. Remote monitoring of patients by 'internet of things' (IoTs) telecommunication technologies is an extraordinary opportunity to increase effectiveness and reduce costs of health-care systems. A survey on IoTs applications in health-care systems is provided in [19]. Cell phones of new generations, Wi-Fi networks, personal digital assistants (PDAs), and digital and RFID technologies can be connected by IoTs infrastructures. Patients and medical staff as well as medical devices, intelligent wheelchairs, wireless sensors and mobile robots can now support the de-hospitalization process with expected reduction in both health-care costs and human errors, mainly for long term care. Logistic processes in hospital structures and health-care systems are strongly affected by new 'intelligent' products and ICTs infrastructures.

Logistics of health-care systems requires IE competence [9].

Work Force Aging

The phenomenon of population aging affects the ageing of work force, phenomenon of great interest for industry as it strongly affects work force availability.

New safety devices, working time models, and ergonomic standards are required due to workers' aging. In [2] learning frameworks for ageing workers perspectives from comparative cross-national research on ageing, learning, and working in European Countries, are provided.

Industrial and academic research are required to investigate the influence of workforce aging in formulating new working time models and job rotation scheduling solutions. Finding optimal trade-off between production targets and ergonomic constraints is a field of interest for European industry. IE research and education is the natural cultural reference to properly address the issue. In [1] the problem of finding working time models while respecting ergonomics constraints is formulated and solved for upper limbs work-related musculoskeletal disorders in case of high repetitive, low-load technical actions in manual assembly lines. Models proposed refer to current ergonomic standards (ISO 11228-3:2007). New standards and operations management models are required in the view of production lines staffed with workers with higher average age.

Field investigations on the effect of workers aging on production performance were carried out at the BMW' plant in Dingolfing, Bavaria [13]. A pilot production line was staffed with workers with an average age of 47 years, in this way simulating the effects of the higher average age, as expected by 2017, on the line performance; this was compared with performance of the same line staffed with workers having an average age of 39 years. Workers were involved in the evaluation process providing suggestions for physical and organizational changes to be implemented in the '2017 production line'.

1.2.2 Sustainable Urban Development

'Smart city' is a common definition of an urban area were smart solutions are implemented to support a sustainable urban development. IE competence is required to provide solutions to many urban systems like energy and water storage and dispatching, urban mobility, waste collection, waste and wastewater treatments systems. The world potential market is of interest for the European industry. As an example, the sector of automatic waste separation is expected to grow by the factor five by 2020 [4]. Technology of waste separation could have a significant impact on the municipal waste collection service of a smart city and European industry has a dominant market share. Internet of Things (IoTs) telecommunication technologies connect decentralized electronic devices with application in urban services. RFID technologies are being adopted for monitoring waste production rate of citizens. ICTs are currently applied also to municipal utilities (gas, water) and urban security systems. A urban control center allow remote ICT-based monitoring of the City 'performance'.

The abovementioned technology and management issues required IE competence. Under this new perspective, at least three sustainability issues pertaining to IE area are becoming prominent as they represent opportunities of industry growth consistent with the EU '2020 strategy':

- Future Energy Networks;
- Urban/Industrial Symbiosis;
- Global Security.

Future Energy Networks

Most of today's energy infrastructures are approaching their expected life. However, an increasing demand for energy, mainly produced by fossil resources, is expressed by both industrial and civil users. Furthermore, according to the International Energy Agency [8], more than 60 % of energy demand is concentrated in the cities; about 75 % of EU population lives in urban areas responsible for the 80 % of energy consumptions and global warming gas emissions [18]. Urban sprawl phenomenon enlarged the city area causing extensions of linear energy infrastructures with an increase of investments and operations costs. A re-thinking of urban development is under way.

Pursuing a new development model of urban areas requires a re-thinking of current energy networks. Up-grading available energy networks with higher performance of unchanged energy transport technology (e.g. higher electric voltage or higher gas pressure) could result as inappropriate to face with the future sustainable energy supply.

Following a Greenfield approach, a vision of future energy networks has been proposed by academic (ETH, RWTH of Aachen University, TU Delft University) and industrial partners (ABB, Areva, Siemens) [5, 11].

Energy networks of the future are conceived as constituted by two major macro-units: the energy hub and the energy interconnector.

The first unit transforms multiple input energy carriers (fossil/renewable energy sources) into multiple output energy carriers like electrical, chemical, thermal (heating/cooling), and mechanical (e.g. compressed air) which have to be stored and dispatched. The second unit, the energy interconnector, integrate different output energy carriers in one underground device. The intermittency and uncertainty of renewable energy sources, like wind and solar, as well as of energy demand and price require appropriate small scale storage solutions [12]. The energy hub and the energy interconnector macro units tend to satisfy energy demand at district (Industrial, rural or urban) level, in this way reducing the extension of energy transport infrastructures. Benefits of a de-centralized layout of infrastructures can significantly reduce investments in infrastructures for energy transportation on long distance. First applications are expected in the very next future within the 'smart cities' initiatives across the EU.

Urban and Industrial Symbiosis

Geographic proximity of urban and industrial districts could lead to ecological and economic solutions for a sustainable urban development.

Recyclable waste (e.g. paper, plastic and organic waste) originated both in a urban and in industrial districts can be utilized by a contiguous industrial district for recycling, re-use or waste—to energy transformation. The phenomenon, known as 'Urban/Industrial Symbiosis', is being receiving wide attention.

The eco-town program in Japan is an outstanding example of symbiosis. The program seeks to maximize economic and environmental performance by integrating commercial, urban, and industrial waste streams in industrial applications [6, 20]. The program identified 26 eco-towns for a global investment of 1.65 billion USD in 61 innovative recycling projects and 107 new recycling facilities.

Future energy networks as well as urban and industrial symbiosis rely on the same strategic view of the cities of the future aiming at reducing or eliminating transportation of energy and materials on long distance enabling technologies for 'on-site' transformation, at the district level.

Technologies involved and management capability require a systemic and inter-disciplinary cultural approach which is inside the background of IE.

Global Security

Global security is a societal challenge which is referred to both unintentional (e.g. fires and natural disasters) and intentional events (e.g. terroristic attacks).

Industrial sites along with sensible infrastructures (e.g. ports, airports, railway stations) are traditionally targets subject to risk analysis and accident management.

However, at current, a sustainable urban development is also referred to the security of the Cities. Here, increasing attention is being paid by security of cultural heritages assets (e.g. museums, churches, monuments, and archaeological sites) to prevent and contrast the effects of potential natural disasters or intentional threats.

Global security represents a very sensible worldwide social issue as well as an opportunity of growth for industry producing sophisticated high technology-based security systems. Safety and Security engineering approaches are currently adopted in designing such complex systems. IE competence requires integration between social/psychology sciences with engineering disciplines.

2 A Preliminary Survey on IE Competence in EU

A wide range of competence can be referred to the IE area. An assessment of IE subjects in the EU education system is provided in [15]. The spectrum of IE subjects is wide and varies from country to country; the assessment is far to be considered as complete and more investigations are still needed.

A major referenced definition of IE is provided by the Institute of Industrial Engineers (IIE) [http://www.iienet2.org/]:

> Industrial Engineering is concerned with the design, improvement, and installation of integrated systems of men, materials, equipment and energy. It draws upon specialized knowledge and skill in the mathematical, physical and social sciences together with the principles and methods of engineering analysis and design to specify, predict, and evaluate the results to be obtained from such systems.

A more recent definition of IE is proposed in [10] by a research group involved in the IESE project (Industrial Engineering Standards in Europe) [http://www.iestandards.eu/]:

> The branch of engineering that engages in the study of how to describe, evaluate, design, modify, control and improve the performance of complex systems, viewed over time and within their relative context.

Definitions are similar: the latter focuses on peculiarity of IE to design and manage complex and dynamic systems capable to interact with the environment.

Major question is whether IE university curricula fulfill industry needs. The question has been addressed by the IESE project involving six European Countries (Denmark, Germany, Iceland, Ireland, Netherlands, and Sweden).

Analysis of educational programmes offered in these countries was carried out and analyzed industry needs. To assess the educational gaps, the survey considered the Industrial Engineering Educational Programme (IEEP) prescribed and adopted by the International Labour Organization (ILO). The model is widely accepted in Europe and in many US Universities.

According to the IEEP model, the five main areas of IE are:

1. IE Base (e.g. work measurement, processes, logistics, organization).
2. Operations Research (e.g. Optimization, Simulation and Network Models;
3. Human Factors Engineering (Ergonomics, Behavioral Science,…).
4. Management Systems (e.g. TQM, Project/Contract Management, Health/Safety Management, HRM, Business Ethics, Cross Cultural Management).
5. Manufacturing Systems Engineering (Production, Maintenance, Automation).

In the EU, the IEEP model is considered not fully fulfilling industry expectations. Industry seems to be more interested to knowledge people able to understand as a company works in general and less interested in very specific types of job. Industry requires capability in solving problems and carry out tasks.

Furthermore, the IESE survey on around 50 EU companies identifies two major IE educational gaps in the following areas:

1. Innovation and Technology: it deals with a fast technological development (e.g. nano/bio-technology) by information/manufacturing technologies.
2. Environment and sustainability: it deals with policies, legislation, energy standards (EN 16001), corporate energy policies, energy management and auditing, sustainable technologies (solar, wind, wave…), combined heat and power, building management systems, lighting, HVAC.

Finally, human factors engineering competence need to be empowered.

Details on the survey can be found in [10].

Both areas synthesize the need to exploit innovation in manufacturing and services within the general paradigm of a Sustainable Development. Major limits of academic higher education in updating curricula are recognized:

- Research personnel recruiting: universities seem to engage candidates with the same professional profile;
- Priority of research and 'publish or perish' policy;
- Changing study regulations and national accreditation is time consuming.

It should be observed how the gap analysis between university programmes and industry needs provided by the IESE project needs to be integrated by further investigations on all levels of academic higher education curricula (from bachelor to PhD), involving more Countries in the EHEA.

3 A New Framework of Competence for Industrial Engineering

Following the context analysis (Sect. 1) as well as the preliminary survey on current IE competence and industry educational needs (Sect. 2), a contribution to define a framework for updating IE competence is proposed. The analysis will be carried out with reference to the framework on industrial and societal challenges introduced in the context analysis (Fig. 2). The framework is far to be considered as exhaustive. The purpose is to promote a wide debate involving both academia and industry.

With reference to industrial challenges, IE competence requiring to be empowered in current curricula refers to the wide area of 'Technology, Innovation and Competitiveness'. Competence in developing new products, processes, and services should be empowered. To this concern, deployment capability of fundamental research into KETs-based productions and services would be of value for industry growth. Knowledge and Technology Management, Project Financing, Management of Public-Private Partnerships, Project Management of Innovation Projects, are subjects of interest for industry.

IE competence should be exploited also to face with societal challenges referable to Healthily Aging Society and Sustainable Urban Development.

3.1 IE in the Healthily Aging Society

Health-care systems are at current of great research interest for IE. More and more industrial engineers are employed in public-private organizations.

Logistics of hospital infrastructures and processes is being managed by industrial engineering techniques to point out criticalities like deficit of human resources,

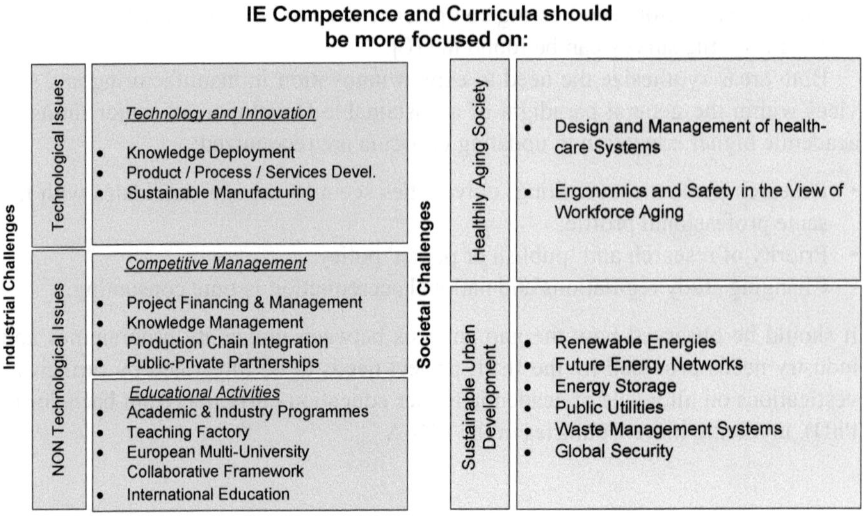

Fig. 2 IE competence to face industrial and societal challenges in the EU

communication errors, faults in technical and professional behavior, human errors (wrong operating of procedures, commission/omission errors, inadequate medical prescriptions), inadequate maintenance or unavailability of medical devices. Ergonomics and safety concerns of workforce aging are relatively new educational and research topics of great interest for IE. Ergonomics and Operations Management should be more integrated disciplines.

3.2 IE for a Sustainable Urban Development

Renewable energies transformation and energy networks design and management are in the cradle of IE competence. Here, new energy storages and networks (electrical, mechanical, chemical, and mechanical), in contexts characterized by randomness of both energy availability (renewable sources) and users' demand are complex systems required to be properly designed and managed. The same competence is required in managing urban utilities and related facilities (gas, water, electricity) as well as public/private services like waste and waste-water management systems.

3.3 Improving the IE Education System in Europe

High qualified educational paths have to be designed and implemented not only for academic purposes but also for industry. Networking for learning/training/working exchange in collaborative workplace and experiential learning approaches are required to innovate IE education. Staff recruiting university policies should also be finalized to bridge educational gaps (besides research needs).

Furthermore, IE education in the EU should aim at a unifying educational framework, meanwhile respecting IE culture and tradition of each Country. To pursue such a goal, a more in-depth survey of IE curricula in the EHEA is needed at different levels of education. Following the example of the 'Advanced Manufacturing Partnership' initiative, launched by Mr. President Obama [16], an 'European Multi-University Collaborative Network' in the form of a 'European School of Industrial Engineering', should be proposed. The initiative should be promoted as a partnership between Universities of the EHEA and Industry interested in educating talented, leadership-prone, young people and industry workforce: empowering human capital is the major driver for Europe to recovery competitiveness worldwide and effectively face with the grand challenges of Europe.

4 Summary and Conclusions

Tracing possible trajectories of future IE competence in the EU is a complex task for three basic reasons:

IE competence is called to face with major EU industrial and societal challenges: further abilities are required to improve the already wide spectrum of IE;

Updating IE curricula as well as teaching/learning tools and practice can be carried under a common European perception and perspective about the role of education;

A complete survey analysis on IE curricula and education methods in the EU is still missing: to look at the future education would require a reliable perception of current situation.

The paper addressed main IE competence to tackle industrial and societal challenges which the EU is called to face with.

The context analysis and the preliminary survey on IE education in the EU outlined major findings on required IE competence:

1. The need of a higher capability of industrial engineers to exploit innovation into new product and process. KETs exploitation is one of major limit to industry growth in the EU.
2. Education and scientific activities are mainly focused on industrial challenges.
3. IE is called also to tackle societal challenges. Healthily Aging Society and Sustainable Urban Development are wide fields of investigations. Safety and ergonomics concerns in the view of workforce aging as well as security engineering require a full involvement of IE education. Health-care and long care systems design and management need of theoretical tools and best practice which are, by nature, in the cradle of IE.

4. Industrial and societal challenges represent a significant opportunity of growth for industry.
5. IE education system is required to increase the availability of skilled human capital to speed up product and process innovation and support industry growth.
6. University and Industry are called to bridge the gap between IE curricula and industry needs. Learning factory and experiential learning approaches are required. An European Multi University Collaborative Network, with industry involvement, is needed to define curricula, educational materials and best practice. Tackling grand challenges of the EU will strongly depend on willingness and capability of universities and industry to share and drive innovation processes in research and education.
7. An 'European School of Industrial Engineering', promoted as a partnership between high qualified Schools of IE and Industry, can coordinate efforts to educate classes of talented, leadership-prone human capital.

References

1. Boenzi F, Digiesi S, Mossa G, Mummolo G, Romano VA (2013) Optimal break and job rotation schedules of high repetitive—low load, manual tasks in assembly lines: an ocra—based approach. 7th IFAC Conference on Manufacturing Modelling Management and Control. MIM 2013, Petersburg, 19–21 Sept. 2013
2. European centre for the development of vocational training (2010), Working and ageing. Emerging theories and empirical perspectives. Publication Office of the European Union
3. EU Commission (2012) Communication from the Commission to the European Parliament, The Council, The European Economic and Social Committee and the Committee of the Regions. A European strategy for key enabling technologies—a bridge to growth and jobs. COM (2012) 341
4. EU Commission (2012) Industrial revolution brings industry back to Europe. Memo 12:749
5. Geidl M, Koeppel G, Favre-Perrod P, Klockl B, Andersson G, Frohlich K (2007) The energy hub—a powerful concept for future energy systems. Third Annual Carnegie Mellon Conference on the Electricity Industry , 13–14 March 2007
6. Geng Y, Tsuyoshi F, Chen X (2010) Evaluation of innovative municipal solid waste management through urban symbiosis: a case study of Kawasaki. J Clean Prod 18:993–1000
7. HLG (2011) KET working document, mastering and deploying key enabling technologies (KETs), building the bridge to pass across the KETs 'Valley of Death' for Future European Innovation and Competitiveness. http://ec.europa.eu/enterprise/sectors/ict/files/kets/hlg-working-document_en.pdf, access date: 9 May, 2014
8. IEA (2012) Urban energy policy design. http://www.iea.org/newsroomandevents/news/2012/june/name,28079,en.html, access date: 9 May, 2014
9. Ignone G, Mossa G, Mummolo G, Pilolli R, Ranieri L (2013) Increasing public healthcare network performance by de-hospitalization: a patient pathway perspective. Strateg Outsourcing 6:85–107. ISSN: 1753-8297
10. Jensson P, Byrune T, Nolan D, Norgaard B, Rokkjaer O, Schinner HD, Appold W, Polman T, Schut A, Bayard O, Areskoug M (2011) Industrial engineering standards in Europe—industry needs versus education. 1st world engineering education flash week. Lisbon 27–30 Sept. Eds Bernardino J. and Quadrado J.C.

11. Kienzle F, Favre-Perrod P, Arnold M, Andersson G (2008) Multi-energy delivery infrastructures for the future. International Conference on Infrastructure Systems. The NL, Rotterdam
12. Kienzle F, Ahcin P, Andersson A (2011) Valuing investments in multi-energy conversion, storage, and demand-side management systems under uncertainty. IEEE Trans Sustain Energy 2(2):194–202
13. Loch CH, Sting FJ, Bauer N, Mauermann H (March, 2010) How BMW is defusing the demographic time bomb. Harward Bus Rev 88(3):99–104
14. Mohammed S, Amiratb Y, Rifaic H (2012) Lower-limb movement assistance through wearable robots: state of the art and challenges. Adv Robot 26(1–2):1–22
15. Mummolo C, Mangialardi L, Kim JH (2013) Quantifying dynamic characteristics of human walking for comprehensive gait cycle. Journal of Biomechanical Engineering vol. 135, n. 9, pp. 091006/1-091006/10
16. Paci AM, Lalle C, Chiacchio MS (2013) Education for innovation: trends, collaborations and views. J Intell Manuf 24:487–493
17. Rechel B, Grundy E, Robine JM, Cylus J, Mackenbach JP, Knai C, McKee MA, 2013) Aging in the European Union. www.thelancet.com, vol 381, Issue 9874, pp. 1312–1322
18. Tajani A (Oct, 2012) Le città come incubatori di innovazione: politiche e tecnologie al servizio dei cittadini. Mobility Tech, Milan
19. Turcu CE, Turcu CO (2013) Internet of things as key enabler for sustainable healthcare delivery. Proced—Soc Behav Sci 73:251–256
20. Van Berkel R, Fujita T, Hashimoto S, Geng Y (2009) Industrial and urban symbiosis in Japan: analysis of the eco-town program 1997–2006. J Environ Manage 90(3):1544–1556

A "Collaborative Me" Crossroad: Individual Beliefs and the Adoption of Corporate Blogs

Pedro Fernández-Cardador, Ángel Hernández-García and Santiago Iglesias-Pradas

Abstract Collaboration has become an essential process to improve business performance. Firms need to enhance innovation, and this is where 2.0 collaborative tools are expected to play a key role. Corporate weblogs are one of the main 2.0 tools that have raised more interest among managers, due to their ease of use and potential to bring together employees and partners who have to collaborate in order to achieve a common goal. However, individuals' reactions to the use of such new systems may differ, which in turn might lead to rejection of corporate blogs. Thus, the objective of this exploratory research is to study the influence of individuals' beliefs in the adoption of corporate weblogs; more specifically, the factors discussed in this research are self-efficacy, personal outcomes expectations and anxiety. In order to assess predictive ability for the exploratory research model, we have developed a theory grounded model, which has been validated with data from 70 employees from the Information Technology department of a large industrial Spain-based company. Findings from the results show that perceived usefulness is predicted by anxiety and personal outcomes expectations, perceived ease of use is predicted by blog anxiety, and behavioral intention to use corporate blogs is predicted by perceived usefulness.

Keywords Corporate blog · Technology acceptance · Self-efficacy · Personal outcomes expectations · Computer anxiety

P. Fernández-Cardador (✉)
Universidad Politécnica de Madrid, Madrid, Spain
e-mail: pedro.fernandez.cardador@alumnos.upm.es

Á. Hernández-García · S. Iglesias-Pradas
Departamento de Ingeniería de Organización, Administración de Empresas y Estadística, Universidad Politécnica de Madrid, Av. Complutense 30, 28040, Madrid, Spain
e-mail: angel.hernandez@upm.es

S. Iglesias-Pradas
e-mail: s.iglesias@upm.es

C. Hernández et al. (eds.), *Managing Complexity*, Lecture Notes in Management and Industrial Engineering, DOI 10.1007/978-3-319-04705-8_2,

1 Introduction

Collaboration is a fundamental process in knowledge creation and sharing which takes place through interaction among individuals. Collaboration is also one of the main drivers of change in organizations, and especially in knowledge-oriented organizations [27]. According to Prahalad and Ramaswamy [26], there is a strong link between collaboration within a company and firm's profitability, and this also holds true when considering collaboration with other companies. However, adoption of collaborative technologies is not happening as fast as expected [24].

There are five major groups of characteristics influencing adoption of collaborative tools; these characteristics are related to the technology, individual, group, task, and context [4], respectively. In the mid-seventies, many technology adoption processes were failing due to fear of new technologies [16] and doubts of employees about their own ability to use a new technology and obtain the expected results in return [7]. In organizational contexts, these behaviors are user-specific, as different users have different needs depending on their role [4].

Collaborative systems in corporate settings have undergone a radical change with the introduction of 2.0 tools, such as wikis and weblogs [6]. Blogs allow readers and writers to express their opinions, exchange different points of view and add new complementary knowledge [30]. In the early days of weblogging, their use was limited to personal journals; but later on, blogs began to receive significant attention as useful knowledge sharing tools [5] due to their potential to facilitate exchange of knowledge among people [17] with a significant cost reduction [31].

The fact that the beliefs which lead to adoption and rejection of technologies highly depend on intrinsic characteristics of the individual has not been previously addressed in the context of corporate blog adoption. Therefore, the main objective of this exploratory research is to study the influence of individual characteristics—namely, self-efficacy, personal outcomes expectations and anxiety—in the adoption of 2.0 collaborative tools and, more specifically, in the adoption and use of corporate blogs.

The remainder of this chapter is organized as follows: Section 2 presents the theoretical framework and research hypotheses; research methodology, procedures and measures are shown in Section 3; Section 4 covers the data analysis technique and empirical results for the study, which are followed by a discussion of results and implications for theory and practice in Section 5.

2 Theoretical Framework and Research Hypotheses

The Technology Acceptance Model (TAM) [9] explains how and why individuals adopt and use a technology—in this case, corporate blogs. According to Davis, an individual's behavioral intention towards the use of a new technology is the best predictor of actual use. In TAM, behavioral intention is influenced by two beliefs: **perceived usefulness (PU)** and **perceived ease of use (PEOU)**; in this study, PU and PEOU refer to the users' perception of their own performance when using corporate blogs and the degree to which a user believes that using a corporate blog is free of effort [15], respectively. Applying TAM relations to the case of corporate blogs, we find that perceived ease of use (*H1*) and perceived usefulness (*H2*) positively predict intention to use corporate blogs, and also that perceived ease of use positively predicts perceived usefulness (*H3*).

Anxiety is a generalized emotional distress [21] experienced by an individual. According to Bandura [2], anxiety appears when individuals try to carry out behaviors they do not feel competent to perform. There are two kinds of anxiety: trait-based, i.e. personality anxiety, and anxiety associated to a specific situation; computer anxiety falls into the second category, and may be defined as an irrational "generalized emotional distress experienced by an individual" when using or considering the use of computers [16]. Based on this concept, we have defined **blog anxiety** as the anxiety experienced by individuals when they perceive themselves to be underperforming at using corporate blogs. Previous studies found that computer anxiety causes computer use avoidance [7]; following this rationale, we posit that blogging anxiety negatively predicts perceived usefulness (*H4*) and perceived ease of use (*H5*).

Personal outcomes expectations refer to prospective rewards and/or improvements of perception of an individual's image by other members of the organization [8]. This concept is based on the Social Cognitive Theory [3], which states that an individual will engage in a behavior if he expects some kind of reward after performing it. We have adapted this concept taking into consideration that users may post in corporate blogs when they expect some pleasure such as enjoyment, organizational recognition or improvement of their image in return. Previous studies about outcomes expectations are inconclusive; for instance, Papadopoulos et al. [25] found that people will continue to share information on the Internet if they expect praise or rewards and Kankanhalli et al. [18] found a positive relation between outcomes expectations and intention to use knowledge sharing systems, while Lu and Hsiao [20] found that this relation was non-significant in the case of blogs. Given this lack of consensus, we will assess the nature of this relation in the case of corporate blogs, and therefore posit that personal outcomes expectations positively predict perceived usefulness (*H6*) and perceived ease of use (*H7*).

Self-efficacy is a human regulatory mechanism which affects individuals' judgments about their ability to perform a given task [2]. We will adapt this concept to our research context by defining **blogging self-efficacy** as the self-confidence in one's ability to collaborate using corporate blogs. Blogging self-efficacy emerges thus as a barrier for corporate blog adoption; i.e., if users believe that they are not able to use blogs, they will most likely be reluctant to collaborate with their community via corporate blogs.

Prior research has found that computer self-efficacy may be an antecedent of perceived ease of use [1]. Moreover, previous studies have found that computer self-efficacy may increase personal outcomes expectations [7] and perceived usefulness [19]. Empirical studies have also found a negative correlation between self-efficacy and anxiety [29] and other research has established that increasing levels of

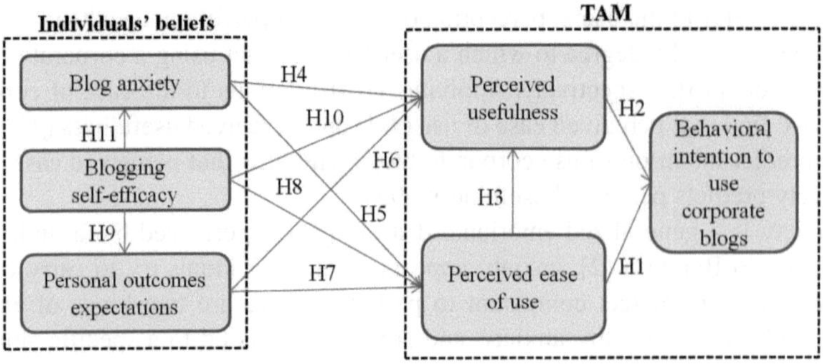

Fig. 1 Research model

self-efficacy cause anxiety reduction [10]. Taking all this into account, we find that blogging self-efficacy positively predicts perceived ease of use (*H8*), personal outcomes expectations (*H9*) and perceived usefulness (*H10*), and negatively predicts blogging anxiety (*H11*).

Following the hypotheses presented in this section, the complete research model for this study is presented in Fig. 1.

3 Research Methodology

For this study, we have selected a sample of employees from a Spain-based large industrial company where 2.0 collaborative tools are currently being deployed as part of a plan to establish a new collaborative culture. Data were gathered via online survey during January and February 2013. Ninety employees from the IT department were invited to participate in this study, and 70 valid answers—70 % from male and 30 % from female respondents—were collected, for a total response rate of 77.8 %. The Likert-7 scales used have been validated in prior literature, and were adapted to the context of corporate weblogs: blog-anxiety scales were adapted from Heinssen et al. [14]; self-efficacy and personal outcomes expectations were taken from Kankanhalli et al. [18]; perceived ease of use and perceived usefulness were based on Davis [9]; scales for intention to use corporate blogs were adapted from Oum and Han [23].

4 Data Analysis

We used a partial least squares (PLS) technique in order to assess the structural model and the software used was SmartPLS 2.0M3 [28]. Since this was an exploratory research focused on prediction, PLS was chosen instead of covariance-based

Table 1 Item reliability and convergent validity

Constructs	Item	Outer loadings	Composite reliability	AVE
Blog anxiety	BBA01	0.809		
	BBA03	0.700	0.806	0.581
	BBA04	0.774		
Blogging self-efficacy	BSE01	0.934	0.920	0.852
	BSE03	0.913		
Personal outcomes expectations	BPO03	0.884	0.879	0.784
	BPO04	0.887		
Perceived usefulness	BPU02	0.760	0.867	0.619
	BPU03	0.789		
	BPU04	0.812		
	BPU05	0.785		
Perceived ease of use	BPE01	0.726	0.893	0.737
	BPE03	0.901		
	BPE04	0.934		
Behavioral intention to use blogs	BBI01	0.883	0.926	0.758
	BBI02	0.835		
	BBI03	0.862		
	BBI04	0.901		

Significance level $p < 0.001$ for all items

structural equation modeling [13]. Besides, PLS does not need strict assumptions of sample size or measurement scales and allows using smaller sample sets.

Item reliability was evaluated by observing the standardized loadings of latent variable indicators—all the indicators were defined as reflective. Indicators with loadings not exceeding the ideal cutoff level of 0.7 [22] were dropped for subsequent analysis. Item reliability results are shown in Table 1. A bootstrap resampling procedure was used to test the stability of the estimates, with significance values of $p < 0.001$ for all cases. To ensure convergent validity we calculated the constructs' composite reliability and average variance extracted (AVE). Values were higher than 0.81 and 0.60 respectively (see Table 1), well over the acceptable threshold values of 0.7 [12] and 0.5 [11]. Discriminant validity was confirmed upon Fornell and Larcker's [11] recommendation, as the square root of AVE was greater than bivariate correlations between each construct and the rest of constructs.

The results related to the prediction of behavioral intention to use corporate blogs showed that perceived usefulness had significant and large effect supporting H2 but perceived ease of use did not exert a significant effect and thus, H1 was not supported. Moreover, perceived ease of use had no significant effect on perceived usefulness, leading to the rejection of H3.

Perceived usefulness was predicted by blog anxiety and personal outcomes expectations supporting H4 and H6 respectively, but not by self-efficacy, thus not supporting H10. On the other hand, perceived ease of use was only predicted by

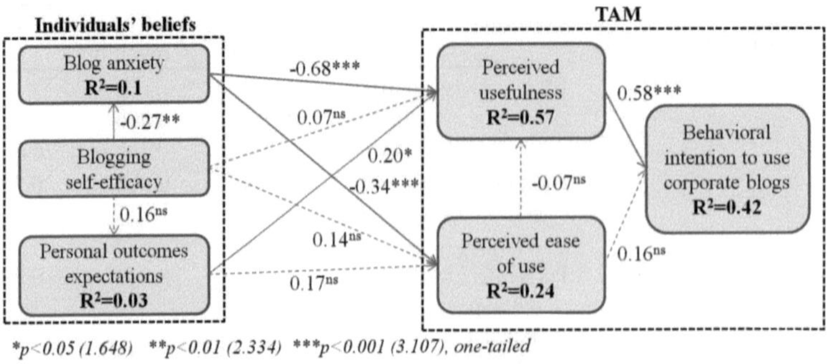

Fig. 2 PLS research results

blog anxiety—supporting H5—but not by blog self-efficacy and personal outcomes expectations—leading to rejection of H7 and H8.

Blogging anxiety was significantly predicted by blog self-efficacy—supporting H11, and blog self-efficacy did not exert a significant influence on personal outcomes expectations—and then H9 was rejected. As for variance explained, the model is able to explain 42 % of behavioral intention to use corporate blogs, a value which stresses the relevance of individual beliefs in the process of corporate blog adoption. We also calculated the Stone-Geisser (Q^2) values to test predictive relevance, and all values were positive, confirming predictive relevance. Figure 2 summarizes the structural model analysis results.

5 Discussion

This study emphasizes the fundamental role played by anxiety in the adoption of corporate blogs. From the results, when the most common communication and collaboration tools, such as e-mail, are replaced, employees may believe they will not be able to catch up on new tools or to achieve an optimal use of their functionalities, which may lead to a lower perceived usefulness, resistance and, ultimately, to rejection. On the other hand, fear of making fatal errors, loss of information when using corporate blogs, or fear of blogs itself, were not found to be a significant cause of anxiety, and nor did they exert significant influence on perceived usefulness.

As for blogging self-efficacy, the results—consistent with previous studies—reveal a negative relation with anxiety, but only explaining 10 % of variance. Some of the rest of variance might be explained by the traditional concept of self-efficacy, which was omitted in this research. Therefore, the study of its effect on anxiety is recommended for subsequent research.

We have also found that users are not expecting monetary rewards or promotions in exchange for the use of blogs, but they do perceive that corporate blogs may be useful inasmuch as they may lead to improvements in their image and prestige within the organization.

Finally, the high value of R^2 for perceived usefulness suggests that the perceived usefulness of corporate blogs for collaboration is mainly determined by individuals' beliefs, stressing out the need to consider these factors in adoption processes. In addition, perceived usefulness is the main driver of corporate blog adoption, as expected, but perceived ease of use had no significant influence on behavioral intention to use corporate blogs; a possible explanation for this finding is that the sample was selected from IT-oriented professionals with high experience on the use of collaboration tools; however, this finding should be confirmed by further research.

References

1. Agarwal R, Karahanna E (2000) Time flies when you're having fun: cognitive absorption and beliefs about information technology usage. MIS Q 24(4):665–694
2. Bandura A (1977) Self-efficacy: toward a unifying theory of behavioral change. Psychol Rev 84(2):191–215
3. Bandura A (1989) Social cognitive theory. Annals of child development, six theories of child development, vol 6. JAI, Greenwich, pp 1–60
4. Brown S, Dennis AR, Venkatesh V (2010) Predicting collaboration technology use: integrating technology adoption and collaboration research. J Manage Inf Syst 27(2):9–54
5. Chai S, Kim M (2010) What makes bloggers share knowledge? An investigation on the role of trust. Int J Inf Manage 30(5):408–415
6. Chen D, Hu N, Liu L (2007) Corporate blogging and firm performance: an empirical study. 2007 International Conference on Wireless Communications. Networking and Mobile Computing, pp 6152–6155
7. Compeau DR, Higgins CA (1995) Computer self-efficacy: development of a measure and initial test. MIS Q 19(2):189–211
8. Compeau DR, Higgins CA, Huff SL (1999) Social cognitive theory and individual reactions to computing technology: a longitudinal study. MIS Q 23(2):145–158
9. Davis FD (1989) Perceived usefulness, perceived ease of use, and user acceptance of information technology. MIS Q 13(3):319–340
10. Dwyer KK, Fus DA (2002) Perceptions of communication competence, self-efficacy, and trait communication apprehension: is there an impact on basic course success? Commun Res Rep 19(1):29–37
11. Fornell C, Larcker DF (1981) Evaluating structural equation models with unobservable variables and measurement errors. J Mark Res 19(1):39–50
12. Hair J F, Anderson RE, Tatham RL, Black WC (1998) Multivariate data analysis. Prentice Hall, Englewood Cliffs
13. Hair JF, Christian M, Sarstedt M (2011) PLS-SEM: indeed a silver bullet. J Mark Theory Pract 19(2):139–151
14. Heinssen RK, Glass CR, Knight LA (1987) Assessing computer anxiety: development and validation of the computer anxiety rating scale. Comput Hum Behav 3:49–59
15. Hsu C, Lin J (2008) Acceptance of blog usage: the roles of technology acceptance, social influence and knowledge sharing motivation. Inf Manage 45(1):65–74
16. Igbaria M, Iivari J (1995) The effects of self-efficacy on computer usage. Omega 23(6):587–605
17. Jackson A, Yates J, Orlikowski W (2007). Corporate blogging: building community through persistent digital talk. 40th Hawaii International Conference on System Sciences, pp 1530–1605
18. Kankanhalli A, Tan BCY, Wei K (2005) Contributing knowledge to electronic knowledge repositories: an empirical investigation. MIS Q 29(1):113–143

19. Liu X (2010) Online posting anxiety: impacts on blogging. Chin J Commun 3(2):202–222
20. Lu H-P, Hsiao K-L (2007) Understanding intention to continuously share information on weblogs. Internet Res 17(4):345–361
21. Nietzel MT, Berstein DA, Russel RL (1988) Assessment of anxiety and fear. In: Bellack AS, Herson M (eds) Behavioral assessment, a practical handbook, 3rd edn. Pergamon, Toronto
22. Nunnally JG (1978) Psychometric theory. McGraw Hill, New York
23. Oum S, Han D (2011) An empirical study of the determinants of the intention to participate in user-created contents (UCC) services. Expert Syst Appl 38(12):15110–15121
24. Palen L, Grudin J (2002) Discretionary adoption of group support software. In: Munkvold BE (ed) Implementing collaboration technologies in industry: case examples and lessons learned. Springer, London
25. Papadopoulos T, Stamati T, Nopparuch P (2013) Exploring the determinants of knowledge sharing via employee weblogs. Int J Inf Manage 33(1):133–146
26. Prahalad CK, Ramaswamy V (2001) The collaboration continuum: understand the full goals and complexity of collaboration before moving forward. Information Week, November (22)
27. Riemer K, Frößler F (2007) Introducing real-time collaboration systems: development of a conceptual scheme and research directions. Commun Assoc Inf Syst 20:204–225
28. Ringle CM, Wende S, Will A (2005) SmartPLS 2.0 (beta). www.smartpls.de
29. Sam H, Othman A, Nordin Z (2005) Computer self-efficacy, computer anxiety, and attitudes toward the internet: a study among undergraduates. Unimas Educ Technol Soc 8(4):205–219
30. Wattal S, Racherla P, Mandviwalla M (2009) Employee adoption of corporate blogs: a quantitative analysis. System Sciences, 2009. HICSS '09. 42nd Hawaii International Conference, 1–10
31. Yu T-K, Lu L-C, Liu T-F (2010) Exploring factors that influence knowledge sharing behavior via weblogs. Comput Hum Behav 26(1):32–41

Towards Strategic Project Management

Víctor Hermano and Natalia Martín-Cruz

Abstract This paper focuses on the application of the Dynamic Capabilities framework to Project Management discipline. A review of project management literature illustrates the new project and project managers' conceptualization and the shift towards a more strategic perspective. Commonalities and overlaps between project management and dynamic capabilities approach are highlighted both from a theoretical and professional point of view. The findings indicate the closeness between project management and dynamic capabilities by creating an integrative framework useful both for top and project managers. Besides, we show the potential benefits of the application of strategic management theories to Project Management.

Keywords Project management · Strategic management · Theoretical framework · Dynamic capabilities approach

1 Introduction

Since the 1950s, when it was recognized as a distinct discipline, project management has become a widespread practice in modern organizations [25]. In today's turbulent and dynamic environments, traditional elements of business success—owning tangible assets, controlling costs, maintaining quality- are necessary but appears to be insufficient to obtain sustainable abnormal results [19]. In order to compete in this environment, organizations are increasingly shifting towards more-flexible project-based structures instead of rigid traditional ones [14, 25]. Based on the report "New challenges for a land on expedition", developed by the Deutsche Bank, in 2020, the "project economy" delivers 15 % of value creation in Germany [7]. Moreover, project-based organizations (PBOs) have been considered to obtain better results, especially in unstable and dynamic environments [8, 12, 21].

V. Hermano (✉) · N. Martín-Cruz
INSISOC Research Group, Department of Business Management and Marketing,
Business School of Universidad de Valladolid, Pso del Cauce S/N, 47011, Valladolid, Spain
e-mail: vhermano@eco.uva.es

N. Martín-Cruz
e-mail: ambiela@eco.uva.es

C. Hernández et al. (eds.), *Managing Complexity*, Lecture Notes in Management and Industrial Engineering, DOI 10.1007/978-3-319-04705-8_3,

This "projectification" process comes along with a paradigm shift in project definition. As Turner and Müller [23] claim, the traditional conceptualization of project as an endeavor or task is unable to fully gather all roles, functions and limits of projects, therefore, this definition is just not enough. Nowadays, projects are conceived as organizations' best drivers for change [21, 23]. Even more, some authors perceive projects as powerful strategic weapons initiated to create economic value and thus, being source of sustainable competitive advantages [13, 17]. Therefore, projects are conceived as central element for implementing companies' intended strategy [13], and are nowadays defined as agencies for change, and even, as temporary organizations [23, 24]. Therefore, a paradigm shift seems to be happening and projects are becoming vectors of organizations' strategy [21].

Although project management theory has been frequently recognized as predominantly implicit and practitioner-driving, in the last years, project management academics have increased their efforts in a new attempt to create a strong theoretical framework for project management discipline. Specifically, some authors claim that a cross-fertilization among strategy and project management could create great opportunities to enrich analysis and techniques in both disciplines [5, 13]. However, although project and strategic management have developed parallel, there is a gap between these disciplines both in theory and in practice [5, 13, 21, 23]. A review of strategic management literature shows that Project Management does not appear as a keyword for papers nor for conferences [21]. Project management as a research topic is confined to specific project management journals such as the PMI Project Management Journal, or the IPMA International Journal of Project Management.

This paper works, first, on the conceptualization of projects as temporary organizations, and the importance of PBOs as flexible organizational forms well prepared for competing in today's turbulent environment. Therefore, the objective of this paper is to close the gap between project and strategic management. Specifically, we claim that project management constitutes a dynamic capability which helps PBOs to adapt to rapidly changing environments. We seek to show the potential convergence between dynamic capabilities and project management by providing a framework for the consideration of project management as a dynamic capability both from a theoretical and a professional point of view. Thus, we shed light on the need to find the fundamentals that allow for an explanation of PBOs as a superior form of organization.

2 From Project Management to Project-Based Organizations

Traditionally, organizations formed temporal project teams inside their functional structure in order to deal with specially challenging operations (e.g. new products development; commissioning of a new procedure; etc.) or to respond to rapidly changing market conditions [9]. Projects were understood as temporal endeavours undertook to achieve a particular objective. Classic project management literature was only concerned for the critical success factors in project operations, while in a more professional side, the concerns focused on professionalizing project management techniques [12]. All these efforts were oriented to secure single project success. However, in a more and more complex and uncertain environment, undertaking successful projects has been recognized as necessary but no longer sufficient strategy to secure long term and sustainable performance [12].

Thus the focus has been shifted from individual projects to portfolios, and a new area of practice, known as Project Portfolio Management, has appeared [12]. The key point in project portfolio management is not to secure single project performance but to create structures able to optimize the use of resources among many projects and thus, allowing organizations to efficiently coordinate collections of projects [21]. Furthermore, nowadays project management literature has adopted an even more strategic perspective since the focus has moved from project portfolio management to the organizational level -PBOs [12]. Scholars claim that organizations carrying out their core operations mainly in project form -PBOs- could obtain better results in dynamic environments [12, 21].

Although profusely cited, the concept of PBO, also cited as project-oriented company, project-based firm, and p-form organization; is not without controversy. There is one literature stream emerging from Hobday's seminal paper [6] that identifies PBOs as those organizations executing projects in industries focused on complex industrial products and systems. These PBOs are intrinsically innovative as they continually reorganize their structure around the demands of each project and to changing client needs. Thus, these PBOs are very flexible and able to cope with project uncertainties and risks [6]. However, these PBOs suffer when company projects are not independent from each other. Specifically, these PBOs are weak in coordinating cross-project resources and promoting organizational knowledge and learning. Moreover, these PBOs could promote individual project objectives over the overall organization goals.

Scholars have recently suggested that only PBOs capable of developing project capabilities would we able to overcome the dichotomy among the individual nature of projects and the long-term and stable objectives of the overall organization's strategy [12]. Project capabilities are defined as the internal ability of a PBO to create lasting performance based on multiple short term projects [2]. Following Hobday [6], a PBO is one in which the project is the primary unit for production organization, innovation, and competition, and project management is the main source for capabilities and knowledge formation. For the purpose of this paper, we define PBOs through a translation of Hobday's [6] ideas into the dynamic capabilities approach. Unlike earlier approaches (e.g. Porter's competitive forces framework, or the resource-based view) which are static in nature, from the dynamic capabilities approach organizations obtain superior performance not just because of their assets endowment but through their ability for reconfiguring their competences to address rapidly changing environments—dynamic capabilities [20]. Drawing on the dynamic capabilities approach, we define PBOs as those where project capabilities shape not just project management processes but all internal and external competences of the organization. In a PBO, project (dynamic) capability represents the mechanism for integrating, building and reconfiguring organizational competences, so project (dynamic) capability constitutes the main source for a competitive advantage.

3 Project Management and Dynamic Capabilities: Implications from a Professional Perspective

The cross-fertilization between dynamic capabilities approach and project management can be extended to a professional perspective to give theoretical meaning to the guidelines offered by the different global project management standards such as the well-known PMBOK®. In the following, we interpret the PMBOK®'s processes and indications throughout the dynamic capabilities theoretical framework. We focus on the main components and routines of dynamic capabilities and the guidelines provided by the PMBOK® for dealing with them.

PMBOK® defines project management as the application of knowledge, skills, tools and techniques to project activities to meet the project requirements [16]. Furthermore, it recognizes that project management is accomplished through the appropriate application and integration of 42 processes. The observation of PMBOK®'s definition of project management as a collection of processes reminds to the definition of organizational capabilities given by Winter [26]—collection of routines that confers a set of decision options for producing significant outputs[1]. Although there are some authors claiming that the temporary nature of projects makes it difficult to develop routines so making also difficult to develop dynamic capabilities [6], recent research has shown that PBOs have viable alternatives both individual (e.g. managers of competencies) and collaborative (e.g. industry-wide social networks) by which they can create routines and distribute social learning [1]. Therefore, and based on the work of Winter [26] we consider organizational project capabilities as collection of routines that can be clarified through the study of the 42 processes (routines) contained in the PMBOK®.

Regarding the components of dynamic capabilities and for the purpose of this paper, we assume the framework established by Teece et al. [20] in their seminal paper where dynamic capabilities possess three different roles -coordination/integration, learning, and reconfiguration/transformation [20]. The first one, coordination/integration refers to the tasks managers perform for coordinating and integrating activities inside the firm and also for the coordination of external activities and technologies [20]. The PMBOK® possess a whole chapter (Chap. 4) dedicated to project integration management in which the integration role is defined as the processes (routines) and activities needed to identify, define, combine, unify and coordinate the various processes and project management activities within the Project Management Process Group [16]. Processes in that chapter detail how to develop the project management plan, how to direct and manage project execution or how to monitor and control project work. Regarding the second role, learning, is defined as a process (routine) by which repetition and experimentation enable tasks to be performed better and quicker [20]. By learning, firms recognize dysfunctional routines and prevent strategic blind spots. The learning role is a prevalent concept in

[1] Zollo and Winter [27] define routines as stable patterns of behavior that characterize organizational reactions to variegated, internal, or external stimuli (p. 340).

the PMBOK®. Practitioners claim that after developing a project, the implementing organization and actors must have learned something, and the PMBOK® includes that idea by including document lessons learned as a task to perform during project closing. Finally, the reconfiguration/transformation role refers to the need to reconfigure the organizational asset structure to address environmental changes [20]. The PMBOK® establishes the concepts of *rolling wave of planning* and *progressive elaboration* as two principles when developing project planning. Those two concepts refer to the process of making the project plan in successive waves as the project proceeds and later details become clearer. Moreover, PMBOK® suggests to create change requests as an output of almost all of its 42 processes. These change requests refer to modifications in project procedures, policies or documents that are requested by the project team due to issues found while project work is being performed. PMBOK® also advise project managers to perform the task called integrated changed control by which project managers have to review all change requests, and manage changes to deliverables, organizational process assets, and project documents and plans.

Teece et al. [20] recognizes that both the firm asset endowment—position—and the strategic decision history—path—affect firm's processes and so firm's dynamic capabilities. In this sense, the PMBOK® offers guidance for dealing with both position and path of the firm. In almost all of its 42 processes (routines), the PMBOK® include two inputs called enterprise environmental factors and organizational process assets which refer to firm's heritage and firm's asset endowment respectively.

Finally, an important aspect of dynamic capabilities is the commonalities/firm-specific duality. Dynamic capabilities are not exactly the same across firms even though they present common features and can be understood as best practices [3]. This duality is recognized by project management practitioners. On the one hand, the PMBOK® identifies the processes (routines) configuring the subset of the project management body of knowledge generally recognized as good practices. However, the PMBOK® itself notifies that the knowledge it describes, should not always be applied uniformly to all projects. Delving into this idea, the Project Management Institute has developed extensions of the PMBOK® like the construction extension of the PMBOK® or the government extension to the PMBOK® to provide more information on specific project types or industry information.

4 Conclusions, Implications, and Direction for Future Research

This paper highlights that the new conceptualization of both projects and project managers is imbued with strategic essence. Project management literature focus on achieving success for the lonely project is being substituted by papers claiming that projects should be understood in relation to previous projects, future plans and the environment surrounding the implementing organization [4].

Drawing on the dynamic capabilities approach we consider a PBO as an organization where project capabilities shape not just project management processes but all internal and external competences of the organization. By considering project management as a dynamic capability we offer an explanation of the better results PBOs are considered to obtain in dynamic and unstable environments [12, 21]. Thus, we show that dynamic capabilities and project management as well as project and top managers are much closer than what professionals of both disciplines believe. We claim that project management dynamic capability constitutes a source of competitive advantage for PBOs.

From the project management point of view, the cross-fertilization first provides a strong theoretical framework, the dynamic capabilities approach, for the project management discipline. A strong theoretical foundation is recognized as missing for the project management discipline [15, 23] and as one of the most important obstacles for the project management progress [11]. Thus, we strengthen the incipient project management theoretical framework with the explanatory power and theoretical foundations of the dynamic capabilities approach.

We also claim that dynamic capabilities foundations could be useful to face project management problems such as frequent project failures or slow rate of methodological renewal [10, 11]. Specifically, the issue of project failure has been studied through the traditional CSFs—typically attached to the iron triangle, and through single project case studies focused on technical aspects. However, recent studies have demanded to focus on managerial aspects instead of technical ones, since for many cases, the root of failure is constituted by issues such as the decision-making process [18].

Finally, this paper suggests wide-ranging opportunities for future research. First of all, taking the dynamic capabilities as research background scholars could research the conditions under which the project management processes become dynamic capabilities. Secondly, once recognized that project management could be understood as a dynamic capability, the next step is to look empirically at the benefits that project management dynamic capabilities provide to the organizations possessing and developing it and their role as source of competitive advantages. One possible stream of research, following the path initiated by Thomas and Mullaly [22], could focus on the measurement of the value created by project management application.

References

1. Cacciatori E (2008) Memory objects in project environments: storing, retrieving and adapting learning in project-based firms. Res Policy 37(9):1591–1601
2. Davies A, Brady T (2000) Organisational capabilities and learning in complex product systems: towards repeatable solutions. Res Policy 29(7–8):931–953
3. Eisenhardt KM, Martin JA (2000) Dynamic capabilities: what are they? Strateg Manage J 21(10–11):1105–1121
4. Engwall M (2003) No project is an island: linking projects to history and context. Res Policy 32(5):789–808

5. Grundy T (1998) Strategy implementation and project management. Int J Proj Manage 16(1):43–50
6. Hobday M (2000) The project-based organisation: an ideal form for managing complex products and systems? Res Policy 29(7–8):871–893
7. Hofmann J, Rollwagen I, Schneider S (2007) Deutschland 2020—new challenges for a land on expedition, current issues. Deutsche Bank Research, Frankfurt a. M.
8. Huemann M, Keegan A, Turner JR (2007) Human resource management in the project-oriented company: a review. Int J Proj Manage 25(3):315–323
9. Irja H (2006) Project management effectiveness in project-oriented business organizations. Int J Proj Manage 24(3):216–225
10. Kharbanda OP, Pinto JK (1996) What made Gertie Gallop? Lessons from project failures. Van Nostrand Reinhold, New York
11. Koskela L, Howell G (2002) The underlying theory of project management is obsolete. In: Conference proceedings of the 2002 PMI conference, Seatle
12. Melkonian T, Picq T (2011) Building project capabilities in PBOs: lessons from the french special forces. Int J Proj Manage 29(4):455–467
13. Meskendahl S (2010) The influence of business strategy on project portfolio management and its success—a conceptual framework. Int J Proj Manage 28(8):807–817
14. Nohria N (1996) From the M-Form to the N-Form:Taking Stock of Changes in the Large Industrial Corporation, London Business School Working Paper SLRP WP16/1996
15. Pollack J (2007) The changing paradigms of project management. Int J Proj Manage 25(3):266–274
16. Project Management Institute (2008) A guide to the project management body of knowledge, 4th edn. Project Management Institute, USA
17. Shenhar AJ, Dvir D, Levy O, Maltz AC (2001) Project success: a multidimensional strategic concept. Long Range Plann 34(6):699–725
18. Shepherd DA, Patzelt H, Wolfe M (2011) Moving forward from project failure: negative emotions, affective commitment, and learning from the experience. Acad Manage J 54(6):1229–1259
19. Teece DJ (2007) Explicating dynamic capabilities: the nature and microfoundations of (sustainable) enterprise performance. Strateg Manage J 28(13):1319–1350
20. Teece DJ, Pisano G, Shuen A (1997) Dynamic capabilities and strategic management. Strateg Manage J 18(7):509–533
21. Thiry M, Deguire M (2007) Recent developments in project-based organisations. Int J Proj Manage 25(7):649–658
22. Thomas J, Mullaly M (2008) Researching the value of project management. Project Management Institute. Newtown Square, PA
23. Turner JR, Müller R (2003) On the nature of the project as a temporary organization. Int J Proj Manage 21(1):1–8
24. van Donk DP, Molloy E (2008) From organising as projects to projects as organisations. Int J Proj Manage 26(2):129–137
25. Whittington R, Pettigrew A, Peck S, Fenton E, Conyon M (1999) Change and complementarities in the new competitive landscape: a European panel study, 1992–1996. Organ Sci 10(5):583–600
26. Winter SG (2003) Understanding dynamic capabilities. Strateg Manage J 24(10):991–995
27. Zollo M, Winter SG (2002) Deliberate learning and the evolution of dynamic capabilities. Organ sci 13:339–351

The Impact of Innovation Management Techniques on Radical Innovation: An Empirical Study

Juan Ignacio Igartua, Nekane Errasti and Jaione Ganzarain

Abstract While research in innovation management has provided many insights into specific aspects of innovation, the encompassing problems confronting general managers, especially managers of small and medium-size firms, have been overlooked in the development of innovation management techniques and tools. This paper analyses the way innovation management techniques (IMTs) influence innovation in firms. Specifically, this paper focuses on studying the role of IMTs in radical innovation. To this end, we propose a specific model of analysis, tested in a sample of more than 500 Spanish companies. Research results highlight that different sets of IMTs relate to radical and incremental innovation in different ways, and that therefore companies seeking radical innovation look for certain IMTs rather than others. This empirical study will help managers and practitioners to understand the role of IMTs in structuring radical innovation strategy, as well as researchers to focus on the role of such IMTs in innovation.

Keywords Innovation · Radical Innovation · Innovation Management Techniques · IMTs

1 Introduction

The need to understand innovation appears to be widespread, at business level. Some researchers have developed studies regarding the measurement of innovative performance in enterprises [18], using instruments such as the Community Innovation Survey instrument (CIS) trying to discover the factors that influence that result [3]. On the other hand, other scholars have investigated the role of innovation management and the analysis of its impact on innovation and innovation performance of firms [2, 24, 26], including the emphasis on the role of systems and tools [9].

Finally another incipient research approach has been orientated to analyse the role of techniques and tools for managing innovation [14].This approach highlights innovation as a fundamental process in organization performance [12, 25], a process that requires setting up a well-organized and well-run standardized set of tools [16].

J. I. Igartua (✉) · N. Errasti · J. Ganzarain
Departamento de Mecánica y Producción Industrial, Escuela Politécnica Superior, Mondragon Unibertsitatea, C/ Loramendi 4, 20500 Mondragon, Spain
e-mail: jigartua@mondragon.edu

C. Hernández et al. (eds.), *Managing Complexity*, Lecture Notes in Management and Industrial Engineering, DOI 10.1007/978-3-319-04705-8_4,

With this in mind, the aim of this paper focuses on understanding the influence of innovation management techniques (IMTs) on innovation in firms. Specifically, this paper focuses on studying the role of IMTs in radical innovation. Thus, the main purpose of the study reported in this paper is to understand how companies' implementation of IMTs, affects innovation (product, service, process, and other kinds of innovations) and the specific role of certain IMTs when looking for radical product or service innovation [19]. The objective of this paper is to understand whether IMTs play a significant role in innovation and the achievement of radical innovations.

After a brief introduction to innovation management techniques (IMTs), we develop the methodology used in this study. Subsequently we show some empirical results of the investigation, including a conclusions section.

2 Innovation Management Techniques (IMTs)

The need to manage the innovation process and context, demands that managers make effective and timely decisions based on multiple functions, inputs and disciplines [6]; and therefore, management tools and techniques are needed to support these complex decisions [22]. Brady et al. [5] define a management tool as "a document, framework, procedure, system or method that enables a company to achieve or clarify an objective" (p. 418).

Innovation management techniques (IMTs) can be defined as the range of tools, techniques and methodologies intended to support the process of innovation and help companies to meet new market challenges in a systematic way [22]. Chiesa and Masella [9], in their audit model of the process of technological innovation, identified the effective use of appropriate tools and systems as one of three facilitators of innovation processes, in conjunction with the deployment of human and financial resources and the leadership and direction of senior management.

An investigation conducted in Europe [11] affirmed that IMTs allow a company to combine technology and business strategy, fostering increased employee participation, and concluded that there is insufficient awareness of the variety and range of IMTs available, as well as the potential benefits of their use.

More recently, Hidalgo and Albors [14] argue that IMTs are critical to increasing competitiveness, showing that proper application of IMTs facilitates a company's ability to introduce appropriate new technologies in products or processes, as well as the necessary changes to the organization.

As regards the existing IMTs, several authors such as Phaal et al. [22] have worked towards the development of a catalogue of tools, as well as a series of research programs. In this direction, some works have tried to summarize the existing set of techniques [14], an approach followed by other researchers that focused on the role of certain tools [1, 10, 16,15, 17, 21, 27], or empirical studies [13, 4, 8].

Thus, the present research focuses on the role of 17 groups of IMTs identified in the literature (1. creative development, 2. technology management, 3. strategic management, 4. people management, 5. business intelligence, 6. management innovative project, 7. development of new products-services, 8. techniques and practices for collaboration and networking, 9. design management, 10. knowledge management, 11. new business development, 12. financial resource management, 13. industrial property rights management, 14. production management, 15. marketing, 16. organizational practices, and 17. process improvement) [14, 23].

3 Research Methodology

The research was conducted through a survey targeted at business managers, similar to other research studies conducted in the field of innovation [20]. The research is based on a survey focused on innovation management where top managers, from more than five hundred companies over a defined universe of six thousand companies, were asked to answer a structured questionnaire from December 2008 till April 2009.

The instruments developed for the measurement of innovation (product-service and process) were based on variables used in the literature and on the Community Innovation Survey [7]. On the other hand, the measurement of the implementation of IMTs in companies was developed through a scale of 53 items from the identification of the aforementioned 17 IMT groups identified in the literature (Cronbach's alpha for this scale was 0.948).

The gathered data has been analysed using SPSS16 and statistical methods such as T Student Test, ANOVA and regression analysis (simple linear and multiple linear). Due to the fact that the sample meets the sampling criteria needed to ensure its representativeness, the implications of the study are directly extrapolated to the entire study population.

4 Results

In order to examine whether there are significant differences between the implementation of IMTs in companies and their innovation results, a Student's t-test comparison of two means was developed. The results of this test are summarized in Table 1.

Consequently, in all cases the t-statistic takes a critical level of bilateral significance lower than the critical value of 0.005 thereby rejecting the null hypothesis of equality of means, and therefore concluding that the use of IMTs in companies that innovate is higher compared to those companies that do not.

It was also considered that, when analyzing the use of IMTs by companies, these could be classified into four groups, depending on their product-service and process innovations results (Table 2).

Table 1 IMTs use related to Innovation (Product, service, process and other innovations)

		N	Mean	Std. D.	Std. Err. mean			*N*	Mean	Std. D.	Std. Err. mean
Prod	Yes	362	2.4700	0.67465	0.03546	Proc	Yes	378	2.4235	0.71159	0.03660
Inno	No	186	1.7004	0.63001	0.04619	Inno	No	176	1.7732	0.67540	0.05091
Serv	Yes	327	2.4267	0.68541	0.03790	Other	Yes	409	2.4319	0.69673	0.03445
Inno	No	221	1.8741	0.73550	0.04948	Inno	No	148	1.6291	0.62068	0.05102

Table 2 Companies classification based on innovation

Prod-Serv Inno	Process Inno	Group	*N*
YES	NO	1	85
YES	YES	2	332
NO	YES	3	44
NO	NO	4	93

Thus, when analyzing the use of IMTs on these four groups the statistical results based on ANOVA analysis (with a significance of 0.000 lower than the critical value of 0.005), show that for all the IMT groups (except the group of techniques related to production management: Just-in-time, ERP, Lean Management), the mean use of IMTs is higher for those companies that innovate in product-service and process altogether, than for those that only innovate in product, process or do not innovate at all (Fig. 1).

Finally, when analyzing the use of IMTs related to the innovation radicalness, two simple linear regression studies were developed, one for the radical innovation of product-services and the other for the incremental one (see Table 3). The models take a very high R (0.596) for radical innovation and an also high R (0.641) for incremental innovation; indicating that 35.5 % of the variability of performance in radical innovation of product-services depends on the use of IMTs, as opposed to 41.1 % in the case of incremental innovation. In addition, the F statistic shows for both regressions a value below the critical level (Sig 0.05), so it can be argued that variables are linearly related.

Besides, and in order to identify the IMT groups that are most related to the radical innovation of product-services and to the incremental one, two multiple linear regression analyses were carried out introducing variables step by step until the models were validated after five steps and six steps respectively (see Table 4 and Table 5).

So, the model of radical innovation was tested in five steps, after which the proposed model included a constant, and the variables referring the use of IMTs related to networking, economic and financial aspects, creativity techniques, techniques related to industrial property management, as well as those related to business intelligence and technological foresight.

In contrast, the model of incremental innovation was tested in six steps, after which the proposed model included a constant, and the variables referring the use

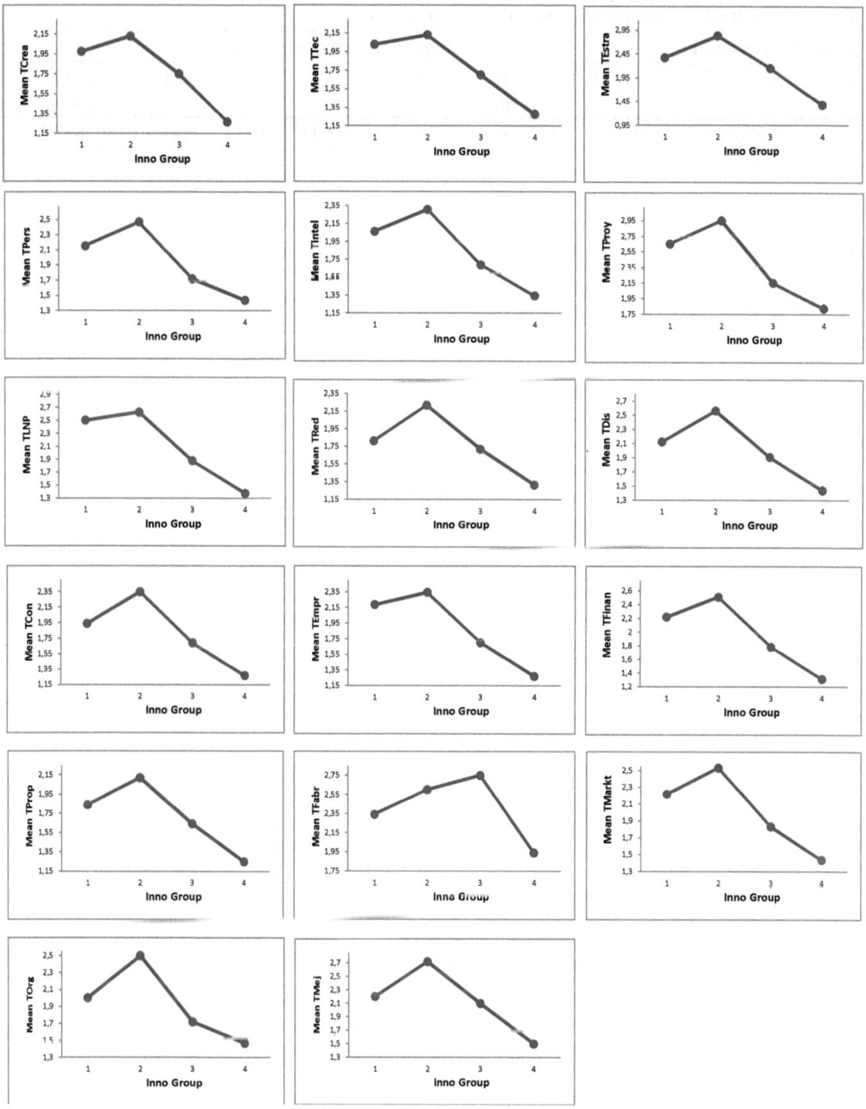

Fig. 1 Mean use of IMTs by companies' innovation activity classification

of IMTs related to new business development (business plan, transfer mechanisms, and spin-offs), new product development techniques, technology management, project management, industrial property management, and those concerning the management of networking activities.

Table 3 Companies classification based on innovation

Radical innovation					Incremental innovation				
Model	R	R^2	Adj. R^2	Std. Err. of estimate	Model	R	R^2	Adj. R^2	Std. Err of estimate
1	0.596[a]	0.355	0.354	0.77389	1	0.641[a]	0.411	0.410	0.78352

[a] Predictors: (Constant), IMTs

Table 4 Multiple linear regression (step-by-step) relating IMTs and Radical Innovation

Radical innovation: model summary[b]									
Model	R	R^2	Adj. R^2	Std.Err. of estimate	Change statistics				
					R Square change	F change	df1	df2	Sig. F. change
5	0.606[a]	0.368	0.361	0w.75135	0.008	6.581	1	508	0.011

[a] Predictors: (Constant), TRed, TFinan, TCrea, TProp, TIntel
[b] Dependent variable: InnoRadical

Table 5 Multiple linear regression (step-by-step) relating IMTs and Incremental Innovation

Incremental innovation: model summary[f]									
Model	R	R^2	Adj. R^2	Std.Err. of estimate	Change statistics				
					R Square change	F change	df1	df2	Sig. F. change
6	0.647[f]	0.418	0.411	0.77402	0.008	7.226	1	504	0.007

[f] Predictors: (Constant), TEmpr, TLNP, TTec, TProy, TProp, TRed
[g] Dependent variable: InnoIncremental

5 Discussion and Conclusions

The main purpose of the article was to identify the link between business innovation and its radicalness and innovation management techniques (IMTs) implemented by companies. The IMTs measurement was based on a set of 17 groups of techniques taken from the literature, while the innovation measurement was based on pre-existing instruments.

Thus, based on the extended set of data and using statistical methods (Student's t-test, linear regression and multiple linear regressions), the research has underlined the importance of IMTs and their differential role in the achievement of different kinds of innovations (product-service and process).

When analyzing innovation in companies, results indicate that the variability of performance in innovation depends on the implementation of IMTs, which underlines the importance of management techniques, coinciding with previous researchers [4, 8, 10, 13, 14, 16, 22, 23] .

Furthermore, the multiple linear regression analysis carried out also stress the role of certain IMTs for the development of radical versus incremental innovations. Thus, networking (open innovation and collaboration), as well as financing, creativity techniques, IPR management and business intelligence (technology watch) seem to be techniques that are important for the development of radical innovations, while new business development techniques, new product development techniques, technology management and project management seem to influence incremental innovation of product or services.

The limitations of this paper result from the research model and the variables used. Further research and analysis would provide more detailed relationships. On the other hand, the contributions of this study must be interpreted with a degree of caution since it has focused on a regional context, which may have certain characteristics that can affect the findings.

References

1. Aagaard A (2012) The contribution of innovation strategy development and implementation in active facilitation of pharmaceutical front end innovation. Sys Pract Act Res 25:457–477
2. Adams R, Bessant J, Phelps R (2006) Innovation management measurement: a review. Int J Manage Rev 8:21–47
3. Arundel A, Hollanders H (2006) Searching the forest for the trees: "Missing" indicators of innovation. 2006 Trend Chart Methodology Report. MERIT–Maastricht Economic Research Institute on Innovation and Technology
4. Barge-Gil A, Nieto MJ, Santamaria L (2011) Hidden innovators: the role of non-RD activities. Technol Anal Strate Manage 23:415–432
5. Brady T, Rush H, Hobday M, Davies A, Probert D, Banerjee S (1997) Tools for technology management: an academic perspective. Technovation 17:417–426
6. Brown D (1997) Innovation management tools: a review of selected methodologies, Luxembourg, European Commission: Directorate-General XIII Telecommunications, Information Market and Exploitation of Research
7. CORDIS (2008) Community innovation survey [Online]. http://cordis.europa.eu/innovation-smes/src/cis.htm. Accessed Aug 2013
8. Chai SNC, Sun HY, Lau A KW (2010) The impact of innovation management techniques on product innovation performance: an empirical study. ICMIT, Singapore. pp 432–437
9. Chiesa V, Masella C (1996) Searching for an effective measure of R & D performance. Manage Decis 34:49–57
10. D'alvano L, Hidalgo A (2012) Innovation management techniques and development degree of innovation process in service organizations. R&D Manage 42:60–70
11. European Commission (2000) Promoting innovation management techniques in Europe
12. Galanakis K (2006) Innovation process. Make sense using systems thinking. Technovation 26:1222–1232
13. Graner M, Mißler-Behr M (2012) The use of methods in new product development-a review of empirical literature. Int J Prod Dev 16:158–184
14. Hidalgo A, Albors J (2008) Innovation management techniques and tools: a review from theory and practice. R&D Manage 38:113–127
15. Huang F, Rice J (2012) Openness in product and process innovation. Int J Innov Manage 16(4):1250020-1-1250020-24
16. Igartua JI, Garrigós JA, Hervas-Oliver JL (2010) How innovation management techniques support an open innovation strategy. Res Technol Manage 53:41–52
17. Lichtenthaler U (2011) Implementation Steps For Successful Out-Licensing. Res Technol Manage 54:47–53
18. Mancebo Fernández NR, Valls Pasola J (2005) El comportamiento innovador de la empresa industrial. Un modelo de análisis a partir de la encuesta del INE

19. Mcdermott CM, O'connor GC (2002) Managing radical innovation: an overview of emergent strategy issues. J Prod Innov Manage 19:424–438
20. O'regan N, Ghobadian A, Sims M (2006) Fast tracking innovation in manufacturing SMEs. Technovation 26:251–261
21. Padilla-Meléndez A, Garrido-Moreno A (2012) Open innovation in universities: what motivates researchers to engage in knowledge transfer exchanges? Int J Entrep Behav Res 18:417–439
22. Phaal R, Farrukh C, Probert D (2006a) Technology management tools: generalization, integration and configuration. Int J Innov Technol Manage 3:321–339
23. Phaal R, Farrukh CJP, Probert DR (2006b) Technology management tools: concept, development and application. Technovation 26:336–344
24. Prajogo DI, Ahmed PK (2006) Relationships between innovation stimulus, innovation capacity, and innovation performance. R&D Manage 36:499–515
25. Raymond L, St-Pierre J (2010) R & D as a determinant of innovation in manufacturing SMEs: an attempt at empirical clarification. Technovation 30:48–56
26. Rigby D, Bilodeau B (2007) Management tools and trends 2007. Boston, MA. Bain Co 35(5):9–16
27. Tipu SAA (2012) Open innovation process in developing-country manufacturing organisations: extending the Stage-Gate model. Int J Bus Innov Res 6:355–378

The Logical Framework Approach and Worker Commitment

Carmen Jaca, Luis Paipa-Galeano, Elisabeth Viles, Ricardo Mateo and Javier Santos

Abstract The main objective of this paper is to present the logical framework approach (LFA) as a tool that prepares companies and workers to implement continuous improvement programs. This methodology encourages worker participation in different steps in order to reach consensus in the organization. In addition, the application of LFA improves the capabilities of workers in areas such as participatory analysis, problem analysis and objectives analysis. These capabilities are necessary in any continuous improvement program. The paper also presents the results of applying LFA in two different companies.

Keywords Participation · LFA · Improvement · Organizational behaviour

1 Introduction

Many companies believe that continuous improvement programs will enable them to survive in today's climate by improving their performance and results (Prado-Prado 2009). Most current systems are based on the principles of the Toyota

C. Jaca (✉) · E. Viles · J. Santos
Department of Industrial Management, School of Industrial Engineers, TECNUN, Universidad de Navarra, Paseo Manuel Lardizábal 13, 20018, San Sebastian, Spain
e-mail: cjaca@tecnun.es

E. Viles
e-mail: eviles@tecnun.es

J. Santos
e-mail: jsantos@tecnun.es

L. Paipa-Galeano
Universidad de La Sabana, Campus del Puente del Común, Km.7, Autopista Norte de Bogotá, Chía, Cundinamarca, Colombia
e-mail: luis.paipa@unisabana.edu.co

R. Mateo
School of Economics and Business Administration, University of Navarra, Pamplona, Spain
e-mail: rmateo@unav.es

C. Hernández et al. (eds.), *Managing Complexity*, Lecture Notes in Management and Industrial Engineering, DOI 10.1007/978-3-319-04705-8_5,

Production System, where one of the core principles is to facilitate the participation of workers and promote their autonomy [6, 11, 17].

In recent years, many authors have suggested that the key to the different methodologies and tools originated in Japan is that they are based on the participation and commitment of employees through training and behavioural change [4, 22]. However, it was found that the application of improvement tools in the West focuses on operative aspects, leaving aside the aspects that have to do with change management and people [7, 10, 12, 15, 18]. Organizations commonly implement continuous improvement systems from top to bottom, where worker motivation and participation is usually considered to result from those improvement programs. However, it is necessary to have a strategy to support behavioural change in people [14].

In fact, getting employees to be involved in and committed to Western companies is an unresolved matter. According to the prestigious consulting firm Gallup, about 70 % of U.S. workers are not committed to their work or are "actively disengaged" from their work, meaning they are emotionally disconnected from their workplaces and are less likely to be productive [9]. As a result, most of the workers' potential is wasted and with it, the opportunity to improve business results. These are the core principles of the continuous improvement methodologies: the autonomy of individuals and their participation in improvement through their own opinions and ideas.

The objective of this paper is to present a tool used in donor organizations as a way to increase the involvement and motivation of employees in continuous improvement processes. Using this tool, which is called the logical framework approach, the organization focuses first on the analysis of environmental comfort and the development of the habits of the participants, and then improves processes and operations.

2 Environmental Comfort as Motivational Factor

Over the past 35 years there has been a profusion of theories linking the workplace with levels of job satisfaction and worker motivation and stress [2, 24]. Several studies have shown that aspects such as spatial organization, architectural details and environmental conditions (order, cleanliness, ambient conditions and resources, spatial organization, architectonic details, and view or visual access from the work area), are associated with motivation, stress, performance and even social interaction at work [3, 8, 16, 23].

As a result, some authors have coined the term environmental comfort, which links the psychological aspects of workers' environmental likes and dislikes with concrete outcome measures, such as improved task performance, and with organizational productivity through workspace support for work-related tasks. Environmental comfort includes three categories: physical, functional and psychological comfort. Together these categories make the work environment stimulate workers so they will perform better as they carry out their tasks [24].

In recent decades, several studies have attempted to explain the factors that affect the success or failure of organizational change. Kristin Piderit [20], for example, proposes a new way of understanding employee response to changes. The author maintains the idea that any change process needs both top-down and bottom-up work. Meanwhile, Hodgson [13] proposes that the development of an organization or a change in its strategy involves, even partially, the development of habits that are agreed upon by employees. The same author also stresses that the psychological mechanism of forming habits is something much more specific than what is commonly denoted as "organizational culture". Thus, he suggests the importance of focusing on processes of habit development as a way to address organizational changes more successfully.

In order to establish a starting point for sustainable continuous improvements, it is necessary to first focus on identifying the needs and interests of individuals with regard to the workplace (environmental comfort) and then focus on improving processes and operations. To facilitate this process, we present a methodology, the logical framework approach, which is useful for promoting logical thinking and checking internal logic. The method also encourages people to consider what their expectations are, while also improving communication between people who are involved in the change [1].

3 The Logical Framework Approach Methodology

The logical framework approach (LFA) is a widespread methodology, particularly in donor-assisted projects in developing countries [1, 5]. This methodology proposes a procedure for the planning of a development project. The different steps used by the LFA promote the participation of the different parties involved, based on the identification of problems in order to reach the proposed solutions. Thus, the methodology encourages participants to identify different visions regarding their particular interests, directives and resources that can be in favour of or against a proposal for a solution. Employees are then able to consider at the same time how the problems are perceived and the expected solution or results. It is therefore ideal to create a shared vision from all stakeholders via consensus building [19].

This methodology has been adapted in order to encourage participation and consensus, as mentioned above. For this purpose, the steps have been tailored to focus on identifying and solving problems and concerns related to environmental comfort. Basically the adapted methodology applies steps adapted from LFA, following the sequence suggested by the methodology, which are the following (see Fig. 1):

- Participatory analysis: Environmental comfort is analysed by the different people involved (workers, staff, supervisors, mid-level managers and directors, among others), according to Vischer's model.
- Problem analysis: This consists of identifying the cause-effect relationships between the major problems found in the participation analysis step, thorough the use of a 'problem tree'. The objective of this phase is to be able to reach the root of the problems.

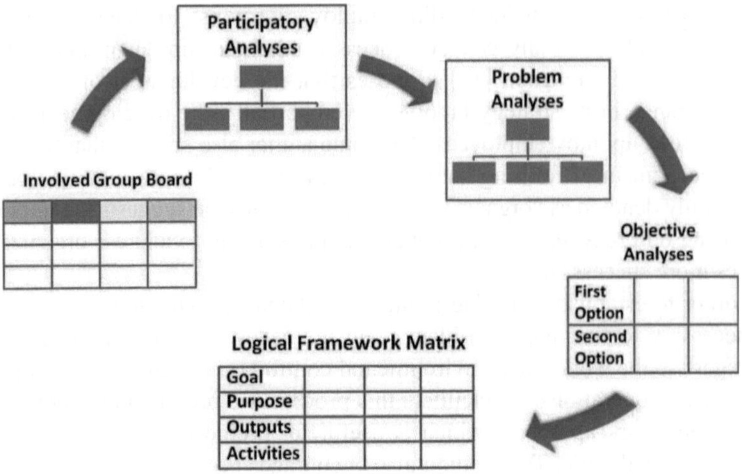

Fig. 1 The logical framework approach methodology. (Adapted from [19])

- Objectives analysis: In the objectives analysis the problem tree is transformed into a tree of objectives (future solutions of the problems) and analysed. So, the group is able to identify the future desired situations in each of the comfort categories and identify the means-ends relationships that allow the desired situation to be reached.
- Alternatives analysis: The purpose is to identify possible alternative options, assess their feasibility and agree upon the future desired solution. This is an exercise in creativity and idea generation.
- Project planning: This consists of preparing and presenting the activities, resources and costs necessary to develop the proposal(s) for solving the problems. All the information is condensed into a single Logical Framework Matrix or LFM

4 Application of the Logical Framework Approach in two Different Companies

The LFA was applied as part of an improvement program in two different companies as a first step toward making people aware of order and cleanliness in their workplaces. The characteristics of both companies are shown in Table 1 below.

The application of LFA allowed us to analyse situations of discomfort that were not critical but that could be demotivating factors for employees. Such circumstances can act as a mental barrier for workers, creating a negative attitude towards organizational changes, sometimes even unconsciously. The LFA methodology brought to light these discomfort issues, and thus the employees were heard and valued by their managers. In turn, the workers' attitude relaxed and they displayed a more

Table 1 Characteristics of the companies

Company	Activity	Location	Turnover	N° employees	Participants
A	Manufacturing	Spain	15M €	120	24
B	Food and agriculture	Colombia	10M €	70	70

Table 2 Example of improvements resulting from the application of LFA in Company A

Area	Improvements proposed by employees
Maintenance	Improvements related to cleaning: Improved coolant leaks, cleaning program establishment and renovation of certain elements Repairs proposed: cranes, damaged glass in machines, heating system in plant Review of preventive maintenance guidelines: cranes, breakdowns reports
Machines	Improvements related to order: Painting of floor space, relocation of current materials, establishment of loading and unloading area Ergonomic study
Health and Safety	Management of safety: Safety audits, improve the use of PPE, create incidents collection system, reactivate the committee to study incidents and accidents, be educated about occupational risks Elements of prevention: Install emergency lighting, create per position plan, close engine cowling
Training	Create a competency management system, create a system of incentives for improvement proposals, create interdisciplinary teams for improvement
Communication	Create shift sheet, train people in assertiveness and social skills, implement boards with production planning
Housekeeping	Put spacers in lockers, perform housekeeping study in shifts, distribute lockers

positive attitude toward listening. In both situations, the application of LFA led to improvements related to environmental comfort, as evidenced in Table 2 and 3, which includes some of the proposed improvements from the implementation of the Logical Framework Approach. The methodology not only encouraged employees to analyse the improvement opportunities, but also to propose and participate in the implementation of those improvements.

As the above tables show, the proposed improvements will not only impact matters directly related to the environmental comfort of workers. This methodology has led to proposals and actions related to the productivity and performance in both companies.

Table 3 Example of improvements resulting from the application of LFA in Company B

Area	Improvements proposed by employees
Logistics	Pre-order programming, programme conveyors at different times, systematic order platform
	Google map for location of farms
Processes	Create and communicate manual for office functions, worker training by function, define organizational chart
	Define processes and participants with systematic template, modify maintenance emergency procedures
Human Resources	Professional Development Plan, worker performance evaluation plan
	Open lines of credit supporting vehicle replacement, promote social activities
	Social area for visitors
Corporate image	Promote the dissemination of corporate image with employees via incentives Establish routine exterior maintenance
Maintenance	Improved parts store, control maintenance plans, maintenance training
Health and Safety	Periodic training in risk prevention

5 Conclusions

The application of logical framework approach combined with Vischer's comfort model is designed to promote the participation of people by attending first to their needs and then to the overall improvement of the process. We argue that the employee's attitude is influenced by his perception of his work, and this perception acts as an input for personal participation and motivation. The proposed methodology is based on organizational change management, which is founded on continuous improvement tools, and organizational behaviour principles, which is proposed as a way of addressing improvement in an organization. Once the employees have had the opportunity to analyse and improve the aspects related to their comfort, the program focuses on improving processes and operations.

The following benefits are expected from applying LFA:

a. Putting the organization on the path of continuous improvement
b. Increased employee morale by improving environmental comfort in its three dimensions: physical, functional and psychological.
c. Reduction of absenteeism and staff turnover by improving environmental comfort in all three dimensions.
d. Increased involvement in the organization by engaging employees in the identification and resolution of problems from building consensus and a shared vision.
e. An increase in productivity through the development of training programs oriented toward changing people's behaviour.

The application of this methodology in both companies shows that LFA is a useful tool for fostering the commitment of workers by aligning their needs and complaints with those of the company where they work.

References

1. Aune JB (2000) Logical framework approach and PRA—mutually exclusive or complementary tools for project planning? Dev Pract 10(5):687–690
2. Björklund C (2001) Work motivation—studies of its determinants and outcomes. Stockholm: Stockholm School of Economics, EFI, the Economic Research Institute
3. Burke RJ (1988) Sources of managerial and professional stress in large organisations. In: Cooper CL, Payne R (eds) Causes, coping and consequences of stress at work. Wiley, Chichester
4. Dahlgaard-Park SM (2011) The quality movement: where are you going? Total Qual Manage 22(5):493–516
5. Dale R (2003) The logical framework: an easy escape, a straitjacket, or a useful planning tool? Dev Pract 13(1):57–70
6. Delbridge R, Lowe J, Oliver N (2000) Shopfloor responsibilities under lean teamworking. Hum Relat 53(11):1459–1479
7. Douglas A (2002) Improving manufacturing performance, Quality Congress. ASQ's Annual Quality Congress Proceedings, (56): 725–32.
8. Fairbrother K, Warn J (2003) Workplace dimensions, stress and job satisfaction. J Manage Psychol 18(1):8–21
9. Gallup (2011) Majority of American workers not engaged in their jobs. http://www.gallup.com/poll/150383/majority-american-workers-not-engaged-jobs.aspx%60. Accessed 12 Dec. 2012
10. Gapp R, Fisher R, Kobayashi K (2008) Implementing 5S within a Japanese context: an integrated management system. Manage Decis 46(4):565–579
11. Hirano H (1989) JIT factory revolution. Productivity Press, Cambridge.
12. Ho SKM (1998) 5-S practice: a new tool for industrial management. Industr Manage Data Syst 98(2):55–62
13. Hodgson GM (2007) Institutions and individuals: interaction and evolution. Organ Stud 28(1):95–116
14. Jaca C, Santos J, Errasti A, Viles E (2012) Lean thinking with improvement teams in retail distribution. Total Qual Manage Bus Excell 23(4):449–465
15. Kobayashi K, Fisher R, Gapp R (2008) Business improvement strategy or useful tool? Analysis of the application of the 5S concept in Japan, the UK and the US. Total Qual Manage Bus Excell 19(3):245–262
16. Leong CS, Furnham A (1996) The moderating effect of organizational commitment on the occupational stress outcome relationship. Hum Relat 49(10):1345–1363. (1996), 49(10):1345–1363
17. Liker JK (2004) The Toyota way: 14 management principles from the world's greatest manufacturer. McGraw Hill, New York
18. Magaña-Campos J, Aspinwall E (2003) Comparative study of Western and Japanese improvement systems. Total Quality Management and Business Excellence, 14(4), 423–436.
19. NORAD (1999) The Logical Framework Approach (LFA): handbook for objectives-oriented planning: NORAD. Norwegian Agency for Development Cooperation
20. Piderit SK (2000) Rethinking resistance and recognizing ambivalence: a multidimensional view of attitudes toward an organizational change. Acad Manage Rev 25(4):783–794
21. Prado-Prado JC (2009) Continuous improvement in the supply chain. Total Qual Manage Bus Excell 20(3):301–309
22. Suárez-Barraza MF, Ramis-Pujol J, Heras MA (2010) Reflecting upon management systems: content analysis and synthesis. 1(2):64–86
23. Sullivan SE, Bhagat RS (1992) Organizational stress, job-satisfaction and job-performance—where do we go from here? J Manage 18(2):353–374
24. Vischer JC (2007) The effects of the physical environment on job performance: towards a theoretical model of workspace stress. Stress Health 23(3):175–184

New Technologies and Entrepreneurial Intention

Natalia Martin-Cruz and Ana Isabel Rodriguez-Escudero

Abstract New technologies are powerful tools to create, disseminate, articulate, and exploit knowledge. Entrepreneurs use these technologies to promote the creation of new ventures. However, recent studies demonstrate that new technologies are not sufficient to enhance the process of venture creation. We use the fundamentals of the theory of planned behavior to understand the impact of new technologies on entrepreneurial intention. Empirical literature related to university students shows that entrepreneurial intention is dependent on attitudes toward entrepreneurship, social norms, and self-efficacy. We therefore evaluate an empirical model in a sample of students enrolled in the 2012–2013 academic year in the University of Valladolid (Spain).

Keywords Entrepreneurship · Entrepreneurial intentions · New technologies · Theory of planned behavior

1 Introduction

New technologies are powerful tools to create, disseminate, articulate, and exploit knowledge. Recently, entrepreneurs are using these technologies to promote the creation of new ventures [8]. However, our understanding of the effect of new technologies on entrepreneurial intention is still lacking. We use the fundamentals of the theory of planned behavior to examine the impact of new technologies on entrepreneurial intention. In particular, we focus on two specific objectives: (i) verify the potential of the planned behavior theory based on the three dimensions—attitude toward entrepreneurship, social norms, and self-efficacy—and some control variables, and (ii) verify the impact of a positive attitude toward new technologies on entrepreneurial intention.

N. Martin-Cruz (✉)
INSISOC Research Group, Department of Business Management and Marketing,
Business School of Universidad de Valladolid, Pso del Cauce S/N, 47011, Valladolid, Spain
e-mail: ambiela@eco.uva.es

A. I. Rodriguez-Escudero
Facultad de Ciencias Económicas y Empresariales, Dpto. de Organización de Empresas y CIM,
Universidad de Valladolid, Pso del Cauce S/N, 47011, Valladolid, Spain
e-mail: ana@eco.uva.es

C. Hernández et al. (eds.), *Managing Complexity*, Lecture Notes in Management and Industrial Engineering, DOI 10.1007/978-3-319-04705-8_6,

To reach those objectives, we use two models. The first model includes only the entrepreneur's attitude toward new technologies. In the second model, we add as the upper echelon characteristics the entrepreneur's personal features including education, personality, and family business background [4, 6, 13, 16, 23]. Once we control all those variables, we find that attitude toward new technologies maintains its significance and thus this variable is important for identifying individuals with a higher entrepreneurial intention. In its practical application, this observation allows specific programs to be targeted to those individuals from public and private institutions.

2 Theoretical Background

The theory of planned behavior [1] has been applied to nearly all voluntary behaviors, and it provides quite good results in very diverse fields [2, 15]. Accordingly, a narrow relation exists between the intention to be an entrepreneur and the effective outcome. In other words, intention is the fundamental element in explaining behavior.

Following this theory, three main elements constitute the explanatory variables of intention toward entrepreneurship: self-efficacy, social norms, and attitude toward entrepreneurship. Self-efficacy, which can also be described as entrepreneurial capabilities, is defined as the perception of the ease or difficulty in the fulfilment of the behavior of interest, namely, the individual's sense of capacity regarding the fulfilment of firm creations behaviors. Social norms measure the social value attributed to entrepreneurial behavior. Finally, attitude toward entrepreneurship refers to the degree to which the individual holds a positive or negative personal valuation about being an entrepreneur.

In addition to these three variables, the literature has identified additional features that may also explain entrepreneurial intention. Prior studies widely consider demographic characteristics (gender, social class), education (training, languages, academic performance), the existence of an entrepreneurial tradition in the family, and personal traits [11, 17, 18, 20]. Although planned behavior theory plays a prominent role in recent entrepreneurship research, other theoretical models have also been used to characterize and explain why some individuals become entrepreneurs. The great person theory focuses on individual intuition and the entrepreneur's unique values and attitudes, such as the need for self-fulfilment. The psychological literature argues that entrepreneurs have a higher propensity for risk-taking, and the management literature emphasizes entrepreneurs' capabilities for innovation, organizing resources, and leadership [5].

Given this discussion, we include in our empirical analysis variables that allow us to verify which variables have a stronger influence on the intention to start-up a business. Fig. 1 shows the proposed model.

We include new technologies in the model by considering the attitude of the individual toward those new technologies. In fact, institutional theory acknowledges

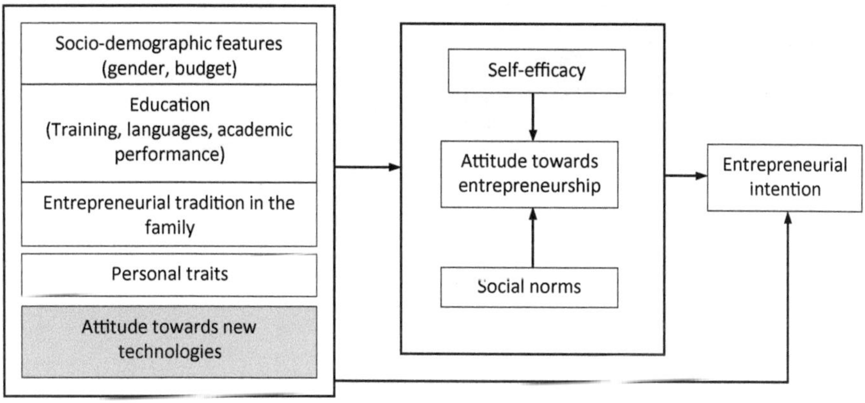

Fig. 1 Model of the planned behavior theory considering the impact of new technologies on entrepreneurial intention

that certain new technologies come to be adopted widely, whereas other, equally plausible, alternative technologies languish [21, 22]. Munir and Phillips [19] attempt to disentangle the role of entrepreneurial institutions regarding the choice of technology. Using observations from discourse analysis, they find that technologies can become a type of institution through processes of social construction. They provide a very useful foundation for the development of an institutional theory of technology that strengthens both technology and innovation research and institutional theory [19]. In addition, the accumulated tacit knowledge and culture of the entrepreneur are essential resources for the creation of wealth from research commercialization leading to technological innovation [9]. We posit that the development of a positive attitude toward new technologies is the starting point to initiate the entrepreneurial career as means to develop new technologies.

3 Method

We evaluate the model presented in Fig. 1 in a sample of students enrolled in the 2012–2013 academic year in the University of Valladolid (Spain). We collected information by means of a questionnaire during the period of February–March 2013. A total of 183 complete questionnaires were obtained. We focus on college students because this group contains the largest proportion of entrepreneurs in the European Union [3]. In fact, the European Union is particularly interested in promoting entrepreneurship and policies as evidenced by the discussion of the Programme for the Competitiveness of Enterprises and SMEs (COSME) 2014–2020.

We use the Entrepreneurial Intention Questionnaire (EIQ) designed by [14] to measure the variables. Whenever possible, items are built as 7-point Likert-type scales. The exceptions to this norm are gender, social class, and entrepreneurial traditional in the family, which are dummy variables. The questionnaire is available from the authors on request.

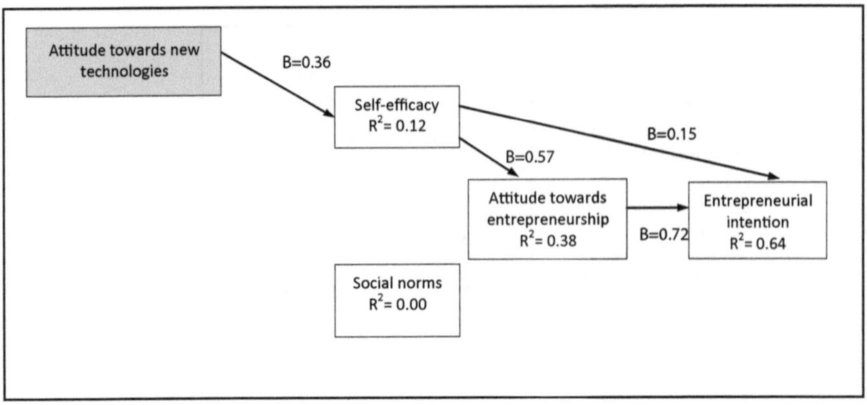

Fig. 2 Model of the planned behavior theory considering the impact of new technologies on entrepreneurial intention

We use a structural equation modeling analysis to test the research hypothesis. Between the two alternative structural equation modeling approaches, we select partial least squares mainly because our variables are not normally distributed and because of the formative nature of some of the measures used in this research. Our model includes both latent (measured with reflective indicators) and emergent (formative) constructs.

4 Results

Figures 2 and 3 summarize the results of the partial least squares analysis performed to test two structural models.

In particular, Fig. 2 shows the results of a model that includes favorable attitude toward new technologies. Figure 3 presents the same model but includes the additional antecedents emphasized by the literature. Specifically, the figures show the standardized path coefficients (B) with the values of the R^2s of the dependent variables. Because traditional parametric tests are inappropriate when no assumption is made about the distribution of the observed variables, we determine the level of statistical significance of the coefficients of both the measurement and the structural models through a bootstrap resampling procedure (500 subsamples were randomly generated).

The estimation of the first model (Fig. 2) shows that a positive attitude toward new technologies favors the feeling of self-efficacy (B=0.36). Those last variables has an effect on the attitude toward entrepreneurship (B=0.57) and on the entrepreneurial intention (B=0.15).

The first model (Fig. 2) does not include the variables that the literature considers to have an effect on entrepreneurial intention. However, when we include those variables in the model (Fig. 3), the positive attitude toward new technologies re-

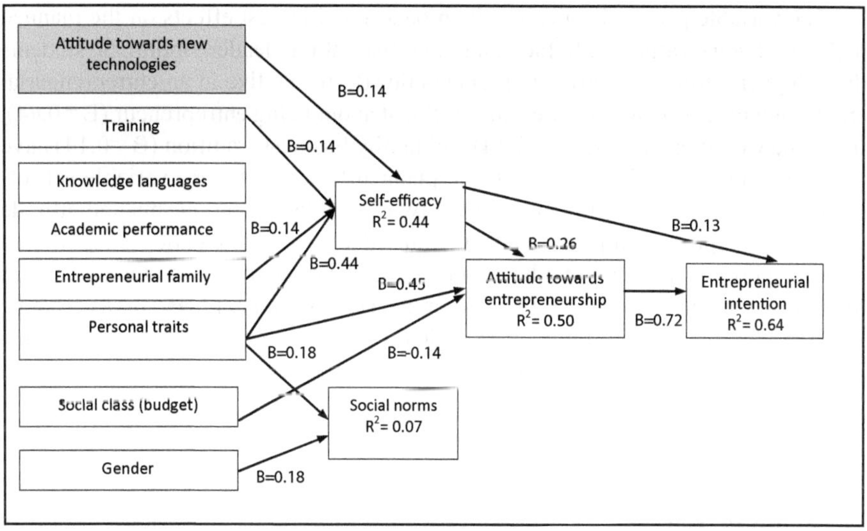

Fig. 3 Model of the planned behavior theory considering the impact of new technologies and other antecedents on entrepreneurial intention

main significant and positive (B = 0.14) along with the effects on the attitude toward entrepreneurship (B = 0.26) and on the entrepreneurial intention (B = 0.13). This attitude of the students is important in the entrepreneurial intention, even though some of its effect is captured by the students' personal traits as we observe in the reduction of the size of the coefficient from the first model (B = 0.36) to the second model (B = 0.14). Consequently, we claim that the attitude toward new technologies may help to identify those students with a higher entrepreneurial intention. Our objective is therefore accomplished.

Other interesting results from the analysis are related to the variables of the planned behavioral theory. In fact, attitude toward entrepreneurship is positively related to entrepreneurial intention (B = 0.72). In addition, self-efficacy has a positive effect on entrepreneurial intention (B = 0.15) that is smaller than the effect of attitude. This result means that students with higher capabilities to create and sustain a firm over time have a higher entrepreneurial intention. This relation may be independent of the attitude toward entrepreneurship. One possible explanation for these results may be that some students consider being entrepreneur as an alternative to find a job in the marketplace. Finally, social norms do have not a significant effect on entrepreneurial intention. The personal traits of students contribute to creating social norms (B = 0.18), such as risk aversion or searching for opportunities. However, this effect is not persistent on the relation between social norms and entrepreneurial intention. Therefore, the intention to be entrepreneur is not related to the way that the students' environment considers the entrepreneurial activity. This result is consistent with previous literature; that is, variables of the planned behavioral theory have an effect on entrepreneurial intention, even though our data do not show a strong effect of the social norms.

The variable personal traits has the broadest and largest effects on the planned behavior theory variables. In fact, outgoing, risk-taking, leader, optimistic students have higher attitude toward entrepreneurship (B=0.45), live in an entrepreneurial environment (B=0.18), and are self-confident about being entrepreneur (B=0.44). Both the variables training (B=0.14) and family business tradition (B=0.14) have an impact on self-efficacy. In fact, the planned behavior theory variable self-efficacy is more influenced than the other two variables (social norms and attitude towards entrepreneurship) by students' individual features. Gender is significant on social norms, meaning that being women is positively related to the perception that those students have about the entrepreneurial environment (B=0.18). That is, women feel that the environment in which they live is more entrepreneurial oriented. Finally, social class has a small effect on attitude toward entrepreneurship (B=0.14).

5 Conclusion

Attitude and perception about new technologies have been considered in different streams of research (e.g., agricultural technology implementation, e-banking; technology adoption). The results of the empirical literature show that when new using or adopting a new technology individuals are driven by their attitudes toward this technology.

We find a similar pattern in university students; their intentions to be entrepreneur are driven by their attitude toward new technologies. For instance, they consider themselves to be on the cutting edge of innovation or experts in new technologies. This relationship between attitudes and technology is important for our understanding of the drivers of entrepreneurial intention and for designing specific training programs to promote this intention in students at the university level [7].

Our empirical model, which is based on the theory of planned behavior, provides a valid explanation for entrepreneurial intentions. Our results are robust to previous literature [1, 10, 12,]. That is, previous teaching practices that have been applied in universities based on this theoretical approach can be still used to promote entrepreneurial intentions.

Our study may be extended by adding other factors, such as the social networks in which students participate and to which they contribute. In fact, social networks are considered a powerful tool to create, disseminate, articulate, and exploit knowledge. Recently, some networks are being used to promote the creation of new ventures [8]. The inclusion of social network variables may be of interest to promote entrepreneurial intentions in students at the university level.

References

1. Azjen I (1991) Theory of planned behavior. Organ Behav Hum Decis Processes 50(2): 179–211
2. Azjen I (2011) The theory of planned behaviour: reactions and reflections. Psychol Health 26(9):1113–1127
3. Centre for Enterprise and Economic Development Research (2000) Young entrepreneurs, women entrepreneurs, co-entrepreneurs and ethnic minority entrepreneurs in the European Union and Central and Eastern Europe. Final Report to the European Commission, DG Enterprise, Middlesex University Business School, UK
4. Choy CS, Kuppusamy J, Jusoh M (2005) Entrepreneurial careers among business graduates: match-making using theory of planned behavior. Int J Entrep 9:67–90
5. Cunningham JB, Lischeron J (1991) Defining entrepreneurship. J Small Bus Manage 29(1):45–61
6. Engle LE, Dimitriadi N, Gavidia JV, Schlaegel C, Delanoe S, Alvarado I et al (2010) Entrepreneurial intent. Int J Entrep Behav Res 16(1):35–57
7. Fayolle A, Gailly B, Lassas-Clerc N (2006) Assessing the impact of entrepreneurship education programmes: a new methodology. J Eur Ind Train 30(9):701–720
8. Field JA, Elbert DJ, Moser SB (2012) The use of social media in building interest in wellness on a college campus. Am J Bus Educ 5(5):515–524
9. Hindle K, Yencken J (2004) Public research commercialisation, entrepreneurship and new technology based firms: an integrated model. Technovation 24(10):793–2003
10. Kautonena T, Van Gelderenb M Tornikoskic ET (2013) Predicting entrepreneurial behaviour: a test of the theory of planned behaviour. Appl Econ 45(6):697–707
11. Kor YY, Mahoney JT, Michael SC (2007) Resources, capabilities and entrepreneurial perceptions. J Manage Stud 44(7):1187–1212
12. Krueger NF Jr, Carsrud AL (1993) Entrepreneurial intentions: applying the theory of planned behaviour. Entrepreneurship theory and practice. Entrep Reg Dev 5(4):315–330
13. Li W (2007) Ethnic entrepreneurship: studying Chinese and Indian students in the United States. J Dev Entrep 12(4):449–466
14. Liñan F, Chen YW (2009) Development and cross-functional application of a specific instrument to measure entrepreneurial intentions. Entrep Theory Pract 33(3):593–617
15. Liñán F, Rodríguez-Cohard JC, Rueda JM (2011) Factors affecting entrepreneurial intention levels: a role for education. Int Entrep Manage J 7(2):195–218
16. Martin N, Rodriguez AI, Hernangomez J, Saboia F (2009) The effect of entrepreneurship education programmes on satisfaction with innovation behaviour and performance. J Eur Ind Train 33(3):198–214
17. Matthews CH, Moser SB (1996) A longitudinal investigation of the impact of family background and gender on interest in small firm ownership. J Small Bus Manage 34(2):29–43
18. Minniti M, Nardone C (2007) Being in someone else's shoes: the role of gender in nascent entrepreneurship. Small Bus Econ 28(2–3):223–238
19. Munir KA, Phillips N (2005/1684) The birth of the 'Kodak moment': institutional entrepreneurship and the adoption of new technologies. Organ Stud 26(11):1665–1687
20. Rauch A, Frese M (2007) Let's put the person back into entrepreneurship research: a meta-analysis of the relationship between business owners' personality characteristics and business creation and success. Eur J Work Organ Psychol 16(4):353–385
21. Tushman ML, Anderson P (1986) Technological discontinuities and organizational environments. Adm Sci Q 31(3):439–465
22. Utterback JM (1994) Mastering the dynamics of innovation: how companies can seize opportunities in the face of technological change. Harvard Business School, Boston
23. Veciana JM, Aponte M, Urbano D (2005) University students' attitudes towards entrepreneurship: a two countries comparison. Int Entrep Manage J 1(2):165–182

Return of Equity Issues in the Spanish Stock Market from the Investor's Perspective, During the 1997–2012 Period

Javier Parreño, Felipe Ruiz and Félix Roux

Abstract Is it profitable for an investor, from a risk-return perspective, to acquire a stake in a quoted company when a capital increase is announced? This paper analyses the return obtained from the investment in equity issues with cash contribution and pre-emptive rights, aimed at funding corporate activities: acquisitions, investments in new facilities and/or strengthening the balance sheet of the companies undertaking the equity issue. During the 16 years covered by the study, the results show a negative average excess risk-adjusted return of almost 5 %, from the moment that the equity offer is announced until the completion of the preferential subscription period. To obtain this excess return, the difference between the nominal Internal Rate of Return (IRR) and the expected return, using the CAPM, is computed for each equity issue. The intention behind this method is to eliminate the effects of time and any other possible effect on the stock price during the period of the analysis. The results from this article are consistent with the Pecking Order theory for the Spanish Stock Market also six months after the preferential subscription period. However, there is a positive return after three months.

Keywords Equity · Issues · Returns · Pecking · Order

1 Capital Structure and the Pecking Order Theory

The theory of capital structure has been widely studied by researchers and professionals, since it corresponds to one of the key financial decisions from the point of view of both academics and practitioners. The modern starting point of this theory was Modigliani and Miller's 1958 paper which stated that, in perfect markets,

J. Parreño (✉) · F. Ruiz · F. Roux
Escuela Técnica Superior de Ingenieros Industriales, Universidad Politécnica de Madrid, José Gutiérrez Abascal, 2, 28006, Madrid, España
e-mail: j.parreno@alumnos.upm.es

F. Ruiz
e-mail: fruiz@etsii.upm.es

F. Roux
e-mail: froux@etsii.upm.es

C. Hernández et al. (eds.), *Managing Complexity*, Lecture Notes in Management and Industrial Engineering, DOI 10.1007/978-3-319-04705-8_7,

financing decisions with either external or internal resources are irrelevant to the market value of companies. From this starting point, additional hypotheses to make the model more realistic (influence of taxes, market behaviour, etc.) have been included, resulting in more complete and explanatory theories to explain the financial preferences of companies.

The Pecking Order theory of financing choices was originated some 25 years ago by Myers and Majluf [6], who stated that companies' funding preferences are as follows: (i) firms prefer internal financing, (ii) they adjust their dividend policies to their investment opportunities; (iii) when external finance is required, they start with debt, followed by convertible securities and, as a last resort, using equity issues.

The rationale for the Pecking Order goes as follows: (a) the investor knows that the management of the company's information is superior to her/his; (b) the investor also knows that the management can time the equity issue so that it does not take place when the stock price, according to the management's superior information, is undervalued; (c) therefore, the investor knows that when the management proposes the equity issue, the price of the stock must be either right or overvalued; and (d) a rational investor should then sell her/his shares at the issue price, the consequence being a drop in the price of the shares when the issue is announced. Fama and French [4] found evidence that the Pecking Order affects in varying degrees depending on the way of issuing equity: there are equity issues, like rights issues, with lower asymmetric information problems.

2 Financing Companies in the Spanish Stock Market Through Equity Issues with Cash Contribution and Pre-Emptive Rights

This study provides new evidence of the Pecking Order theory within the Spanish market. Numerous previous works have shown the hierarchy of financing resources through different methodologies like, without claiming to be exhaustive: Fernández et al. [5], Pastor-Llorca and Poveda [7], Sánchez-Vidal and Martín Ugedo [8] and Aybar et al. [2].

As explained by Sánchez-Vidal and Martín-Ugedo [8], Myers and Majluf [6] analyses the North American stock market, where quoted companies mainly make equity issues on a firm basis, so they usually restrict pre-emptive rights for existing shareholders. In the Spanish stock market, equity issues with monetary contributions in which companies restrict pre-emptive rights are made when an investor or an organized block of investors want to acquire a stake in the company, but this doesn't mean that the company is seeking funding for expansion. The only type of equity issues whose aim is always funding corporate growth using the Spanish stock market are those with cash contribution and pre-emptive subscription.

As a consequence, it seems reasonable to analyse for the Spanish market the return obtained by investors in this type of equity issue as a method of confirming the potential negative impact of this corporate funding criteria, guided by all the papers mentioned before and also in the paper published by Asquith and Mullins [1], which investigated the effect on stock prices of equity offerings on the NY Stock Exchange.

To summarize, there are several major practical reasons for the analysis of this type of equity issues: (i) the objective corporate purpose of the equity issue, (ii) The ability of the entire investor community to subscribe the equity issue (both institutional and domestic investors), (iii) greater systematic follow-up by the stock exchange commission (CNMV), (iv) longer periods to subscribe the equity issue, and therefore to analyse it and (v) the possibility of avoiding shareholder dilution.

3 Methodology

The universe of equity issues analysed meet the following criteria:

1. All companies are quoted in the Madrid Stock Exchange Market when the equity issue is made. Initial Public Offerings (IPOs) have not been considered.
2. The period of analysis is 16 years starting from the beginning of 1997. This period involves two cycles of expansion and contraction of the Spanish economy.
3. There is a monetary contribution in every equity issue considered.
4. The Board of the Company recognises the pre-emptive rights of the existing shareholders.
5. The capital increases are registered in the Spanish Stock Market Commission (CNMV).

The methodology used to obtain the investors' return follows three steps:

- Internal Rate of Return of the equity issue 'n' (IRRn) is the IRR obtained by that investor who, once they know of the intention of the company to proceed with an equity issue, finally decides to invest in it. The stock price information has been obtained from Bloomberg. In this sense, it is important to mention that theoretical value of rights has been adjusted by Bloomberg in the stock price history once the pre-emptive subscription period is completed. IRR has been calculated for three different short term periods: (i) once the pre-emptive subscription period is finished, (ii) three months, and (iii) six months after the forementioned period has finished.
  The main issue in calculating the IRR is the date of the first day the investor decides to acquire shares. The day when enough information about the equity issue is published in CNMV has been designated t=0. Usually, most of the information has been given when the Board of the Company announces the total amount of the equity issue (nominal + premium issue). The next step would be to schedule a General Meeting of the Shareholders in order to approve the equity issue and to delegate to the Board the faculty to carry out the issue.
  During the analysis of the return obtained by the investor, all dividends earned by the shareholder have also been considered as dividend cash flow.

$$0 = O_0 + \sum_{j=1}^{i} \frac{Dj}{(1+IRR)^{\frac{(tj-t0)}{365}}} + \frac{Si}{(1+IRP)^{\frac{(ti-t0)}{365}}} \quad (1)$$

Where:

O_0	Outflow. Disbursement for the investor (t=0)
Dj	Dividend received by the investor on the day j
Si	Inflow. Sale of the investment on the day i

- Expected return of the equity issue 'n' (ERn) that the investor should have obtained in each equity issue. It is calculated during the same period of time as the calculated IRRn.
  The expected return has been calculated with the "Capital Asset Pricing Model" CAPM method [3].
  The risk-free rates of return considered are three months and six months interest rates of the Spanish public debt, depending on the period analysed. The market premium has been calculated considering the difference between the real rate of return of IBEX with dividends Index (Bloomberg ticker: IBEX35TR) and the risk-free rate of return.
- Excess risk-return of the equity issue 'n' (ERRn): it is obtained as the difference between the IRR and the expected return.

$$ERRn = IRRn - ERn \quad (2)$$

By extracting the expected return from the IRR, the effect of time or any other possible effect on the stock price during the period is eliminated.

The universe of stocks combined for this study implies 132 out of 137 (96 %) of the total equity issues with cash contribution and pre-emptive right made by 68 companies. According to the sector breakdown provided by the Spanish Stock Exchange (Bolsas y Mercados Españoles, S.A.), 27 % of the equity issues analyzed comes from companies within the financial services and real estate sector, 24 % basic materials, industry and construction, 21 % consumer goods, 14 % technology and telecommunications, 8 % oil and energy and 8 % consumer services.

An average of eight equity issues per year have been fully subscribed; 2001 was the year with the fewest equity issues (3) and in 2009, 13 companies issued new shares (Fig. 1).

In terms of nominal and total value, equity issues fully subscribed during 2010 represented 25 % and 20 % respectively of the total universe analyzed. On the other side, 2001 was the least active year, representing less than 1 % of the total analysis (Fig. 2).

The most active years in terms of corporate financing through these types of equity issues correspond with the most restrictive period on financing held by financial institutions (2008–2012).

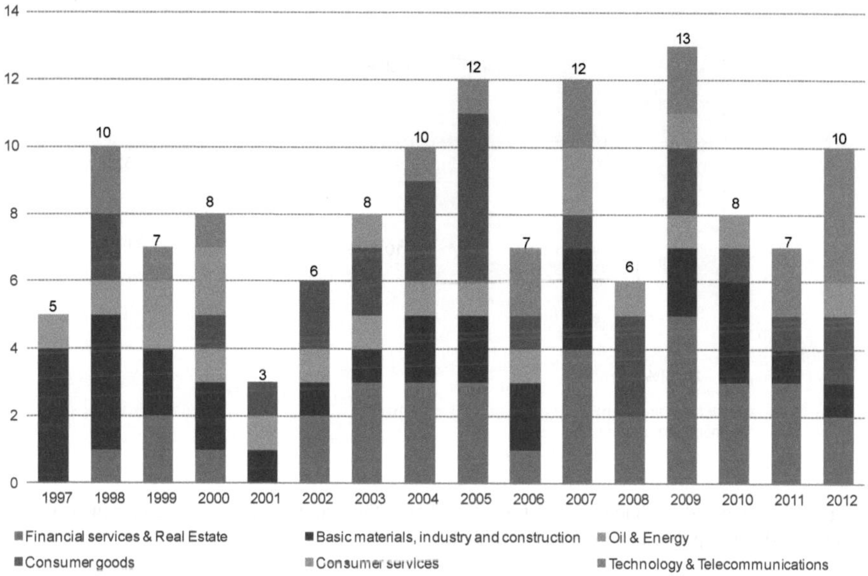

Fig. 1 Equity issues per year considered by the analysis by sector

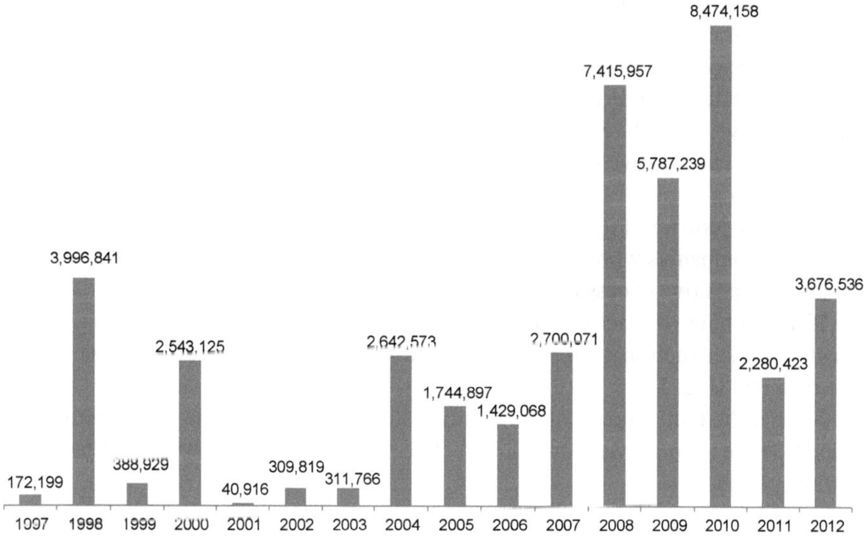

Fig. 2 Total value (nominal + premium) of the equity issues analysed

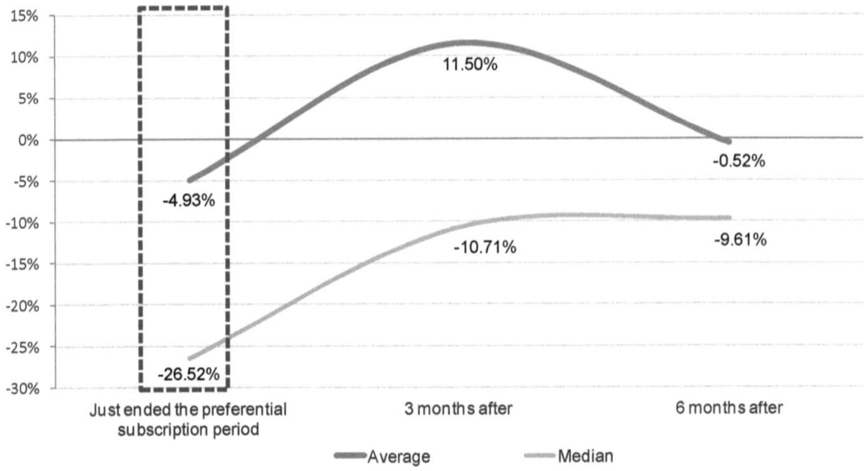

Fig. 3 Average and median excess returns (ERRn) of the equity issues analysed (for those equity issues made during the second semester of 2012, only the first period has been taken into account)

4 Results Obtained

Those investors who have decided to buy shares in a company at the time that it makes public its intention to issue equity in the market have obtained, on average, a negative excess return, both at the end of the preferential subscription period and six months after in terms of the mean and the median. However, the mean after three months shows a positive return. When calculating the median value of the excess returns, one can observe a negative figure for all periods (Fig. 3).

If made a grouping of stocks according to their betas, the following average excess returns shown in Fig. 4 are obtained:

Those companies whose betas are between 0.75 and 0.9 higher than 1.25 obtain an average positive excess return.

If a grouping of stocks according to the total amount funded in the stock exchange market (nominal + premium) is made, the following results shown below are obtained (Fig. 5).

Neither the total amount of the equity issue nor beta affect the excess return obtained. There is no proven correlation between these two parameters and the excess returns obtained.

4.1 *Industry Split*

As shown in Fig. 6, one can observe the main differences between industry weights in terms of average excess returns during the study period: Basic materials, manufacturing and construction is the only sector that has a positive average excess return from the moment that the equity issue is announced until its completion.

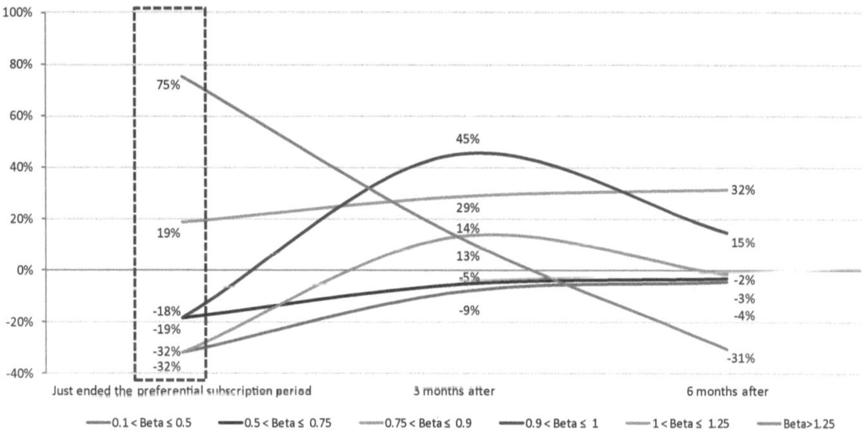

Fig. 4 Average excess returns (ERRn) according to different groups of betas

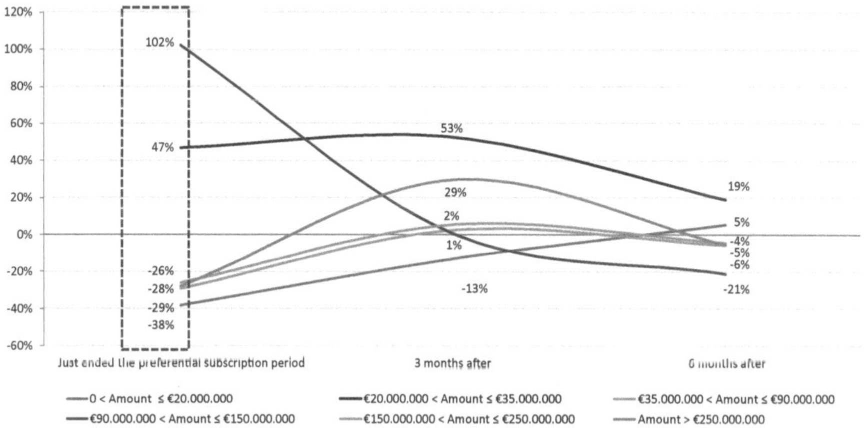

Fig. 5 Average excess returns (ERRn) according to the amount funded in the equity issues

5 Conclusions

The results of this study demonstrate that the announcement of equity issues in the Spanish stock exchange market destined for corporate growth provides a negative excess return for those investors who decide to subscribe them from the moment they are announced. The average excess return obtained by these investors is −5 % and is statistically significant.

Other significant results from this study include: (i) There is no relationship between betas and size of the equity issue and the return obtained for each offering; (ii) according to the sector breakdown made by the Spanish stock market, the

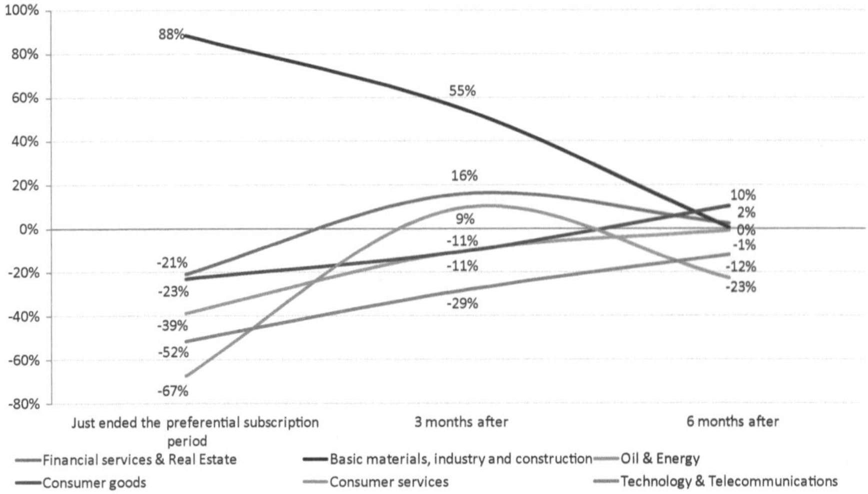

Fig. 6 Average excess returns (ERRn) according to sector

consumer services sector is the most unprofitable sector for investors and the basic materials, industry and construction sector is the only profitable sector.

References

1. Asquith P, Mullins DW Jr (1985) Equity issues and offering dilution. J Financ Econ 15(1–2):61–89
2. Aybar C, Casino A, López J (2001) Preferences hierarchy and Business Strategy in SME's capital structure: approach with panel data. Valencian Institute of Economic Research, Department of Accounting, Valencia University, Spain
3. Campbell JY, Viceira LM (2002) Strategic asset allocation: portfolio choice for long term investors. Clarendon lectures in economics. Oxford university Press, Oxford
4. Fama EF, French KR (2005) Financing decisions: who issues stock? J Financ Econ 76(3):549–582
5. Fernández Ramos MY, de Rojas MC, Zuliani GD (2004) Verification of the pecking order theory: the case of the Spanish companies. Finance and accounting department, University of Valladolid, Spain
6. Myers SC, Majluf NS (1984) Corporate financing and investment decisions when firms have information that investors do not have. J Financ Econ 13:187–222
7. Pastor-Llorca MJ, Poveda F (2004) Capital increases in Spain: an empirical study of equity issues rights. Department of Finance, Accounting and Marketing, University of Alicante, Spain
8. Sánchez-Vidal J, Martín-Ugedo JF (2004) Financing preferences of Spanish firms: new evidence for the theory of hierarchy. Economics analysis working papers, vol 3, N 8, Spain

Sustainable Balanced Scorecard: Practical Application to a Services Company

Alfonso Redondo Castán, José Antonio Pascual Ruano, Angel Manuel Gento Municio and Javier Muñoz Sanz

Abstract We have developed, in partnership with a consulting firm, a tool that provides two basic goals: (1) The integration of the planning of Social Responsibility (CSR) with the strategic and operational planning of the company (BSC), in our case based on a patented management model with 7+1 levels, and (2) Monitoring the management of the CSR according to the activity carried out by the organization. This has allowed us to evaluate/test the quality and adaptability of the developed tool, identifying their strengths and weaknesses, areas for improvement and future lines of research or work.

Keywords Balanced Scorecard (BSC) · Corporate Social Responsibility (CSR) · Strategy · Management

1 Project Justification

Many are the changes that organizations are experiencing at the strategic and tactical levels of management, whether those companies are public/private, large/small, goods/services, etc.. And many organizations are questioning whether it even makes sense to talk about Corporate Social Responsibility (CSR), or whether, by contrast it should be another of their business pillars.

In this line, the crisis (economic/social/values/etc…) presents an opportunity for businesses. Our society highlights the need to build a more sustainable and

A. Redondo Castán (✉) · J. A. Pascual Ruano · A. M. Gento Municio
Dpto. de Organización de Empresas y CIM, Escuela de Ingenierías Industriales, Universidad de Valladolid, Pso del Cauce 59, 47011 Valladolid, Spain
e-mail: redondo@eii.uva.es

J. A. Pascual Ruano
e-mail: pascual@eii.uva.es

A. M. Gento Municio
e-mail: gento@eii.uva.es

J. Muñoz Sanz
Escuela de Ingenierías Industriales, Universidad de Valladolid, Pso del Cauce 59, 47011 Valladolid, Spain
e-mail: msz.javier@gmail.com

C. Hernández et al. (eds.), *Managing Complexity*, Lecture Notes in Management and Industrial Engineering, DOI 10.1007/978-3-319-04705-8_8,

responsible socioeconomic model, aligned with the concepts and vision pursued by CSR: economic models which integrate business development and societal concerns, searching for a balance between renovation and irresponsible consumption of finite natural resources, aligning the interests of all "stakeholders" towards an integral and supportive CSR, etc. By integrating this all, we will achieve a CSR that becomes a competitive element within the business framework, and that is able to generate greater profits for a company.

In this context, two years ago, a service company suggested us to collaborate in developing a model that would allow the integration of the classical Balanced Scorecard (BSC) with CSR: Responsible Scorecard. This project aims to know the impact and the level of implementation and control of an existing CSR, through the development of a tool that allows to strategically manage all CSR indicators.

2 Scope and Planning

According to the objectives stated on the abstract, the project was divided into three phases:

- First, we had to analyze the strategic management model implemented in the company (7 + 1 levels), emphasizing the identification of the most relevant indicators.
- Later, taking as starting point a work by Muñumer [2] and Garcia [1] which we had previously directed, we set in context CSR in firms, from the strategic point of view, and considering the intended development of a sustainable management tool.
- After the previous stages, we proceeded to develop and implement the sustainable management tool, characterized, strategically and from the point of view of CSR, by the business management of the Consulting firm.

In this article we will focus on the latter stage, since it constitutes the central and practical core of the developed work.

3 Development and Implementation of Sustainable Balanced Scorecard

The design/development of the theoretical model, as a previous step before its application in the consulting firm, was carried out in three phases that we summarize below:

3.1 Phase I: Diagnostic

This is a critical and laborious phase, which consists in analyzing and measuring the level of involvement, both for each of the dimensions of the CSR of the company, as well as for the indicators of each of the strategic perspectives (CMI) of the company.

CSR INDICATOR			STRATEGIC PERSPECTIVES			
CODE	DIMENSION	DESCRIPTION	FINANCIAL	STAKEHOLDERS	INTERNAL PROCESSES	LEARNING AND GROWING
EC8	ECONOMIC	Development and impact of infrastructure investments and services provided primarily for public benefit through commercial liabilities, pro bono, or in kind.	EC8			
EC9	ECONOMIC	Understanding and description of significant indirect economic impacts, including the extent of such impacts.	EC9			
EN1	PRODUCT/ SERVICE	Materials used by weight or volume			EN1	
EN2	ENVIRONMENTAL	Proportion of materials used which are valorized			EN2	
EN3	PRODUCT/ SERVICE	Direct energy consumption by primary source.			EN3	
EN4	PRODUCT/ SERVICE	Indirect energy consumption by primary source.			EN4	

Fig. 1 Classification of CSR indicators in strategic perspectives

CSR INDICATOR			STRATEGIC PERSPECTIVES			
CODE	DIMENSION	DESCRIPTION	FINANCIAL	STAKEHOLDERS	INTERNAL PROCESSES	LEARNING AND GROWING
EC8	ECONOMIC	Development and impact of infrastructure investments and services provided primarily for public benefit through commercial liabilities, pro bono, or in kind.	EC8			
EC9	ECONOMIC	Understanding and description of significant indirect economic impacts, including the extent of such impacts.	EC9			
EN1	PRODUCT/ SERVICE	Materials used by weight or volume			EN1	
EN2	ENVIRONMENTAL	Proportion of materials used which are valorized			EN2	
EN3	PRODUCT/ SERVICE	Direct energy consumption by primary source.			EN3	
EN4	PRODUCT/ SERVICE	Indirect energy consumption by primary source.			EN4	

Fig. 2 Analysis of the state of the CSR indicators in the company

In order to carry out an internal scanning, the following steps were performed.

3.1.1 Classification of CSR Indicators Aligned with the Strategic Perspectives

In Fig. 1 we can see the classification of CSR indicators aligned with the strategic perspectives of the company.

3.1.2 Diagnostic/Analysis of the State of the Indicators of the Company

The analysis/screening was performed based on a set of templates specifically developed for that purpose, involving a huge amount of hours and meetings. This check/data capture allows us to know the level of commitment and the implementation status that each company may have with relation to its CSR, both in dimensions as well as in prospects. Figures 2, 3 and 4 shows, in "white": those indicators that do not apply to the company; in "gray": those indicators that apply and cannot be measured; and in "purple": those indicators that are currently applied and measured.

GENERAL DIAGNOSTIC OF THE INDICATORS OF STRATEGIC PERSPECTIVE.										
INDICATORS			FINANCIAL		STAKEHOLDERS		INTERNAL PROCESSES		LEARNING AND GROWING	
Type	nº of indicators	%	nº of indicators	%	nº of indicators	%	nº of indicators	%	nº of indicators	%
Currently measured										
Measurable										
Not applicable										
TOTAL										
IMPROVEMENT RATIO FOR PERSPECTIVE (%)										

Fig. 3 General diagnostic of the CSR indicators by strategic perspectives

GENERAL DIAGNOSTIC OF THE INDICATORS BY CSR DIMENSION.										
INDICATORS			ECONOMIC		ENVIRONMENTAL		PRODUCT / SERVICE		SOCIAL	
Type	nº of indicators	%	nº of indicators	%	nº of indicators	%	nº of indicators	%	nº of indicators	%
Currently measured										
Measurable										
Not applicable										
TOTAL										
IMPROVEMENT RATIO FOR DIMENSION (%)										

Fig. 4 General diagnostic of the indicators by CSR dimension

3.1.3 Identification of Critical Indicators

The preceding step is enriching, but unapproachable as a whole, so it is necessary to make a selection of those indicators which are key to the firm, also named *critical indicators*. That is, those indicators which are vital both for CSR management and for the management of the strategic perspectives.

The tool so designed, based on correlation tables between indicators, allows us not only to select the critical indicators, but also to know their status, or, equivalently, to determine the percentage of improvement both by strategic perspective and by social dimension.

3.2 Phase II: Improvement Plan

Now, the tool allows us to analyze the indicators that could be applied and are not currently measured by the company. Thereby demonstrating the capacity of improving the firm on CSR and its location (perspective/dimension), facilitating thus its Integral management (Fig. 5).

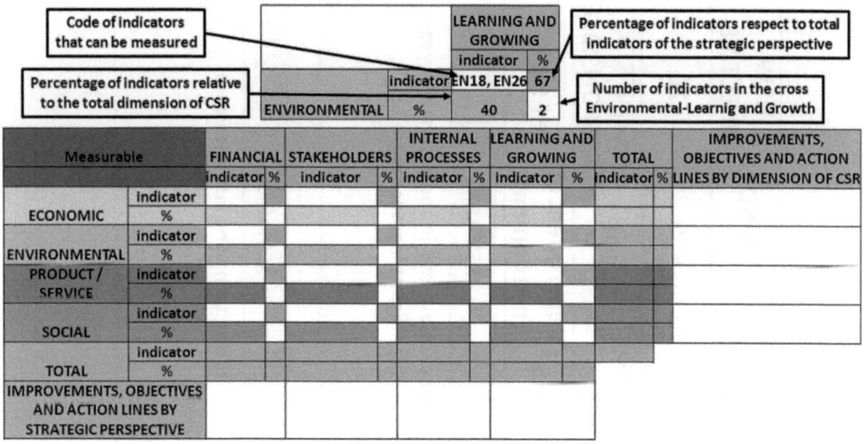

Fig. 5 Improvement Plan

	FINANCIAL	STAKEHOLDERS	INTERNAL PROCESSES	LEARNING AND GROWING	TOTAL	IMPROVEMENTS, OBJECTIVES AND ACTION LINES BY DIMENSION OF CSR
ECONOMIC						
ENVIRONMENTAL				EN18, EN26	2 0 2	100
PRODUCT / SERVICE						
SOCIAL		LA2, ES4.1, ES4.2, ES4.5			4 1 5	80
TOTAL		4 1 5		2 0 2		
IMPROVEMENTS, OBJECTIVES AND ACTION LINES BY STRATEGIC PERSPECTIVE		80		100		

Fig. 6 Table for managing and controlling indicators

3.3 *Phase III: Implementation and Monitoring*

The role of this third and final phase is to monitor and to control all the critical indicators, providing information about their evolution. This will allow us, on Phase IV, its integration with the Operating Plan.

In short, it allows us to know the fraction of strategy goals effectively implemented (the advance on the fulfillment of the objectives of the operational plan that are related to CSR), as well as the advance achieved between different periods.

This last section of the tool uses red and green colors to distinguish between the status of each perspective-dimension at a glance. Those indicators that do not meet the objectives are highlighted in white (see Fig. 6).

This analysis can be extended further, making the tool more effective for management, both by an integration into the company's operating plan, a process that is developed later, as well as by the possible monitoring of each indicator. Figure 7 shows an example of the record sheet that can be performed for each indicator.

INDICATOR	DESCRIPTION	LAST MEASURE	PROPOSED OBJECTIVE	DEGREE OF ACHIEVEMENT		TREND	NEW OBJECTIVE
LA2	Total number of employees and average turnover rate	15/11/2011	24	18	75%	↑	20
INDICATOR	IDENTIFIED RISKS	ACTIONS	ESTIMATED COSTS	ESTIMATED TIME	SCOPE		CONTROL RESPONSIBLE
LA2	Lack of employee motivation	Increase in remuneration for achievement	3.000 €	6 MOTHS	Department of Environment		HF Staff

REMARKS

Little pronounced in the first half of year
Significant increase from July

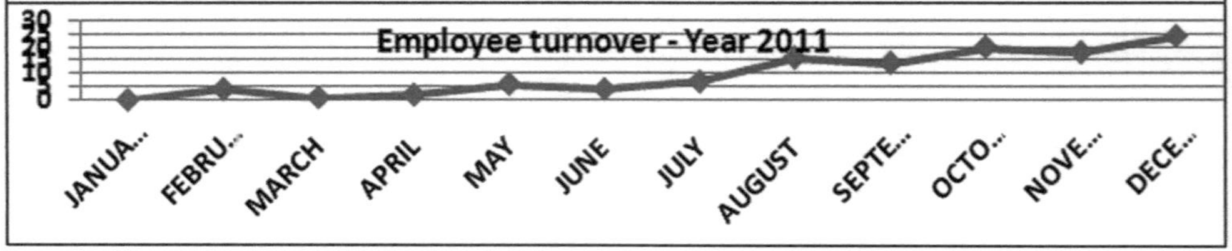

REMARKS

Rising trend the last few years

Fig. 7 Tracking card for indicator LA2

			ASPECTS OF OPERATIONAL PLANNING				
	LEVEL	CONTENT	INNOVATOR ASPECT Aspects that enable the organization to generate a capacity to adapt to opportunities or threats	MANAGER ASPECT Implementation and monitoring of all planned in the innovative aspect	PRODUCTOR ASPECT Results obtained feedback to the innovative aspect. Indicates the process of continuous improvement	TOTAL INDICATORS BY LEVELS	%
CORPORATIVE LEVELS	N7+1	R & D & I	Identification of opportunities, risks and threats EN6, EN7, EN18, EN26	Management of opportunities, risks and threats	Developments in progress	4	4
	N7	SOCIAL	Social presence	Management of alliances ES 2.3, ES 4.1, ES 4.2, ES 4.4, ES 4,5, ES 5.5	Corporate Social Responsibility HR9, ES6.1, EN13, SO7	10	10
	N6	CUSTOMERS, TRADE PROCESS	Commercial development	Management of the customer portfolio PR5	Marketing PR7, PR9, PR6	4	4
	N5	RELATIONSHIPS, STAFF, HR	Planning for the needs of people	HR Management EC5, EC7, LA1, LA2, LA4, LA6, LA9, LA12, LA13, LA14, HR8, ES2.1, ES2.2, ES3.5	People satisfaction EC3, LA3, LA7, HR3, HR4, SO3, ES3.7, LA10, LA11, LA8	24	24
	N4	DIRECTION, IMPROVEMENT GROUPS	Delivery ES 3,3	Internal management LA5, SO1, SO2, SO4, ES1.1, ES3.2, ES3.8	Learning and improvement ES3.1, ES3.4	11	11
	N3	SYSTEM, QUALITY	System EC2	Management of improvement EN1, EN2, EN3, EN4, EN5, EN11, EN12, EN14, EN15, EN25, EN29	Customer satisfaction EN8, EN9, EN10, EN16, EN17, EN19, EN20, EN21, EN22, EN23, EN24, EN27	24	24
	N2	PRODUCTION AND SERVICES	Production Planning	Productive resources management PR1, PR3, HR2, HR5, HR6, HR7, ES5.4	Productivity PR2,PR4	9	9
	N1	RESOURCES AND ASSETS	investment plan EC4	Economic and financial management EC6, ES5.2, EC8, EC9, EN30, HR1	Economic outcome EC1, EN28, SO6, SO8, PR9, 7.1, 7.2, 7.3	15	15
	TOTAL INDICATORS BY ASPECT		7	53	41	101	100
	%		7	52	41	100	

Fig. 8 Integration of CSR indicators with the operational plan of the company

3.4 *Phase IV: Integration with the Operational Plan*

The final process of this tool is the phase of integration with the company's operating plan. It is very versatile, with high adaptability and vital to the establishment of relations with the previous phase of implementation and monitoring.

The aim is its practical application to a "Service Company: Consulting", making evident not only its capacity for measuring and tracking, but also its ability to be adapted when classifying CSR indicators for each management area of the company, whatever the degree of complexity of its operational planning. The studied company uses a complex integrated management system based on a patented model with 7+1 levels. The result is shown in Fig. 8.

After a thorough analysis (available in [3]), the conclusions drawn from its integration into the operational plan are as follows in Fig. 9.

4 Final Syntheses

The proposed model, which has been tested in a consulting firm, but which can be applied/adapted to any other company, allows:

- To perform an internal review of the company with relation to CSR, according to its indicators of interest, and to integrate new indicators. This makes the

	DEGREE OF ACHIEVEMENT OF OBJECTIVES OF CRITICAL INDICATORS FOR EACH LEVEL (%)		
	PROPOSED OBJECTIVE	DEGREE OF ACHIEVEMENT	NEW OBJECTIVE
N7+1	80	100	85
N7	90	100	90
N6	55	40	50
N5	60	25	42
N4	100	100	100
N3	100	100	100
N2	60	100	65
N1	70	100	80

		INNOVATOR ASPECT	MANAGER ASPECT	DIRECTOR ASPECT
DEGREE OF ACHIEVEMENT OF OBJECTIVES OF CRITICAL INDICATORS FOR EACH ASPECT (%)	PROPOSED OBJECTIVE	55	70	80
	DEGREE OF ACHIEVEMENT	100	85	45
	NEW OBJECTIVE	60	65	90

Fig. 9 Achieved goals for critical indicators

management of each individual indicator easier, providing information about monthly and annual evolution, rate of progress towards objectives, trends, new objectives, estimated cost and time, scope, etc.

- To identify the strategic CSR indicators for the company, allowing it to focus their strategic and operational efforts. Controlling the critical indicators for each management area, and its corresponding margin for improvement.
- To assess the monitoring of these indicators as currently undertaken by the company, classifying them according to their CSR dimension or strategic perspective, and checking the improvement ratio for each dimension and perspective.
- To evaluate the CSR by the desired level or perspective: by areas or corporative levels of the operational plan, by strategic lines or CSR dimensions; For strategic or operational decisions, watching their evolution over time; it allows the introduction of new critical or general indicators.
- To diagnose the analysis and evolution of CSR in the company, checking the degree of implementation of the strategy, and, thus, to establish corrective measures.
- A model adaptable to any type of operational planning in order to perform a scanning of the effectiveness of CSR at each management level of the company.

In conclusion, it is a tool that allows the introduction, implementation, analysis and evolution of CSR in a company, facilitating the establishment of operational and strategic CSR lines, and integrating the BSC.

5 Future Lines of Research and Development

Due to the scope of this study, the main future R & D lines are addressed towards an integral automation of the tool, looking for its full integration into the enterprise. It is necessary to create an interface that allows managing the tool in harmony with the rest of applications implemented in the company, so that data treatment can be carried out with the greatest possible traceability and integrity.

References

1. García Vílchez EJ (2010) Desarrollo del modelo de Sostenibilidad Integrado (M.S.I.) para la medida de la gestión sostenible de una industrial de procesos: Aplicación al sector de fabricación de neumáticos. Tesis Doctoral, Universidad de Valladolid
2. Muñumer H (2011) La Responsabilidad Social de la Empresa. Actualización del Sistema de Gestión Ética (SGE21:2008) e Integración de los criterios de la Guía de Responsabilidad Social ISO26000:2010. Proyecto Fin de Carrera, E.I. Industriales, Universidad de Valladolid (Winner of the UVA RSU 2011)
3. Muñoz Sanz J (2012) Cuadro de mando responsable: aplicación práctica en la consultora 1A Consultores. Proyecto Fin de Carrera, E.I. Industriales, Universidad de Valladolid (Awarded with a special mention in the Organization section, in "X Premios Michelin Valladolid a los mejores Proyectos Fin de Carrera sobre Innovación, Calidad Total y Organización" (2012))

Predisposition of Workers to Share Knowledge: An Empirical Study

Lourdes Sáiz, José Ignacio Díez, Miguel Ángel Manzanedo and Ricardo del Olmo

Abstract Our objective is to determine the factors and obstacles that either contribute or complicate knowledge exchange between workers. This research area is new in Spain, as only three references have been found for firms in China and North America. In our approach to this topic, we apply concepts of a sociological, psychological, and motivational nature, which affect knowledge exchange and sharing and allow us to justify the theoretical basis of this study, as well as its purpose and its organizational benefits. A detailed survey was prepared for the empirical research with 21 questions given to a sample of 557 workers from firms in Burgos. Among the positive factors that contribute to knowledge sharing, the results highlight recognition and an appreciation of the worker's contribution, the work environment and reciprocity. The most important barriers are the poor quality of employment contracts, fellow workers that do not wish to learn, and unfair and disloyal behaviour.

Keywords Knowledge management · Knowledge sharing · Assistance and barriers to knowledge exchange · Employment contract · Work environment · Motivation policies

---

L. Sáiz (✉) · J. I. Díez
Department of Civil Engineering, Higher Polytechnic School, University of Burgos, C/ Francisco de Vitoria, S/N, 09006 Burgos, Spain
e-mail: lsaiz@ubu.es

J. I. Díez
e-mail: ignacorreo@hotmail.com

M. Á. Manzanedo · R. del Olmo
Department of Civil Engineering, Higher Polytechnic School, University of Burgos, C/ Villadiego, S/N, 09001 Burgos Spain
e-mail: mmanz@ubu.es

R. del Olmo
e-mail: rdelolmo@ubu.es

C. Hernández et al. (eds.), *Managing Complexity*, Lecture Notes in Management and Industrial Engineering, DOI 10.1007/978-3-319-04705-8_9,

1 Introduction

Acceptable use of knowledge and its management within the firm is increasingly gaining ground in the promotion of the strong points with which the organization wishes to compete [19]. However, there are workers who are more or less committed to exchange, in accordance with their personality, the context and their motivation, be it intrinsic or extrinsic, that is provided by the firm [17, 24]. The firm is, therefore, committed to knowing about and taking action on all behaviour and attitudes that favour the exchange of knowledge [6]. Transfer, sharing and management of knowledge within the firm has been considered a process or a function that contributes to the success of the firm, for which purpose general models for exchange have been designed and developed to support this activity [12, 20]. Nevertheless, there are still people and attitudes within the firm who are contrary to knowledge sharing, which is evident from the barriers to the full application of knowledge management [4]. On occasions, such barriers are raised because of the managers and the management style of the firm. Hence, the directors and those responsible for decision taking are key elements, perhaps the most important, so that all employees find their position and share knowledge in the posts where they can best contribute [7, 26]. In this work, as well as studying the purpose and the elements that influence knowledge exchange between workers, an empirical study is conducted, the novelty of which stands out, which has contrasted the positive factors that are likely to promote knowledge sharing and transfer and the barriers that are likely to prevent such activities. The results and conclusions that are reached will help us progress in this area of research and will provide information to firm directors, so that they take more informed decisions that will enable the effective application of knowledge management.

2 Knowledge Exchange Within the Firm

Knowledge sharing and exchange in the workplace is a complex task with all too many links and consequences. Its fundamental objective is to transfer knowledge from its generation to the different places in the firm that need it; it therefore affects organizational survival, prosperity and growth, improving its capability to respond and to take action and making it more competitive. Knowledge management, of which this exchange forms part, bases its function on competitive advantage that is gained with the improvement and the quality of the knowledge of people that help to create new knowledge, in a permanent cycle [9]. Efficient knowledge management means that workers lose their fear of sharing knowledge with other fellow workers [3], in an environment where the system or management style provides and favours the trust and the reciprocity needed for workers to perform their knowledge exchange [23].

The exchange of knowledge has beneficial effects for workers, given that learning is quicker, workers learn what is necessary and gain further experience. Even though it may not be completely assured, because there are other factors with an influence, it also has a beneficial effect on the performance of employees [13]. The creation of an environment of trust with acceptable behaviour improves the exchange of knowledge, acting on the personal emotions, minimizing destructive emotional conflict and assisting a work environment based on mutual trust. All of this is supported by distribution of knowledge and information that motivates exchange and mutual learning [25].

Nevertheless, tacit knowledge is the most difficult to capture and to transfer, because it is non-explicit knowledge or because it can be intentionally concealed by the person who holds it. This type of knowledge might imply a point of negotiation or a guarantee to remain in the firm, because of the exclusivity that it represents, which may even be used as a token for exchange [27].

Despite the beneficial effects of the exchange of knowledge, on occasions, workers deliberately hide and even deny the existence of their knowledge [4], or raise seemingly insurmountable barriers that appear in various forms [14]. In some situations, when the worker does not share knowledge, it is out of a fear of losing the status that the firm has and by which it is valued. In these cases, if the work environment is not conducive to knowledge exchange, the majority of workers will be fearful of doing so [23]. This fear brings to the surface the feeling of being replaceable in the firm, when the exercise of sharing means that the worker loses any exclusive idea of possessing knowledge [23]. In this context, a possible consequence, in the worker's view, is redundancy, with all of its very significant psychological and social consequences [8, 15, 22].

The disastrous economic situation at present means that redundancy is often determined by the shorter employment record of the worker as much as by criteria based on any corresponding compensation for loss of employment [10]. It generates an uneasy situation in which the youngest or the most recently incorporated workers are unjustly disadvantaged [16] and not those who contribute least knowledge. Together with this, the practice of early retirement may imply the loss of more expert workers, who will be replaced by others at a lower level. This retirement is encouraged for purely pecuniary reasons, as the new workers are on considerably lower salaries, with no consideration of the knowledge that it contributes [5, 18].

Incentives and rewards, valuable allies for the correct application of knowledge management, should therefore be offered to alleviate the consequences of these practices in the workplace [21]. With an acceptable design, a supportive and beneficial work environment may be created for knowledge exchange and transfer. Organizations may provide various rewards in multiple forms, such as salary increases, bonuses, stable contracts, opportunities for promotion and other types of benefits for the workers [1].

Bearing in mind cultural, national and other factors linked to where the incentive is created, it has been demonstrated that the organization is, on occasions, more effective in the formation and teaching of its employees, when there is an incentive that rewards that attitude towards learning [11]. However, research also exists that

Table 1 Research data sheet

Sample population	557 workers
Sectors of economic activity	Automobiles, services, energy, agriculture, and livestock
Field	Burgos and province (Spain)
Date of survey	June—October 2012
Design and technical direction	J. I. Díez, L. Sáiz, M. A. Manzanedo
Sources for the preparation of the questionnaire	[2, 4, 14]

proves that the organizational response can have a negative effect on the attitude of the individual towards the exchange of knowledge [1].

3 Empirical Study: Methodology, Justification for the Study, Sample and Questionnaire

The objective of this study is to survey the views and perceptions of workers towards knowledge exchange and sharing within the firm. In other words, the aim is not only to establish whether workers share knowledge in a natural way in their jobs in the firm, but also to establish which, in their opinion, are the drawbacks and barriers that complicate that exchange. The justification for this study is based on the need to know some of the aspects that affect knowledge exchange, from the work environment, the management style, the personality of the individual and the social and interpersonal relations that are found in the workplace.

The sample consists of 557 workers from different firms in the city of Burgos (northwest Spain) and the province of the same name, found in the automobile, services, energy and agricultural and livestock sectors. Further data that are representative of the sample are as follows: workers are contracted by firms with over ten workers and have, on average, two years service, for those within the 18–31-year-old age bracket, ten years of service for those within the 31–45-year-old age bracket, and 20 years service for those older than 46 years old.

An "ad-hoc" questionnaire was designed to gather the data, composed of 21 questions and structured into four sections: (a) general data for the identification of the participants; (b) predisposition to exchange knowledge; (c) human relations and work environment; (d) motivation and incentives to encourage exchange. The questions were drafted with the help of two experts, knowledgeable of the reality of these firms and of the challenges and trials that they have faced in the exchange and sharing of their key knowledge.

Having completed the questionnaires, performed the data treatment and the analysis of the information that had been gathered, the results and the conclusions helped us to identify barriers to knowledge sharing and incorporated recommendations of interest on how to increase and to improve the exchange of knowledge in the firm (Table 1).

Table 2 Aspects that support and that prevent knowledge sharing

Aspects that support sharing	(%)	Aspects that prevent sharing	(%)
Recognition and appreciation	24.5	Unstable contract	17.2
Good work environment	14.2	Fellow workers who do not wish to learn	15.2
Reciprocity	16.1	Disloyal fellow workers	14.6
Stable contract	9.7	Lack of recognition and appreciation	14.5
Remuneration	8.9	Bad work environment	13.3
Labour responsibility	8.4	Bad relation with fellow workers	9.6
Business organization	7.1	Unfairness of salary or in decision-making	7.3
Comunication with management	4.7	Pressure or bullying	4.2
Fair firm	3.3	Badly organized firm	4.1
Job security	3.1		

4 Results and Conclusions

This study has summarised its results, in particular, those that refer to situations or aspects that support knowledge sharing within the firm and equally those that prevent it or complicate it. Moreover, it reflects the predisposition of workers to share knowledge and the incentives that encourage behaviour that will support knowledge exchange within the firm. The descriptive results are presented as frequencies and percentages, to facilitate a clear understanding (Table 2).

With respect to the predisposition of the worker to share knowledge, the results revealed that 80.3 % would be willing to teach others, provided that the people they teach do not appropriate their ideas, and 40.4 % would not share it with fellow workers unfairly assessed as better than them. If workers considered themselves poorly valued or underused, which implies a lack of motivation, they would not share their knowledge. Moreover, if the acquisition of such knowledge had implied a significant effort, 29.4 % of individuals would not share it. Reciprocity in the transmission of knowledge emerged as an essential element for knowledge exchange, given that 38.8 % would not transfer their knowledge if others did not do likewise.

The leading incentives for knowledge transfer were, for 72 % of interviewees, an organizational structure that promotes and facilitates sharing knowledge and know-how; the existence of a stable contract in 65 % of cases; and 34 % responded positively to the idea of additional remuneration for passing knowledge on to others.

We may deduce from these results that the study has revealed some unknown factors in relation to the exchange and the sharing of knowledge. By means of an empirical study that used primary information (from 557 workers in different firms), we tested the elements and opportunities that supported knowledge transfer in the workplace, as well as its obstacles and limitations. The most prominent aspects are recognition and appreciation of the worker by the firm, up to the point that when this recognition is lacking, a highly demotivating situation is generated, not only for knowledge sharing, but also to be efficient and to contribute value to the function in question. Another relevant element is a work environment oriented towards knowledge sharing, as in situations of conflict, there is a lack of clarity in the assignation of tasks and objectives and little or no consideration of the worker, which is demotivating and provokes passive attitudes with little or no involvement in improvements made to the job. Mutual reciprocity is a further element that favours exchange and notably influences human behaviour, it being a quality that affects, in a significant way, not only the quantity, but also the quality of valuable knowledge exchange.

In turn, the poor quality of employment contracts are found among the barriers that, more than any other, complicate knowledge sharing between workers. A direct tie between the quality of the contract and knowledge exchange has been confirmed; fellow workers unwilling to learn or showing unjust and disloyal behaviour represent an added problem given that it is an intentional attempt to render the exchange inoperative, either by the negation to acquire more knowledge or by not using what has been acquired.

This study has also served to establish certain factors that influence the predisposition of the worker to share knowledge, which are maintenance of knowledge ownership, fair valuation of the work carried out by everybody, and the fact that others share and interact with their knowledge. The motivating elements or most prominent incentives to achieve efficient exchange reside in an appropriate organizational structure, stable contracts and specific remunerative complements for those who share knowledge.

References

1. Bock GW, Zmud RW, Kim YG et al (2005) Behavioral intention formation in knowledge sharing: examining the roles of extrinsic motivators, social-psychological forces, and organizational climate. MIS Q 29(1):87–111
2. Bures V (2003) Cultural barriers in knowledge sharing. E+M Ekon Manage Lib 6(special):57–62
3. Cabrera Á, Cabrera EF (2002) Knowledge-sharing dilemmas. Organ Stud 23(5):687–710
4. Connelly CE, Zweig D, Webster J et al (2012) Knowledge hiding in organizations. J Organ Behav 33(1):64–88
5. Cremer H, Lozachmeur J-M, Pestieau P (2009) Use and misuse of unemployment benefits for early retirement. Eur J Polit Econ 25(2):174–185
6. Chen S-S, Chuang Y-W, Chen P-Y (2012) Behavioral intention formation in knowledge sharing: examining the roles of KMS quality, KMS self-efficacy, and organizational climate. Knowl Based Syst 31(0):106–118
7. Chow WS, Chan LS (2008) Social network, social trust and shared goals in organizational knowledge sharing. Inf Manage 45(7):458–465
8. Eguchi K (2007) Productivity loss and reinstatement as a legal remedy for unjust dismissal. J Jpn Int Econ 21(1):78–105
9. Erden Z, Von Krogh G, Nonaka I (2008) The quality of group tacit knowledge. J Strateg Inf Syst 17(1):4–18
10. Galdón-Sánchez JE, Güell M (2003) Dismissal conflicts and unemployment. Eur Econ Rev 47(2):323–335
11. Hall H (2001) Input-friendliness: motivating knowledge sharing across intranets. J Inf Sci 27(3):139–146
12. Hedlund G (1994) A model of knowledge management and the N-form corporation. Strateg Manage J 15(S2):73–90

13. Hsu IC (2006) Enhancing employee tendencies to share knowledge-Case studies of nine companies in Taiwan. Int J Inf Manage 26(4):326–338
14. Lin C, Wu J-C, Yen DC (2012) Exploring barriers to knowledge flow at different knowledge management maturity stages. Inf Manage 49(1):10–23
15. López BSA (2007) Efectos individuales del despido y la resilencia como facilitador en la búsqueda del empleo. Panor Socioecon Redalyc 25(035):168–172
16. Malo MÁ, Cueto B (2012) Biografía laboral, ciclo económico y flujos brutos en el mercado de trabajo español. Panor Soc, 15:43–60
17. Martín CN, Martín PV, Trevilla CC (2009) Influencia de la motivación intrínseca y extrínseca sobre la transmisión de conocimiento. El caso de una organización sin fines de lucro. CIRIEC-España, Revista de Economía Pública, Social y Cooperativa. Redalyc 66(octubre):187–211
18. Messe PJ (2011) Taxation of early retirement windows and delaying retirement: the French experience. Econ Model 28(5):2319–2341
19. Ndofor HA, Levitas E (2004) Signaling the strategic value of knowledge. J Manage 30(5): 685–702
20. Nonaka I, Takeuchi H (1995) The knowledge-creating company. Oxford University Press, Inc., Oxford
21. Quigley NR, Tesluk PE, Locke EA et al (2007) A multilevel investigation of the motivational mechanisms underlying knowledge sharing and performance. Organ Sci 18(1):71–88
22. Raday F (1989) Costs of dismissal: a analysis in community justice and efficiency. Int Rev Law Econ 9(2):181–207
23. Renzl B (2008) Trust in management and knowledge sharing: the mediating effects of fear and knowledge documentation. Omega 36(2):206–220
24. Van Den Hooff B, Huysman M (2009) Managing knowledge sharing: emergent and engineering approaches. Inf Manage 46(1):1–8
25. Xia L, Ya S (2012) Study on knowledge sharing behavior engineering. Syst Eng Procedia 4(0):468–476
26. Yu CP, Chu TH (2007) Exploring knowledge contribution from an OCB perspective. Inf Manage 44(3):321–331
27. Yuqin Z, Guijun W, Zhenqiang B et al (2012) A game between enterprise and employees about the tacit knowledge transfer and sharing. Phys Procedia 24, Part C(0):1789–1795

The M.A.G. Factor. Management, Administration and Guidance. Where and How Much MAG Does Each Project Deserve and Need? A New, Original Assessment and Scoring System

Tom Taylor

Abstract This paper explains in outline a new and original assessment and scoring system to assist project managers in assessing the areas and levels of Management, Administration and Guidance (MAG) that clients may require when being involved in projects. The development of this system arose from consideration of client: project manager relationships on a variety of projects; and in particular the amounts of assistance and help that may need to be provided. A number of possible criteria, circumstances and influences were identified; and these have continued to be refined through project activities and by undertaking exercises with students on post graduate courses. Research and applications are continuing. Contributions, comments and corrections are welcome.

Keywords Clients · Project managers · Help and assistance · Management · Administration and guidance · Score system and analysis

1 Introduction

This short paper seeks to introduce the MAG factor—a new original assessment and scoring system for a whole range of projects (Taylor 2008, 2010 and 2011). A MAG factor assessment will greatly assist addressing the following conundrums:

1. Why is it that some Clients appear to need more help than others with the management, administration and guidance (MAG) for their projects?
2. Why is it that projects which seem to be similar, even identical, in fact require different MAG contributions and workloads?
3. When there is only a limited MAG expertise and resource available where should they be applied most effectively?
4. When there are particular MAG concerns what should be done about them?
5. What type, style and ethos of project management are needed and available?

T. Taylor (✉)
Dashdot & President of Association for Project Management, London, UK
e-mail: ttaylor@dashdot.co.uk

C. Hernández et al. (eds.), *Managing Complexity*, Lecture Notes in Management and Industrial Engineering, DOI 10.1007/978-3-319-04705-8_10,

2 What is MAG? Management, Administration and Guidance

When a client/customer identifies a project or is landed with a situation to which a project approach is most appropriate then they will at some point consider how the project is to be organised. From a client's perspective the project management requirements may be broken down into three groups of: Management, Administration and Guidance—hence MAG.

Management is the management of the project, programme or collection of projects. This will involve determining the project, devising the course to deliver it, selecting the team and driving all aspects of the project forward to achieve its goals. Leadership is involved and the strategic aspects are established and prioritised. Some Clients (or their internal managers) may want and feel able and willing to undertake this role throughout the project. Others may feel that they are in need of MAG help overall, or for some aspects and/or some stages. The question is: "How much Management help in the MAG Factor does the Client require?"

Administration covers the more technical and tactical aspects of projects. Inevitably there is some administration involved in all projects—sometimes a lot—covering secretarial, accountancy, budgeting, payments, arranging events and meetings, keeping records, monitoring, analysing and reporting, etc. Much of this activity and data supports the people and organisations who are managing as above. Some clients have experienced in-house resources who are available and willing to undertake all or most or only some of these tasks. The project activities and administration needs usually grow as the project progresses. The question is: "How much Administration help in the MAG Factor does the client require for this project?"

Guidance even when a client has confidence in their own Management and Administration there may be circumstances when they need Guidance in the definition or delivery aspects of a project. Whilst most clients have reasonable knowledge of the legal system, accountancy, human relations, property, etc. they still retain or go to specialist advisors for guidance. This may extend even to representations by such people but will not necessarily replace the overall management, decision making and ownership which will remain with the client or their designated project or operational representatives. Such guidance if required may be provided by mentors, advisors, gurus, friends, managers—as above, administrators—as above. So the question is "How much Guidance help in the MAG Factor does the Client need?"

3 Timing and Plans of Work

Most projects can be broken down into discrete stages. For example for construction/building projects there are usually three key stages:

- *Feasibility*: when the brief and scheme solution are established, key approvals and funding sought and go ahead received.

- *Pre-Construction*: when the main detail design, procurement and orders are completed.
- *Construction*: when the main works are executed with any remaining design and procurement through to completion—and probably a bit beyond to deal with settlement of accounts and any outstanding issues.

And then there is:

- Post Completion: this stage is vital in securing the original outcomes, benefits and more for the client that justified the resources and efforts in the first place.

Therefore it is possible to undertake a MAG Factor review at the start of each of these stages, as well as any single time on a project when such matters are being addressed.

In addition there are two other circumstances that a MAG Factor review might take place:

- At times of difficulty/problems/crisis—this is when a review might address if the appropriate levels of client contribution and MAG help are being applied and to appropriate aspects of the project.
- At the end of a project as part of the project debrief or lessons learned to ascertain where the pressures and problems occurred and how they were handled.

4 Outputs

From experience, research and application ten key criteria have been identified to establish the MAG factor plus some further other optional criteria.

It is possible to apply these criteria to any project to establish:

- the overall and average MAG Factor scores—in comparison with other projects and benchmarks,
- the aspects on which low scores have been established and how they are to be maintained,
- the aspects on which high scores have been established—and the consequential efforts and attention that are required on these aspects to deal with them or to endeavour to lower their scores
- a common understanding within the client body on the likely MAG needs and solutions.
- a common approach on how naturally limited MAG efforts are to be expended and prioritised.

5 Mag Scoring

The recommended approach to scoring is to use a 1 to 10 approach; with 1 being low and 10 being high; "not applicable", if really true, can score 0/zero/nil.

The scores can be allocated in relation to the resource efforts to be applied to each criteria in a reasonable time period of say a week, month or period for a stage within a plan of work.

In cases where the team cannot agreed on a single score then they can record the range under consideration, move on and return on conclusion of the exercise.

Similarly if some circumstances for a criteria would score high while others would score low then record circumstances, assumptions and scores for both—and place their resolution in the recommendations.

6 MAG Criteria

Ten criteria have been identified as consistent influences which affect the amount and foci of MAG required on projects. The ten selected criteria with a brief description of each are as follows:

1. *Same or Different Sector*

Clients who operate within the same sector as the project will probably need less help. For example a bank might need less help bringing in revised banking regulations but might need more help in setting up Health and Safety arrangements.

2. *First Time Type/Volume*

Clients who are undertaking this type or size of project for the first time will probably need more help.

For example a retailer who has previously acquired and fitted out their shops on an individual basis would need more help for say a national acquisition and makeover of fifty units.

3. *First Time Contract/Procurement*

New, different or complex contract and procurement will probably need more help. For example in understanding and applying the first design and build arrangement or multi packaged contracts or foreign sourced services or products. Similarly assembling suitable component tender lists—with sourcing, assessing and selecting in a new market place or against new requirements will require more effort.

4. *In Occupation/Use*

Clients with premises in occupation or ongoing operations will need more help with their capital projects. Also for example places adjacent to day-to-day use—road widening; airport facilities; railway track, signalling and stations—these all requires more help, than a green field situation.

5. *Individual or Committee Culture (double edged)*

Single headed clients probably need less help then say large complex committees—possibly. Except sometimes the individual client can be quite demanding and/or distant; while the experienced committee can be effective, authoritative and prompt—possibly.

6. *Funding*

Externally funded projects probably need more help to secure monies, satisfy funders, deal with payments, etc. compared with internally, simple or self funding. For example projects with Lottery backing and/or the need for public fundraising will need help with financial expertise and fund raising resourcing.

7. *Own Occupations/Use*

Projects for self occupation probably need more help. Owner occupiers have been known to demonstrate characteristics of being fussy, multi headed, have difficulties making decisions, and wish to change their minds to achieve perfection—while at the same time they may work on projects for others without these features.

8. *In Relationships (double edged)*

Clients in established satisfactory partnering, technical staff employed or other relationships will need less new help—possibly.

However it may be that not all the team members are in such relationships with the client and/or the relationships have become casual and not consistent with the formal agreements or reasonable expectations—possibly.

9. *Stakeholders*

Projects with diverse or multiple stakeholders will probably need more help.

Management, coordination and liaison of stakeholders and participants can be underestimated as soft skills, compared with the other more tangible hard tasks.

10. *Availability (double edged)*

Clients with predominant day jobs and distractions will need more help with their additional projects—or will they?

However some clients with busy schedules can be quite decisive and hands off, whilst clients with time on their hands can become over involved?

11. *Other Criteria*

There may be other circumstances in which clients require more help with the management, administration and guidance (MAG) of their projects which can be individually recognised.

- For example "Health Safety and Welfare" requirements are a high priority and require extra attention in power generation and transmission projects as well as other sectors. Additional help may be required when these are particular issues.

- "Diverse Locations" such as in manufacture at various plants of the components which constitute modern aircraft—compounded by language, culture, time zone differences.
- "Unknowns, Uncertainties and Complexity" cover situations where there are likely to be a greater number of changes than usual on a project; or there are more unknowns at a stage then might reasonably be expected (these circumstances should also be reflected in higher than usual budget contingencies to deal with them as well as aspects of Agile Management and Complex Project Management).

7 Way Forward

The recommended way forward for the first time is on the following lines:

1. Read over and become familiar with the criteria.
2. Select all or some of the case study projects as trials—below.
3. Select a team including the client for a workshop set of exercises which hopefully will include quite experienced and less experienced people working in mixed sets of two or three or individually.
4. Undertake the case studies and complete their MAG Factor scores and prepare the recommendations—discuss the outcomes—use a sample score sheet. [Inevitably there are ranges of interpretations for each case study and how to apply each criteria and as a group. There are no right answers. Some sample model answers are available to aid discussions of the outcomes.]
5. Now the real thing! Provide a brief summary of your project to hand, add some assumptions, including for whole project duration or a stage only, assess the project against the criteria and any other criteria that may be pertinent—calculate the scores and make recommendations for the organisation of the project.
6. Discuss the project, the assumptions, the scores and recommendations to decide what is to be done to clarify any issues, make decisions, implement decisions and set date or circumstances to monitor the MAG criteria and undertake further review(s).
7. It is advisable to identify, consider and record any further assumptions the team may wish to adopt from the outline project description and the application of criteria.

8 Case Studies

The following case studies, all with construction sector content, are offered to test readers' approach to the MAG Factor. Readers can add their own case studies as previous or hypothetical projects with other features or from other sectors.

1. *Road Signs in Blobshire*

To replace all road signs in Blobshire UK with miles and kilometre measurements to new European standards within 18 months

2. *More Housing*

Phase 3B of a housing estate for a further 120 semi-detached and detached two and three bedroom units on previously agricultural land by "Top Ten House Builders Co. Ltd"

3. *Rejuvenated Theatre*

Demolition of 50 % of community theatre premises, rebuild, refurbish, extend as part of urban regeneration with European, regional, local and public funding contributions—to correspond with 100 year anniversary—while continuing theatre productions elsewhere.

4. *Mixed Development at Transport Interchange*

First new, privately-developed mainline railway station in a city centre on contaminated railway land, 800 m from existing station, with specially assembled consortium covering commercial offices, property, construction with design, retail operator, railway company and local authority—with some social residential, some leisure, some public space and link to adjacent separate bus/coach station and tram terminus.

5. *Improved Security to Retail Units*

Following a series of break ins and robberies on security and insurance advice it has been decided to improve security to 1,000 shop units throughout the country and 20 in North America with replacement locks throughout, internal CCTV systems to half and internal or external shutters to about quarter—to be undertaken in evenings and Sundays over a concentrated four week period as soon as possible.

9 Conclusions

In all project situations there is a need for competent, good practice, helpful Management, Administration and Guidance. The reverse is not appropriate or helpful e.g. inexperienced or remote or very light management; or bureaucratic or burdensome or inadequate administration; or inappropriate or mistimed or self-servering guidance.

The MAG Factor is a fairly simple concept to understand. It provides a considered and measured way to deal with a range of issues which otherwise can be vague and difficult—and on which possibly only intuition and good/bad experience would otherwise be used.

However despite the simplicity developing skills in applying and using the MAG Factor approach takes some time and effort—hence the inclusion of case studies and suggestions of a joint workshop approach to bring out understanding and assist applications.

Skills in use will be improved by being organised in the approach, carefully considering criteria (and changing or adding other criteria), by keeping previous score profiles, and observing the influences of refining project descriptions or assumptions.

This approach can also be of assistance to or be combined with risk reviews, benefits management, governance contributions etc.

Persons and organisations who have a structured view concerning Management, Administration and Guidance by using the MAG Factor or similar will be able to provide leadership and efficiency to projects and have advantages over persons who do not have such outlooks or information.

Good Luck with your projects.

References

1. Taylor T (2008) How to select the right project manager. Published by Dashdot, London
2. Taylor T (2010) Sustainability interventions—for managers of projects and programmes—with some serious opportunities, challenges and dilemmas. Published by CEBE and Dashdot, London
3. Taylor T (2011) Leadership in action. Published by Dashdot, London

Information Capability in Basque Country Quality Award Winners

Marta Zárraga Rodríguez and M. Jesús Álvarez

Abstract Given the global environment that companies have to compete in nowadays, changes are so frequent that companies have to adopt a proactive attitude by trying to anticipate those changes. Using quality information while making decisions has become a critical factor for success, and nobody disputes the importance of having this quality information, which comes from the efficient use and management of information. Companies that have such quality information will have a competitive advantage and improve their results. Under the RBV theory, this efficient use and management of information could be considered a capability of a company. The aim of this paper is to explore the degree to which certain companies have developed this information capability. We focused the study on companies committed to Total Quality Management models because, due to the nature of these information intensive models, such companies can be expected to have developed information capability. The findings confirm this fact, although there are still opportunities for improvement.

Keywords RBV · Information capability · Information practices · EFQM

1 Introduction

Nowadays companies compete in a changeable word. It is essential that organizations be aware of trends in order to be ready for new situations and, even better, to anticipate them. In this scenario, having reliable and complete information is a key factor in making the right decisions.

In order to obtain this quality information, the efficient use and management of information is critical, and its presence can be evidenced through a set of observable practices. In this study we present a list of information-related practices based on a literature review. Then we explore whether there are practices common to

---

M. Zárraga Rodríguez (✉)
ISSA, Universidad de Navarra, Cuesta de Aldapeta nº 49, 20009 San Sebastián, Spain
e-mail: mzarraga@unav.es

M. J. Álvarez
TECNUN, Universidad de Navarra, Pº Manuel Lardizabal 13, 20018 San Sebastian, Spain

C. Hernández et al. (eds.), *Managing Complexity*, Lecture Notes in Management and Industrial Engineering, DOI 10.1007/978-3-319-04705-8_11,

companies committed to TQM, which is considered an information intensive management system.

2 Framework and Objectives of the Study

Companies need information when performing almost every activity, and it is a necessary input when making decisions at the operational, tactical and strategic level [7, 10, 15, 16]. Information is a strategic company resource, just as human and financial resources are, but it is not always managedwith the same degree of awareness and structure [15].

Information can be seen as just a resource or as a capability. To the extent that the efficient use and management of information provides a competitive advantage to the company, the efficient use and management of information can be seen as a capability under the Resource Based View (RBV) organizational theory. The RBV organizational theory states that the key to the strategic success of a company lies not only in the environment but also in the resources the organization has and in the strategic use the company makes of them; that is, in its organizational capabilities. A capability is a source of competitive advantage for the company that allows the generation of value and differentiation through the combined use of a series of resources [1, 13].

We consider, therefore, that a company has information capability when the use and management of information is so efficient that it is a source of competitive advantage for the company.

It is difficult to observe capabilities, but we can detect practices that evidence their presence. According to Ashurst et al. [1], practices are described as a set of socially defined ways of doing things in order to achieve an outcome. To the extent that a practice is defined in terms of its outcomes, it is easier to demonstrate the success of the practice. The question becomes: what practices would be related to information capability as we have defined it?

2.1 Information Practices

When defining information practices, we take as a reference the work by Marchand et al. [11]. They point out several information practices that are pooled into three major groups: practices related with the management of the information life cycle, practices related to the integration of information technology into day-to-day business, and practices where behaviours and values assumed by employees in relation to the use and management of information are shown. We also take into account proposals by other authors ([2, 3, 4, 5, 9, 14], among others), and as a result we obtained a list of information practices that serve as evidence of information capability development (Table 1). We also show the expected outcomes of each practice in the table.

The objective of this exploratory study is to know whether or not these practices are common practices in companies. We will focus on companies committed to Total Quality Management (TQM) because TQM is an information intensive management system [12]. In this way, for instance, Fok et al. [6] explored the relationships between TQM and information systems; they state that in order to implement quality systems successfully, an effective information system could be expected to be in place at the company. Thus it seems reasonable to expect that companies that have obtained quality awards have also promoted information capability development and that information practices are actually common practices in these companies.

3 Research Methodology

Our empirical study was carried out in the Basque Country, and participants were selected according to their commitment to quality management. This region in the north of Spain is the European region with the highest density of EFQM awards [8].

The data were collected via questionnaire. Items were presented as statements and respondents had to indicate their agreement on scale of 1 (strongly disagree) to 10 (strongly agree). The statements aim to measure the outcomes of the practices because this allows the success or the presence of the practices to be demonstrated.

Statements were developed from a literature review and the instrument was tested by faculty members, senior managers and Euskalit members in order to make sure the items' meanings were clear and that the questionnaire was easy to answer. (Euskalit is the Basque Foundation for Excellence, a private non-profit organization founded in 1992 by the Department of Industry and Energy of the Basque Government, which supports the policy of promoting the quality of the Basque Government).

The final instrument contained three main sections: outputs of practices related to the management of the information life cycle (that is, the ILCM group), outputs of practices related to the integration of information technology in day-to-day business (the ITI group), and outputs of practices related to behaviours and values assumed by employees and displayed when using information (the AIBVC group).

The questionnaire was administered via a web page, which participants accessed with a link.

The participants were contacted by Euskalit. The link to the questionnaire was sent to 292 companies that had received a quality award and 43 of them responded, which is a response rate of 15 %. The Basque Government provides quality awards according to the points obtained by the companies during external evaluation, which employs the scoring system used by the European Model of Excellence.

Table 1 Information practices

Practice	Outcomes	Code[a]
Collect the information	Compilation of the information needed by stakeholders	ILCM1a
	Compilation of information to collect from the environment and the inner workings	ILCM1b
Sense the information	Process for competitive and technology surveillance	ILCM2a
	Process to anticipate problems with suppliers and partners	ILCM2b
Organize the information	Process to ensure that the information is available	ILCM3
Process the information	Process to transform data into useful information	ILCM4
Maintain the information	Process to have updated databases	ILCM5
Disseminate the information	Process that ensure the distribution and exchange of information	ILCM6
Employ information technology to support daily operations	Daily operations supported by IT	ITI1
Employ information technology for business process support	Management of business processes automated and integrated using IT	ITI2a
	People management supported by IT	ITI2b
	Interaction and relationships with stakeholders strengthened with IT	ITI2c
Employ information technology for innovation support	Development and exchange of new ideas supported by IT	ITI3
Employ information technology for management support	Monitoring and analysis of internal or external business aspects supported by IT	ITI4
Employ information technology for strategy support	Anticipation of possible future scenarios supported by IT	ITI5a
	Competitive and technology surveillance supported by IT	ITI5b
Employ information technology for information sharing support	Exchange of information supported by IT	ITI6a
	Document location automation supported by IT	ITI6b
Prevent manipulating or hiding information	Absence of handling information for personal gain	AIBVC1
Establish formal and reliable sources of information	Embedded formal and reliable sources of information used by organization's members	AIBVC2
Transmit information about the performance of the company to all employees	Information about the performance of the company communicated to all employees	AIBVC3
Exchange of sensitive and non-sensitive information collaboratively	Exchange of sensitive and non-sensitive information collaboratively among team components and between areas	AIBVC4a
	Exchange of sensitive and non-sensitive information in a collaborative way with outside	AIBVC4b
Trust in each other	Failures discussions in an open and constructive manner and without fear of repercussions unfair	AIBVC5
Have concern with obtaining and applying new information	Quickly respond to changes and innovation promoted	AIBVC6

ILCM information life cycle management practice outcomes, *ITI* information technology integration practice outcomes, *AIBVC* assumed information behaviours and values convey practice outcomes

[a] Key to outcome "Code"

4 Results and Discussion

Table 2 presents a summary of the scores given to each statement by the respondents, separated by section. All the means are between 6 and 8.5, which implies that a high level of information practices is commonly held.

Figures 1, 2 and 3 also deal with the results pooled into three groups, showing the level of agreement between the respondents and the presence of practice outcomes.

Figure 1 displays the perceived level of practices associated with the management of the information life cycle (ILCM). Notice that the outcomes related to the practice defined as sense the information (ILCM2a and ILCM2b) are the ones with the lowest scores in this group. Interestingly, these organizations do not perceive the existence of competitive and technological surveillance processes when they are understood to be a key element in setting strategy. On the other hand, one of the outcomes related to the practice defined as collect the information has the highest score (ILMC1a), revealing that these kinds of organizations seems to be aware of the information that employees, customers, suppliers and other stakeholders need and they gather it systematically.

Figure 2 displays the perceived level of practices related to the integration of information technology in day-to-day business. It can be observed that according to the scores these companies don't seem to use information technologies (TICs) as support for competitive and technology surveillance (ITI5b), which is in accordance with the result pointed out in Fig. 1. Moreover, higher scores correspond to the use of TICs to support daily operations, improving individual productivity and the use of TICs to facilitate process monitoring and indicator analysis. The use of TICs is, in general, an increasingly widespread practice in organizations and, in the case of the organizations that are committed to total quality and thus, must monitor the processes having an information system that is supported by new technologies is very helpful.

Figure 3 displays the perceived level of practices that convey behaviours and values assumed by employees in relation to information use. According to the results shown in Fig. 3, the free exchange of sensitive information in a collaborative way with people outside the organization (customers, partners, suppliers, society at large) is the set of practices with the lowest score (AIBVC4b). On the other hand, the higher scores obtained by the AIBVC1, AIBVC2 and AIBVC3 outcomes emphasize that these organizations value people sharing sensitive information rather than manipulating or hiding it, that they have formal and reliable sources of information that their members use and that information about the performance of the company is revealed to all employees in order to influence and direct individual performance and consequently the company's performance.

Table 2 Quantitative results: mean ($\overline{\mathbf{x}}$) and deviation ($\boldsymbol{\sigma}$) of the questionnaire scores

	ILCM1a	ILCM1b	ILCM2a	ILCM2b	ILCM3	ILCM4	ILCM5	ILCM6		
$\overline{\mathbf{x}}$	8,05	7,42	6,47	7,00	7,49	7,40	7,42	7,56	–	–
$\boldsymbol{\sigma}$	1,23	1,59	2,00	1,91	1,45	1,50	1,58	1,39	–	–
	ITI1	*ITI2a*	*ITI2ba*	*ITI2c*	*ITI3*	*ITI4*	*ITI5a*	*ITI5b*	*ITI6a*	*ITI6b*
$\overline{\mathbf{x}}$	8,23	7,98	7,72	7,47	7,30	8,23	7,49	6,37	7,84	7,74
$\boldsymbol{\sigma}$	1,07	1,22	1,49	1,71	1,49	1,00	1,50	1,84	1,19	1,35
	AIBVC1	*AIBVC2*	*AIBVC3*	*AIBVC4a*	*AIBVC4b*	*AIBVC5*	*AIBVC6*	–	–	–
$\overline{\mathbf{x}}$	8,35	8,26	8,47	7,63	6,58	7,07	7,60	–	–	–
$\boldsymbol{\sigma}$	1,54	1,20	1,47	1,48	1,88	1,74	1,40	–	–	–

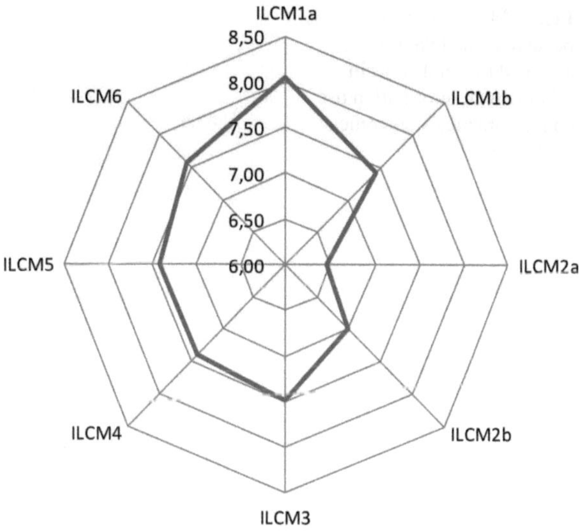

Fig. 1 Mean scores for management of the information life cycle practice outcomes (ILCM)

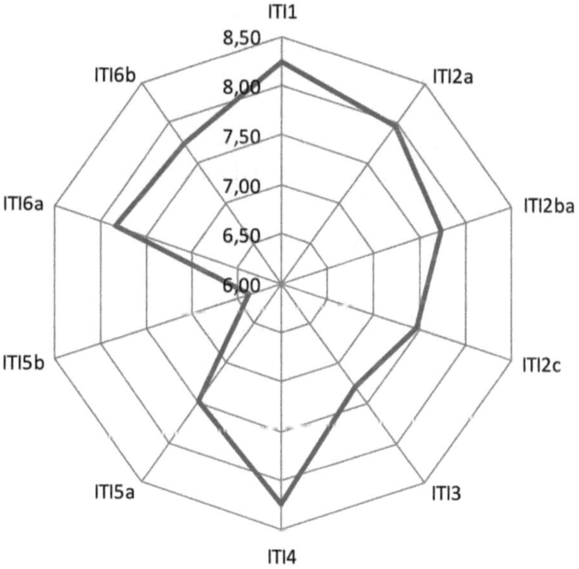

Fig. 2 Mean scores for integration of information technology in day-to-day business practice outcomes (ITI)

5 Conclusions

As a general conclusion, we see that the companies analysed seem to have developed information capability to the extent that many information practices are perceived as common practices; this was to be expected since those companies were committed to Total Quality Management and have even won a quality award.

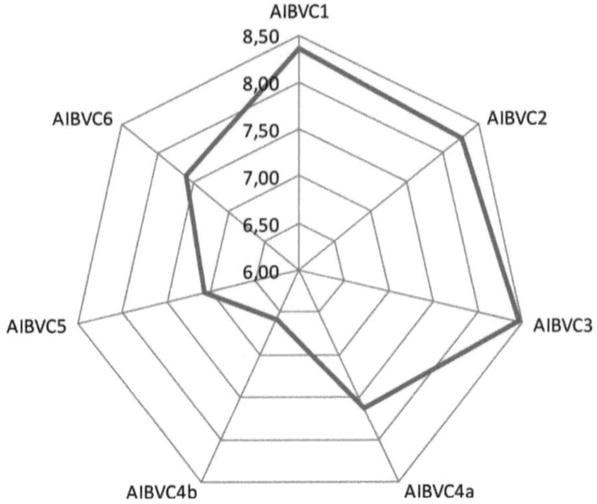

Fig. 3 Mean scores for behaviours and values assumed by employees in relation with information use convey practice's outcomes (AIBVC)

Correctly applying the EFQM model involves establishing processes that must be under control by using the right indicators. The monitoring of the whole performance of the quality system would be allowed by using the appropriate indicators that have to be produced and updated. An information system designed by taking information capability into account would be the most suitable.

The study has allowed areas of improvement to be identified by pointing out practices that, if they were commonly implemented, would improve the use and management of information and make it more efficient, which would be reflected in a company's results.

As for research limitations, the first one is that the questionnaires were answered by managers, which could have introduced a bias. Given that information flows throughout the company and involves everyone's work, it would be better to know the perception of all the employees. The second limitation is the small size of the sample.

References

1. Ashurst C, Doherty NF, Peppard J (2008) Improving the impact of IT development projects: the benefits realization capability model. Eur J Inf Syst 17:352–370
2. Carmichael F, Palacios-Marques D, Gil-Pechuan I (2011) How to create information management capabilities through web 2.0. Serv Ind J 31(10):1613–1625
3. Choo CW, Furness C, Paquette S, van den Berg H, Detlor B, Bergeron P, Heaton L (2006) Working with information: information management and culture in a professional services organization. J Inf Sci 32(6):491–510
4. Chou T, Chan P, Cheng Y, Tsai C (2007) A path model linking organizational knowledge attributes, information processing capabilities, and perceived usability. Inf Manage 44(4):408–417

5. Coltman T, Devinney TM, Midgley DF (2010) Customer relationship management and firm performance. J Inf Technol 26(3):205–219. (This journal article is available at Research Online: http://ro.uow.edu.au/commpapers/768)
6. Fok LY, Fok WM, Hatman SJ (2001) Exploring the relationships between total quality management and information system development. Inf Manage 38:355–371
7. Gorla N, Somers TM, Wong B (2010) Organizational impact of system quality, information quality, and service quality. J Strateg Inf Syst 19:207–228
8. Heras-Saizarbitoria I, Marimon F, Casadesús M (2012) An empirical study of the relationships within the categories of the EFQM model. Total Qual Manage Bus Excell 23(5–6):523–540
9. Hwang Y (2011) Measuring information behaviour performance inside a company: a case study. Inf Res 16(2), paper 480. (Available at http://InformationR.net/ir/16–2/paper480.html)
10. Lin S, Gao J, Koronios A, Chanana V (2007) Developing a data quality framework for asset management in engineering organizations. Int J Inf Qual 1(1):100–126
11. Marchand DA, Kettinger WJ, Rollins JD (2000) Information orientation: people, technology and the bottom line. Sloan Manage Rev 41(4):69–80
12. Matta K, Chen HG, Tama J (1998) The information requirements of total quality management. Total Qual Manage 9(6):445–461
13. Peppard J, Ward J (2004) Beyond strategic information systems: towards an IS capability. J Strategic Inf Syst 13:167–194
14. Sabherwal R, Chan YE (2001) Alignment between business and IS strategies: a study of prospectors, analyzers, and defenders. Inf Syst Res 12(1):11–33
15. Tee SW, Bowen PL, Doyle P, Rohde FH (2007) Factors influencing organizations to improve data quality in their information systems. Acc Financ 47:335–355
16. Wang WT (2012) Evaluating organisational performance during crises: a multi-dimensional framework. Total Qual Manage Bus Excell 23(5–6):673–688

Part II
Production

A Model of Makespan Flow-Shop Scheduling Under Ergonomic Requirements

Sabina Asensio Cuesta, Pedro Gómez Gasquet, Carlos Andrés Romano and María del Mar Eva Alemany Díaz

Abstract This paper address the recent growing interest the industrial environment has put over the healthy work. The ergonomics studies worker's fatigue and muscular stress risks. It shows the need to measure and evaluate risks to improve the efficiency and reduce the costs. Based on a literature review from the scheduling and ergonomics point of view implications are highlighted. The paper presents a MILP mathematical model to minimize makespan in an n-job flow shop problem with sequence dependent setup times considering recovery times.

Keywords Flow-shop · Setup time · Makespan · Ergonomic · Recovery time

1 Approach to an Ergonomic Assessment in Flow-Shop Scheduling

In a wide range of industrial situations companies usually work in flow-shop configurations, where there are m resources in series. Some papers [7, 8, 10] have contributed to reduce the gap between academic and real problems. Each job has to be processed on each one of the m resources (worker or machines). All jobs have to follow the same route, i.e., they have to be processed first on worker 1, then on

S. Asensio Cuesta (✉)
Departamento de Proyectos de Ingeniería, Escuela Técnica Superior de Ingeniería de Diseño, Universitat Politècnica de València Cno, De Vera s/n, 46022 Valencia, Spain
e-mail: sasensio@dpi.upv.es

P. Gómez Gasquet · M. del Mar Eva Alemany Díaz
Centro de Investigación de Gestión e Ingeniería de la Producción (CIGIP), Universitat Politècnica de València, Camino de Vera s/n, 46022 Valencia, Spain
e-mail: pgomez@cigip.upv.es

C. Andrés Romano
Research Group in Reengineering, Operations Management, Group Work and Logistics Excellence, Industriales, Universitat Politècnica de València Cno, De Vera s/n, 46022 Valencia, Spain
e-mail: candres@doe.upv.es

M. del Mar Eva Alemany Díaz
e-mail: mareva@cigip.upv.es

C. Hernández et al. (eds.), *Managing Complexity*, Lecture Notes in Management and Industrial Engineering, DOI 10.1007/978-3-319-04705-8_12,

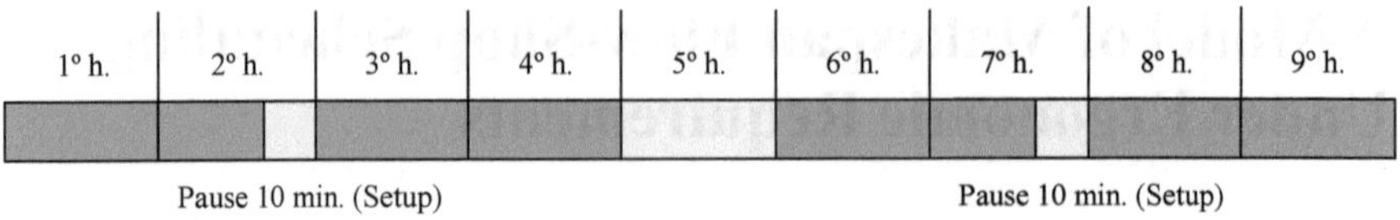

Fig. 1 Working scheduler with a single repetitive task is carried out. (Adapted from [4])

machine 2, and so on. After completion on one resource a job joins the queue at the next resource [3]. Musculoskeletal disorders (MSDs) are associated with maintenance of awkward postures, lifting loads, performing repetitive movements, etc. [1, 2, 5, 9] and stress [6]. In order to determine a quantitative ergonomic risk value in a flow-shop scheduling due to inadequate recovery periods for workers our proposal applied [4] approach. In [4] two methods to assess ergonomic risk level due to inadequate work-rest distribution are described. First method considers the optimal distribution ratio of repetitive tasks and recovery as 50 min work and 10 recovery. Under those criterion, the maximum continued period that can be spent carrying out repetitive task in condition defined as "acceptable" are 50 min. All extra minutes, continue with respect to those periods, which are spent without significant recovery periods, are considered as periods of potential overload. In proposed second method each hour is defined as being "risk-free" or "at risk". An overall risk is determined by the overall number of hours at risk. If the ratio between time work-recovery periods is 5:1 to 6:1, the hour is considered as being risk-free (risk 0). If ratio is between 7:1 and 11:1, the risk assessed is 0.5. If the work-recovery ratio exceeds 11:1, the risk factor is 1, because the ratio is judged as being unsatisfactory. Over an 8-h shift, interrupted by lunch break, but with no other pauses at all, the maximum score of 6 will be counted: in fact, the hour of work which is followed by the lunch break and the last hour of work in the sift, can be considered as "not risk", because they are followed by sufficient recovery periods. Figure 1 illustrate an example of task and breaks schedule and Table 1 shows total work time, rest time and risk values.

In Fig. 1 job A schedule lasts 460 min in that worker has 200 min of recovery and 260 min are spent in conditions of potential overload. In order to illustrate the complexity and the change of paradigm that is proposed, an example is shown in Fig. 2. In the upper part of figure jobs are scheduled without considering recovery periods, in the schedule 1 the sequence is A–B, in the schedule 2 is B–A, where process and setup times are indicated in the figure.

The makespan of permutation sequence B–A is 225 min, the best. But at the bottom, a rest period of 10 min is added every 50 min of work in both sequences. Considering ergonomic aspects the makespan A–B is the best option with 240 min. As it has been mentioned before, as a result of adding rest periods during a processing time sublot appear, job A is divided into 3 sublots A(50), A(30) and A(20) in schedule 1–e (Fig. 2).

Applying [4] methods flow-shop schedules can be assessed to determine ergonomic risk level due to inadequate work-rest distribution. In Fig. 3 that illustrates an example of flow-shop schedule without ergonomic restrictions worker 1 (Wk1)

Table 1 Counting time respectively in good recovery or in potential overload and the risk level due to "lack of recovery" for every hour. (Adapted from [4])

Hour	Work (min)	Recovery	Risk	Minutes spent after recovery	Minutes spent in potential overload
1	60	0 min	1	50	10
2	50	10 min (setup time)	0	–	50
3	60	0 min	1	50	10
4	60	0 min	0	–	60
5	0	60 min (lunch break)	0	–	–
6	60	0 min	1	50	10
7	50	10 min (setup time)	0	–	50
8	60	0 min	1	50	10
9	60	0 min	0	–	60
Total	*460*	*80 min*	*4*	*200*	*260*

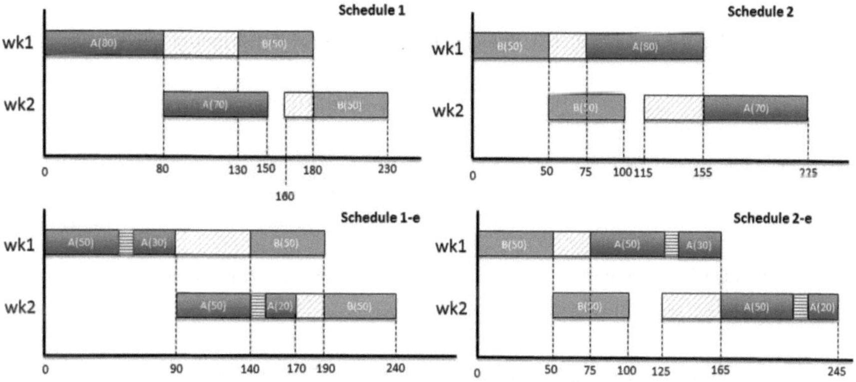

Fig. 2 An example of setup time sequence dependent flow-shop scheduled without consider rest periods (Schedule 1 and 2) and considering ergonomic rest period (Schedule 1–e and 2–e)

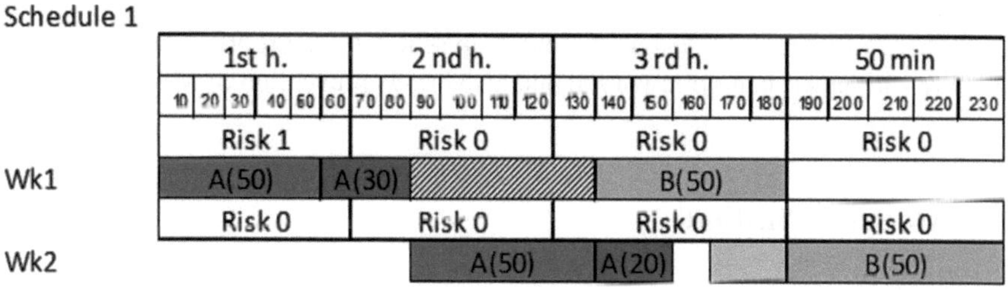

Fig. 3 Scheduler 1 ergonomic risk assessment due to work-rest distribution

spent 30 min in potential overload and the first hour has Risk level 1 (50 min without recovery), worker 2 (Wk2) 20 min with potential overload and there are not hours under risk. While in Fig. 4 after ergonomic restriction added workers are not in potential overload any minute.

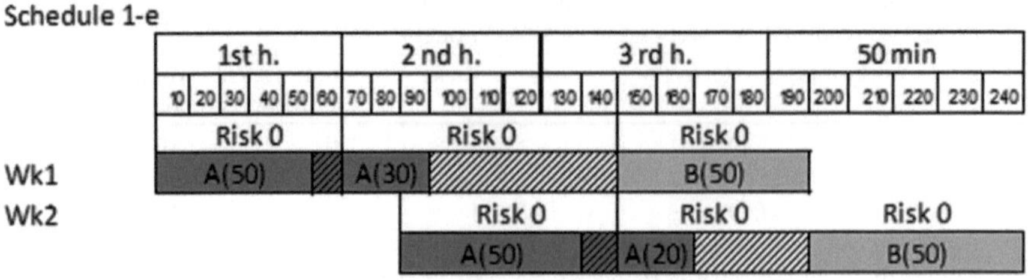

Fig. 4 Scheduler 1–e ergonomic risk assessment due to work-rest distribution

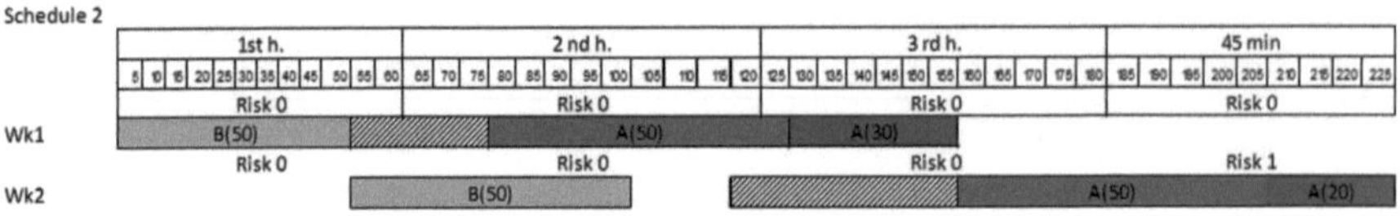

Fig. 5 Scheduler 2–e ergonomic risk assessment due to work-rest distribution

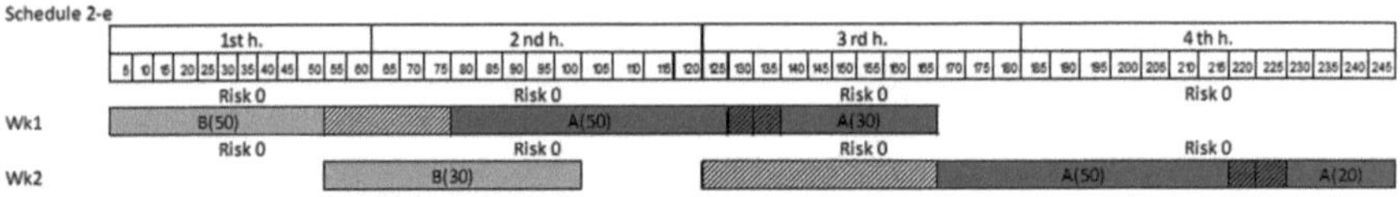

Fig. 6 Scheduler 2–e ergonomic risk assessment due to work-rest distribution

Figure 5 shows a schedule with a makespan of 225 where a worker 1 is under ergonomic risk 30 min because of not adequate rest periods are including and a worker 2 is in the same situation 20 min. However in schedule 2–e (Fig. 6) designed with ergonomic work-rest satisfactory ratio (5:1) workers a not exposed to overload due to lack of recovery period risk.

Thus, consider in the design of flow-shop schedules appropriate breaks is compatible with makespan optimization and also brings prevention of MSDs benefits.

2 Model Definition

In this section, a proposed MILP for a flow-shop scheduling problem is presented. Sequence dependent setup time and operations standby for a recovery time when exceeding the maximum recommend continuous working period considering the ergonomic aspect are the main characteristics of the proposed problem.

Using the classical notation the model will be presented as $n/F/SRW, SDST/C_{max}$. This model consists in a flow-shop (F) of r stages (R), one worker each stage, where n jobs (N) must be processed. The following assumptions are made: (1) all jobs are available at time zero; (2) the processing and setup time of each item is known and deterministic; (3) no preemption is allowed; (4) machines are available at any time; (5) each machine can process at most one job at a time; (6) each job can be processed on one machine at a time; (7) sequence dependent setup times (SDST) are considered; (8) and, job operation could be paused, standby recovery worker (SRW). In the MILP model jobs are divided in sublots if some recovery time is required in order to avoid working periods larger than e_{max}. In this case sublots are separate by a fixed recovery period T_R. The transfer unit between operations is the job, not the sublot. In other case, one sublot by job is considered. Setup times are always considered time for recovery the workers, so setup time are forced to be at least T_R period large. The information will be presented using the following indexes:

i, t index set of jobs {1..n};
l, v index set sublots of *i* job in *r* stage $\{1..Z_i\}$.
r index set of stages on the shop {1..R}

Parameters in the model are the data known beforehand:

Z_i number of units in job *i*
$P_{i,r}$ processing time for one unit of job *i* at stage *r*
$S_{t,i,r}$ setup time for job *i*, preceded by a job *t*, at the stage *r*
T_R recovery time for any worker after a working period
e_{max} ergonomic working time. Maximum working time after a recovery
M a positive number larger than makespan
M' a positive number larger than maximum processing time of any job in any stage + T_R

MILP model determines the following variables:

$X_{l,i,r}$ (integer) number of units in sublot *l* of job *i* at stage *r*
$C_{l,i,r}$ (integer) completion time of sublot *l* of job *i* at stage *r*
$S^X_{t,i,r}$ (integer) extended setup time for job *i*, preceded by a job *t*, at the stage *r*

$$q_{t,i,r} = \begin{cases} 1, \text{ if job } t \text{ is performed before job } i \text{ at stage } r \\ 0, \text{ otherwise} \end{cases}$$

$$\mathbf{h}_{l,v,i,r} = \begin{cases} \mathbf{1, if\ sublot\ } \boldsymbol{l} \mathbf{\ is\ performed\ before\ sublot\ v\ of\ job\ } \boldsymbol{i} \mathbf{\ at\ stage\ } \boldsymbol{r} \\ \mathbf{0, otherwise} \end{cases}$$

$$Y_{l,i,r} = \begin{cases} 1, \text{ if sublot } l \text{ of job } i \text{ at stage } r \text{ } is\ processed\ (X_{l,i,r} > 0) \\ 0, \text{ otherwise} \end{cases}$$

With these notations, the problem can be formulated as the following MILP model. The objective is to minimize makespan (1):

$$\boldsymbol{F.O.\ min\ z = C_{max}} \tag{1}$$

The constraints of the model are presented below in two sets, each representing one type of system constraint. The model is subject to:

Precedence constraints: This set of constrains ensures the processing order of jobs and sublots.

$$C_{max} \geq C_{l,i,r} \quad \forall l,\ \forall i,\ \forall r \tag{2}$$

$$C_{l,i,r} \geq C_{v,t,r} + P_{i,r} * X_{l,i,r} + S^{X}_{t,i,r} + M*\left(q_{t,i,r} - 1\right) \quad \forall l,\ \forall v,\ \forall t,\ \forall i,\ \forall r\ t \neq i \tag{3}$$

$$C_{l,i,r} \geq C_{v,t,r} + P_{i,r} * X_{l,i,r} + S^{X}_{t,i,r} - M*q_{i,t,r} \quad \forall l,\ \forall v,\ \forall t,\ \forall i,\ \forall r\ t \neq i \tag{4}$$

$$C_{l,i,r} \geq C_{v,i,r} + P_{i,r} * X_{l,i,r} + T_{R} * Y_{v,i,r} + M' * (h_{v,l,i,r} - 1) \quad \forall l,\ \forall v,\ \forall i,\ \forall r > 1\ l \neq v \tag{5}$$

$$C_{l,i,r} \geq C_{v,i,r} + P_{i,r} * X_{l,i,r} + T_{R} * Y_{v,i,r} - M' * h_{l,v,i,r} \quad \forall l,\ \forall v,\ \forall i,\ \forall r > 1\ l \neq v \tag{6}$$

$$C_{l,i,r} \geq C_{v,i,r-1} + P_{i,r} * X_{l,i,r} \quad \forall l,\ \forall v,\ \forall i,\ \forall r > 1 \tag{7}$$

$$C_{l,i,r} \geq P_{i,r} * X_{l,i,r} \quad \forall l,\ \forall i,\ \forall r \tag{8}$$

The constraint set (2) determines maximum completion time or makespan. Constraint (3) y (4) ensures that a job cannot start before the previous job at the same stage r has been completely processed. Constraint (5) y (6) ensures that any two sublots of any job are processed simultaneously and between both a recovery time T_R is added. Constraint (7) ensures that any sublot of a job can start in the next stage before all sublots have been completed in the actual stage. Constraint (8) ensures first sublot start time is not negative. Constraints related to sublot sizes and duration of working and recovery periods:

$$\sum_{l=0}^{Li} X_{l,i,r} = Z_{i} \quad \forall i,\ \forall r \tag{9}$$

$$P_{i,r} * X_{l,i,r} \leq e_{max} \quad \forall l,\ \forall i,\ \forall r \tag{10}$$

$$S^{X}_{t,i,r} \geq S_{t,i,r} \quad \forall t,\ \forall i,\ \forall r\ i \neq t \tag{11}$$

$$S^{X}_{t,i,r} \geq T_{R} \quad \forall t,\ \forall i,\ \forall r\ i \neq t \tag{12}$$

$$X_{v,i,r} - Z_i * Y_{l,i,r} \leq M' * \left(1 - h_{l,v,i,r}\right) \quad \forall l,\ \forall v,\ \forall i,\ \forall r\ l \neq v \tag{13}$$

Constraint (9) ensures that all the units are processed for all jobs at all the stages. Constraint (10) ensures that the processing time of each sublot will remain less than ergonomic time for all jobs at all stages. Constraint (11) and (12) ensures setup time considered between two jobs includes at least a recovery time period. And constraint (13) ensures that if a sublot v is processed ($X_{v,\, i,r}=0$) after a sublot l of any job then the value of $Y_{l,\, i,r}$ let add a recovery time T_R in constraint (5) and (6).

3 Conclusions

The proposed model is a suitable tool for designing of shop-flow schedules where the makespan is minimized and, at the same time, ergonomic work-recovery periods are included in order to prevent work related MSDs. An interesting challenge from research point of view, where further research needs to be carried out in order to exploit this technic/concept properly. This approach might improve companies' efficiency it allows us to reflect reality and thus include restrictions on models up often avoided. And we should not forget that these problems, once we introduce SDST combined with lot splitting, are probably NP-hard. The relevance of differentiating between the machine and the operator resource has emphasized. New requirements based on health-related and working conditions have emerged. Although the industrial reality is complex, has shown an example that serves to illustrate some effects on the makespan when considering new restrictions. In the future two paths are open. On one hand, contribute with new methods that should be capable of providing solutions in a reasonable time for realistic problems. On the other hand, our research will introduce and consider more realistic approaches such as Hybrid FS environments (HFS) or HFS/FS and lot streaming combination.

Acknowledgments This research has been carried out in the framework of a project funded by the Spanish Ministry of Economy and Competitiveness (Ref. DPI2011-23597), a Project funded by Consellería de Educación, Cultura y Deportes de la Generalitat valenciana "Programación de la Producción con Partición Ajustable de Lotes en entornos de Planificación mixta Pedido/Stock (PP-PAL-PPS)" (ref. GVA/2013/034" and two projects funded by Universitat Politècnica de Valencia Program for the Support of Research and Development 2011: "Methods and models for operations planning and order management in supply chains characterized by uncertainty in production due to the lack of product uniformity" (Ref. PAID-06-11/1840); "Analysis of the interactions of ergonomic risk factors in the development of work related musculoskeletal disorders and a predictive model design". (Ref. PAID-06-11/2063).

References

1. Asensio-Cuesta S, Diego-Mas JA, Canós-Darós L (2012) A genetic algorithm for the design of job rotation schedules considering ergonomic and competence criteria. Int J Adv Manuf Technol 60:1161–1174
2. Bernard B (1997) Musculoskeletal disorders and workplace factors: a critical review of epidemiological evidence for work-related musculoskeletal disorders of the neck, upper extremity, and low back. National Institute for Occupational Safety and Health (NIOSH), Cincinnati
3. Brucker P (2004) Scheduling algorithms, 4th edn. Springer, Berlin
4. Colombini D, Occhipinti E, Grieco A (2002) Risk assessment and management of repetitive movements and exertions of upper limbs. Elsevier, Oxford
5. Diego-Mas JA, Asensio-Cuesta S, Sanchez-Romero MA, Artacho-Ramirez MA (2009) A multi-criteria genetic algorithm for the generation of job rotation schedules. Intl J Ind Ergon 39(1):23–33
6. Faragher EB, Cass M, Cooper CL (2005) The relationship between job satisfaction and health: a meta-analysis. Occup Environ Med 62:105–112
7. Framinan JM, Gupta JND, Leisten R (2004) A review and classification of heuristics for permutation flow-shop scheduling with makespan objective. J Oper Res Soc 55(12):1243–1255. doi: 10.1057/palgrave.jors.2601784
8. Gomez-Gasquet P, Andres C, Lario FC (2012) An agent-based genetic algorithm for hybrid flowshops with sequence dependent setup times to minimise makespan. Expert Syst Appl 39(9):8095–8107. doi: 10.1016/j.eswa.2012.01.158
9. Neumann WP, Winkel J, Medbo L, Magneberg R, Mathiassen SE (2006) Production system design elements influencing productivity and ergonomics: a case study of parallel and serial flow strategies. Int J Oper Prod Man 26(8):904–923
10. Ruiz R, Maroto C (2005) A comprehensive review and evaluation of permutation flowshop heuristics. Eur J Oper Res 165(2):479–494. doi: 10.1016/j.ejor.2004.04.017

A MILP Event Based Formulation for a Real-world Multimode RCSP with Generalized Temporal Constraints

Tamara Borreguero-Sanchidrián, Álvaro García-Sánchez and Miguel Ortega-Mier

Abstract Scheduling is becoming much more important in every industry. However, the standard RCSP usually does not cover all the characteristics of real world problems. In this work, we present an Event Based MILP formulation for a Multimode Resource Constraint Scheduling Problem of direct application for some industries, as aeronautical assembly lines. Taking as a starting point one of the last MILP formulations for standard RCSP, our contribution is to provide a formulation which covers the multimode case and more general temporal constraints than the ones usually referred to in the literature.

Keywords Scheduling · Multimode · Event based formulation · Temporal constraints · MILP

1 Introduction

Over the last years, the continuous changes on every industry have forced enterprises to explore new manufacturing methods in order to comply with the OTOQOC paradigm (On time, On Quality, On Cost). Production systems based on the Toyota Production System have spread worldwide as a means of reducing waste and optimizing manufacturing processes. The aeronautical industry, since the 1990's has been including the lean techniques into its production systems. Scheduling and line balancing have therefore become two main enablers for lean implementation.

T. Borreguero-Sanchidrián (✉) · Á. García-Sánchez · M. Ortega-Mier
Grupo IOL, Dpto. Ingeniería de Organización, Administración de Empresas y Estadística. ETSII, UPM Madrid, Madrid, Spain
e-mail: tamaraborreguero@alumnos.upm.es

Á. García-Sánchez
e-mail: alvaro.garcia@upm.es

M. Ortega-Mier
e-mail: miguel.ortega.mier@upm.es

C. Hernández et al. (eds.), *Managing Complexity*, Lecture Notes in Management and Industrial Engineering, DOI 10.1007/978-3-319-04705-8_13,

Aeronautical Final Assembly Lines consist on different platforms or stations. Each platform has a fixed team of workers with different skills. The line balancing enables the distribution of the work tasks among the different platforms. Afterwards, the work tasks from each platform must be scheduled in order to complete them within the required Takt Time using the minimum number of operators. We will refer to the scheduling of the tasks as the Aeronautical Platform Scheduling Problem.

This detailed scheduling has the structure of a Resource Constrained Scheduling Problem (RCSP), which was defined by [2] as the allocation of scarce resources to dependent activities over time. It is a NP Hard optimization problem and is actually one of the most intractable classical problems in practice.

There have been a wide range of studies on both heuristic and metaheuristic methods for solving the RCSP, as well as different MILP models. Recently, Koné [4] proposed the use of Event Based Formulations for the RCSP. He provided a benchmark of different models (including MILP and a heuristic) and concluded that event based formulations outperformed the previous MILP models and performed even better than the heuristic for some instances.

However, Koné's Event Based Formulations deal with the standard RCSP, which includes some assumptions that are too restrictive for many practical applications. Therefore, it is of great interest to improve this kind of formulations so that they can be used on industrial applications. On this work, we have developed a new Event Based Formulation that covers the characteristics of the Aeronautical Platform Scheduling Problem. Actually, its main contribution is the allowance of multiple modes per task as well as the use of more general temporal constraints. Furthermore, the alternative objective approach focused on minimizing the cost, is more suitable for nowadays industries where the total Lead Time is usually fixed by the client Takt Time.

Section 2 provides the general classification of the Aeronautical Platform Scheduling problem. Section 3 gives an overview of existing exact formulations for the RCSP. In Sect. 4 we propose the event based multimode formulation with additional temporal constraints. Section 5 contains the results and conclusions of tests performed on several instances and Sect. 6 some conclusions and future research directions.

2 The Aeronautical Platform Scheduling Problem

The RCSP is a combinatorial optimization problem, defined by a 6-tuple (V, p, E, R, B, b), where V is a set of activities, p a vector of processing times per activity, E the set of temporal constraints, R the set of resources, B the resource capacity vector and b the demand matrix (resource consumption per activity) (Koné et al 2011). The objective is to identify a feasible schedule, which assigns a start/completion time (Sj/Cj) to each activity as well as a resource allocation, taking into account the temporal constraints and minimizing the total project lead time.

Until 1999, there was not a common notation for RCSP. [2] proposed a notation based on the extension of the α| β|γ generalized scheme for the machine scheduling literature. In this notation, α refers to the *resource environment*, β to the *activity characteristics* and γ to the *objective function*. According to this notation, the Aeronautical Platform Scheduling Problem is classified as MPSm, σ, r | temp | $\sum c_k \max r_k(S,t)$:

- α = MPSm, σ, r. This stands for a multimode resource constraint project where each activity can be processed in several alternative modes and there exists a set of renewable resources available for each time period during the project execution: *m* being the resources, *σ* the units of each resource available and *r* the maximum number of units of the resources demanded by an activity. For our particular problem, the activities are the work tasks assigned to each platform. The renewable resources are the number of operators (each of them belonging to a particular skill) and the space on the working areas, as platforms are divided into smaller areas where a limited number of operators can work simultaneously. As well as this, each mode for an activity defines a combination of operator skills, number of operators and durations. All the operators assigned to an activity must be from the same skill and the range of possible numbers of allocated operators per tasks is independent from the chosen skill.
- β = *temp*. Among the temporal constrains, there are precedence constraints (task *w'* cannot start until task *w* has been completed), non-parallel constraints (tasks *w* and *w*' cannot be in progress at the same time, but there is no precedence relation between them), and maximal time lags between tasks (task *w'* must start within a maximal time after *w* has been completed).The maximal time lag, if it exists, will be zero for all the pairs of activities. All the temporal constraints are independent from the mode in which a task is executed.
- $\gamma = \sum c_k \max r_k(S,t)$. The objective function is to minimize the resource investment. The total lead time is fixed by the assembly line Takt Time, as stated on Section 1. Therefore, the objective function is to minimize the labor cost of the assembly. The operators, once assigned to a platform stay working on it for all the Takt Time and thus minimizing the labor cost is equivalent to minimizing the maximum number of operators needed throughout the Takt Time.

3 Exact Formulations for the RCSP

Most of the research on RCSP has focused on the core single-mode problem with precedence and resource constraints. In this section, we will review MILP formulations for this core problem. They can be divided on three main groups:

- *Discrete Time Formulations*: In them, the time horizon is divided into time slots. The basic discrete time formulation was proposed by Pritsker (1969). Afterwards, Christofides (1987) proposed the Disaggregated Discrete Time formulation (DDT) that implies a larger number of constraints but, on the other hand, is a tighter formulation and therefore its linear relaxation provides a better LB. The main drawback of discrete time formulations is the increase in the number of variables as the time horizon grows.

- *Continuous time formulation*: In this formulations, the time is represented by continuous variables: Alvarez-Valdés and Tamarit (1993) studied *Forbidden Sets Formulations*which involve a high number of constraints that grows exponentially and cannot be used in practice. *Flow-Based Continuous Time Formulations*, described by Artigues. [1] provide a poor relaxation, compared to discrete time formulations, although they can be preferable to them for instances involving large time scale.
- *Event Based Formulations:* Event Based Formulations for the RCSP where developed by Koné in 2011 [4], from a model introduced by Zapata (2008). These formulations define a series of events which correspond to the start or end of the different activities. They are based on the fact that for the RCSP it always exists an optimal semi-active schedule in which the start time of an activity is either 0 or coincides with the completion time of another activity (Sprecher 1995). Therefore, at most $n+1$ events have to be considered. They have the advantage of not depending on the time horizon, making them especially relevant for long time projects, as is the case.

Among Event Based Formulations, the Start/End Event Based Formulation involves two types of binary variables, x_{we} and y_{we}, that are equal to 1 if task w starts (in the case of x_{we}) or ends (y_{we}) at event e and 0 otherwise.

4 Model Formulation

The Start/End Event Based Formulation (SEE) has been used as a starting point for an extended formulation that copes with the multimode problem and the additional maximal time lag and non-parallel constraints as explained on Sec. 2. The resulting formulation uses four sets: O stands for the operator profiles, W for the work tasks, A for the working areas and E for the events. The model parameters are defined in Table 1.

Due to the new characteristics of our model, we replace the original SEE variables x_{we} and y_{w} with variables x_{weon} and y_{weon}, to be set to 1 if task W starts or ends on event e, using n operators of profile o. As well as this, we define the non-negative variables r^*_{oe} to represent the amount of resource o needed immediately after event e and non-negative variables s^*_{ae} to represent the number of operators working on area o immediately after event e A binary variable $\alpha_{ww'}$ is defined for tasks with non-parallel constraints ($w,w' \in W$ *and* $NONP_{ww'}=1$ *and* $w<w'$) set to 1 if w ends before w' starts and 0 viceversa. A continuous variable $t^i_w \geq 0$ defines the starting time of a task. This will be used for maximal time lag constraints, and defined $\forall w, w' \in W \, and \sum_{w'} MTL_{ww'} + MTL_{w'w} > 0$.

Table 1 Parameters

Parameter	Definition
D_w	Total amount of working hours for task w ∈ W, if assigned only to one operator
G_{nw}	Reduction coefficient to the work task w's makespan when it is done by n operators, $w \in W, MIN_w^{op} < n < MAX_w^{op}$
Pow	1 if task w ∈ W can be done by operators with profile o ∈ O, 0 otherwise
MAX_w^{op}	Maximum number of operators that can work on task w
MIN_w^{op}	Minimum number of operators that can work on task w
$PRE_{ww'}$	1 if the precedence graph includes a precedence relationship between work tasks w and w′, w, w′ ∈ W
$NONP_{ww'}$	1 if the precedence graph includes a non-parallel constraint between work tasks w and w': w, w′ ∈ W and w<w′
$MTL_{ww'}$	1 if the precedence graph includes a maximal time lag constraint between w' and w, w, w′ ∈ W ($MTLww, \leq PREC_{ww'}$)
AR_{aw}	1 if work task w is done on area a, 0 otherwise
CAP_a	Maximum number of operators that can work on area a, a ∈ A
LT	Lead time
M	Big enough number

The continuous non-negative variable t_e represents the time of event e, and the free variable num_o^{op} is used for the total number of operators of profile o needed.

The formulation can be written as follows (domain restrictions omitted):

$$\min.\left(\sum\nolimits_o num_o^{op}\right) \tag{1}$$

Subject to:

$$t_0 = 0 \tag{2}$$

$$t_{e+1} - t_e \geq 0 \forall e \in E <> last(e) \tag{3}$$

$$t_e \leq LT \forall e \in E \tag{4}$$

$$\sum\nolimits_{eon} ey_{weon} - \sum\nolimits_{eon} ex_{weon} \geq 1 \forall w \in W \tag{5}$$

$$\sum\nolimits_{eon} x_{weon} = 1 \forall w \in W \tag{6}$$

$$\sum\nolimits_{eon} y_{weon} = 1 \forall w \in W \tag{7}$$

$$t_f - t_e - \sum\nolimits_{won} D_w G_{nw} x_{weon} + D_w G_{nw}\left(1 - \sum\nolimits_{won} y_{wfon}\right) \geq 0 \forall f > e, w \in W, MIN_w^{op} \leq n \leq MAX_w^{op} \tag{8}$$

$$\sum_{eo} x_{weon} = \sum_{eo} y_{weon} \; \forall w \in W, MIN_w^{op} \le n \le MAX_w^{op} \tag{9}$$

$$\sum_{en} x_{weon} = \sum_{en} y_{weon} \; \forall w \in W, o \in O / P_{ow} = 1 \tag{10}$$

$$\sum_{e''_{on}=0}^{e-1} x_{we''on} + \sum_{e'_{on}=e}^{E} y_{w'e'on} \le 1 \forall \text{ w w}' / PRE_{ww'} = 1, \forall \text{e} \in \text{E} \tag{11}$$

$$t_w^i \ge t_e - M\left(1 - \sum_{on} x_{weon}\right) \forall e \in E, \forall w / \sum_{w'} MTL_{ww'} + MTL_{w'w} > 0 \tag{12}$$

$$t_w^i \ge +M\left(1 - \sum_{on} x_{weon}\right) \forall e \in E, \forall w / \sum_{w'} MTL_{ww'} + MTL_{w'w} > 0 \tag{13}$$

$$t_{w'}^i - t_w^i - \sum_{eon} G_{nw} DUR_w x_{weon} \le 0, \forall w, w' / MTL_{ww'} = 1 \tag{14}$$

$$\sum_{eon} ey_{weon} - \sum_{eon} ex_{w'eon} \le M\left(1 - \alpha_{ww'}\right) \forall w, w' / NONP_{ww'} = 1 \tag{15}$$

$$r_{o0}^* - \sum_{wn} nx_{w0on} = 0 \forall o \in O \tag{16}$$

$$r_{oe}^* - r_{oe-1}^* + \sum_{\substack{w \\ P_{ow=1}}} \left(\sum_n ny_{weon} - \sum_n nx_{weon}\right) = 0 \forall o \in O, e \in E \tag{17}$$

$$r_{oe}^* \le num_o^{op} \forall o \in O, e \in E \tag{18}$$

$$s_{ao}^* - \sum_{won} nx_{w0on} AR_{aw} = 0 \forall a \in A \tag{19}$$

$$s_{ae}^* - s_{ae-1}^* + \sum_w \left(\sum_{on} ny_{weon} AR_{aw} - \sum_{on} nx_{weon} AR_{aw}\right) = 0 \forall a \in A, e \in E \tag{20}$$

$$s_{ae}^* \le CAP_a \forall a \in A, e \in E \tag{21}$$

Equation (1) is the objective function: to minimize the total number of operators. Constraint (2) forces the first event to begin at t=0 and constraint (4) assures that there is no delay in the task completion. The order of the events on time is imposed by constraint (3). Constraint (5) states that the start event of a task must precede its end event. Constraints (6) and (7) limit to one the start and end per work task.

Table 2 Instance characteristics

Set	Precedences	∑MTL	∑NONP	Op. profiles	Areas	No. modes
Set1	6	1	1	2	2	12
Set2	8	1	1	2	2	16
Set3	7	1	1	2	2	17
Set4	7	1	1	2	2	16

Constraint (8) fixes the minimum time difference between the start and the end events to the duration of the task. A single mode for performing the task is imposed by constraints (9) and (10). As for the relations between tasks: (11) is the multimode expression for the precedence constraints. Maximal time lags equal to zero for consecutive events are expressed on constraints (12) to (14). Non-parallel constraints are (15). Resource needs are expressed on equations (16) to (18) in the case of operators and (19) to (21) for the available capacity per area.

5 Results

The computational results were obtained using CPLEX12.4 solver. The tests were carried out on an Intel-Core i7-2630QM processor with 2 GHz and 4 GB RAM, running Windows 7. As the standard PSPLIB instances are not valid for the structure of the problem, four different sets of 8 task instances were used. Table 2 shows the main instance characteristics. Sets 3 and 4 were also extended in order to create instances of up to 11 tasks.

All instances were solved up to optimality, taking times from seconds to fifteen minutes. The solution time grew exponentially with the number of events, even when solving the same set of instances, see Fig. 1. Defining fewer events has also a high impact on the first LP bound, which is tighter. There is always a optimal solution with no more than a number of events equal to $Card(W)+1$. However, all the tested instances had an optimal solution with fewer events than that minimum number.

For each of the eight-task instances, different Lead Times were tested. On average, the solution time also grew as the objective Lead Time approached to the Critical Path Lead Time, see Table 3. Most of the instances require more solution time with the same number of events when new tasks are added. Withal, some were solved faster with more tasks. This shows that in some cases the structure of the problem is more important than the number of tasks itself. The detailed instances and computational results are available on [3].

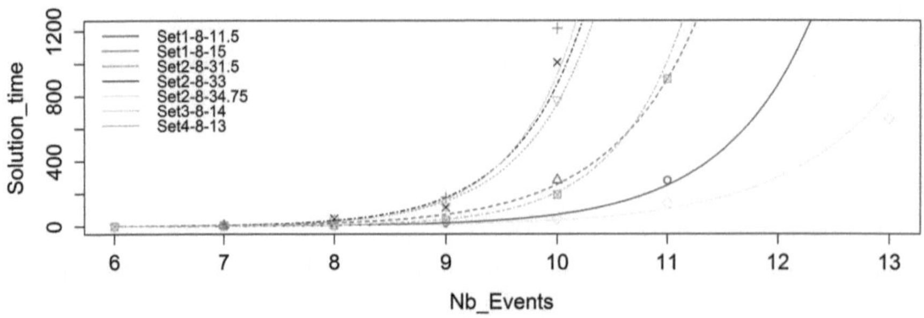

Fig. 1 Solution Time per instance with different number of events

Table 3 Sample solving time for different Lead Times

Instance	LT=31.5	LT=33	LT=34.75	LT=41
Set2_8 Tasks	10.2s	6.65s	1.79s	0.83s

6 Conclusions

This work provides a new MILP formulation for a real case of MRCSP. It is a first insight on the problem and has helped us identify directions for a further research. In order to extend it to bigger instances it is necessary to make a focus on the use of pre-processing to calculate the needed number of events.

As well as this, the solution times have been different for each set of instances, although they had the same task dimension. A characterization of Aeronautical Platform Scheduling Problem instances is required in order to improve the formulation and develop pre-processing techniques suitable for both the formulation and the data sets.

References

1. Artigues C, Michelon P, Reusser S (2003) Insertion techniques for static and dynamic resource-constrained project scheduling, Eur J Oper Res 149:249–267
2. Brucker P, Drexl A, Mhring R, Neumann K, Pesch E (1999) Resource-constrained project scheduling: Notation, classification, models and methods. Eur J Oper Res 112(1):3–11
3. APSP. https://www.dropbox.com/sh/9070da1dz6typ2o/V0jCAtR_8y
4. Koné O, Artigues C, Lopez P, Mongeau M (2011) Event-based MILP models for resource-constrained project scheduling problems. Comput Oper Res 38(1):3–13

NTIGen: A Software for Generating Nissan Based Instances for Time and Space Assembly Line Balancing

Manuel Chica Serrano, Óscar Cordón García, Sergio Damas Arroyo and Joaquín Bautista

Abstract The time and space assembly line balancing problem (TSALBP) is a realistic multiobjective version of assembly line balancing industrial problems involving the joint optimization of conflicting criteria such as the cycle time, the number of stations, and the area of these stations. For this family of problems there is not any repository where researchers and practitioners can obtain realistic problem instances also containing information on mixed products plans. In this contribution we introduce a new TSALBP instance software generator that can produce problem instances having industrial real-like features. This generator is called NTIGen (Nissan TSALBP Instance GENerator) since it is developed from the information and real data of the assembly line and production planning of the Nissan plant of Barcelona. The NTIGen software as well as some benchmark instances are publicly available on Internet and could be used by researchers to carry out general TSALBP experiments and to also discriminate between different assembly line configurations when future demand conditions vary.

Keywords Time and space assembly line balancing · Problem instance generator · Mixed products · Nissan · Optimization

M. Chica Serrano (✉) · Ó. Cordón García · S. Damas Arroyo
European Centre for Soft Computing, Gonzalo Gutiérrez Quirós s/n,
33600 Mieres, Asturias, Spain
e-mail: manuel.chica@softcomputing.es

Ó. Cordón García
e-mail: oscar.cordon@softcomputing.es

S. Damas Arroyo
e-mail: sergio.damas@softcomputing.es

J. Bautista
Dpto. Organización de Empresas, ETSEIB, UPC Catalunya, 08028 Barcelona, Spain
e-mail: joaquin.bautista@upc.edu

C. Hernández et al. (eds.), *Managing Complexity*, Lecture Notes in Management and Industrial Engineering, DOI 10.1007/978-3-319-04705-8_14,

1 Introduction

An assembly line is made up of a number of workstations, arranged either in series or in parallel. Since the manufacturing of a production item is divided into a set of tasks which require an operation time for their execution, a usual and difficult problem, called assembly line balancing (ALB), is to determine how these tasks can be assigned to the stations fulfilling certain restrictions such as precedence relations. The final aim of ALB is to get an optimal assignment of subsets of tasks to the stations of the plant [4]. A well-known family of ALB problems is the simple assembly line balancing problem (SALBP) [2, 10]. The SALBP only considers the assignment of each task to a single station in such a way that all the precedence constraints are satisfied and no station workload time is greater than the line cycle time.

As a result of the observation of the ALB operation in an automotive Nissan plant from Barcelona (Spain), Bautista and Pereira [1] recently proposed a SALBP extension aiming to design a more realistic ALB model. They considered an additional space constraint to get a simplified but closer version to real-world situations, defining the time and space assembly line balancing problem (TSALBP). The TSALBP presents eight variants depending on three optimization criteria: m (the number of stations), c (the cycle time), and A (the area of the stations). The multicriteria nature of the TSALBP favoured the application of multiobjective meta-heuristics such as multiobjective ant colony optimization [5], evolutionary multiobjective optimization [6], and memetic algorithms [7].

However, we noticed the absence of an available dataset with real-like instances for the TSALBP. And what is more, there is not any instance containing mixed product plans when the demand is uncertain. Therefore, we have implemented real-like Nissan TSALBP instance generator software (NTIGen) in order to let researchers to validate their models and methods in a diverse set of TSALBP instances and production plans. The design and implementation of NTIGen is done with the use of real data and industrial features of the Nissan industry plant of Barcelona. The software is freely available on-line to be used for future research works. Using this tool, a set of eight instances has been generated as a benchmark to show the different features of the instances.

The rest of the paper is structured as follows. In Sect. 2 the TSALBP formulation is explained. The description of the NTIGen software is shown in Sects. 3 and 4. A comparison of the eight instances generated by NTIGen is shown in Sect. 5. Some concluding remarks are given in Sect. 6.

2 Time and Space Assembly Line Balancing Problem

The manufacturing of a production item is divided into a set J of n tasks. Each task j requires an operation time for its execution $t_j > 0$ that is determined as a function of the manufacturing technologies and the employed resources. Each station k ($k = 1,$

2,…, m) is assigned to a subset of tasks S_k, called workload. Each task *j* can only be assigned to a single station *k*.

Each task *j* has a set of direct "preceding tasks" P_j which must be accomplished before starting it. These constraints are normally represented by means of an acyclic precedence graph, whose vertices stand for the tasks and where a directed arc *(i, j)* indicates that task *i* must be finished before starting task *j* on the production line. Thus, task *j* cannot be assigned to a station that is ordered before the one where task *i* was assigned. Each station *k* also presents a station workload time $t(S_k)$ that is equal to the sum of the tasks' processing time assigned to the station *k*. SALBP focuses on grouping tasks in workstations by an efficient and coherent way.

In this simplistic model there is a need of introducing space constraints in assembly lines' design based on two main reasons: (a) the length of the workstation is limited in the majority of the situations, and (b) the required tools and components to be assembled should be distributed along the sides of the line. Hence, an area constraint may be considered by associating a required area a_j to each task *j* and an available area A_k to each station *k* that, for the sake of simplicity, we shall assume it to be identical for every station and equal to $A = max_k = 1,2,\dots,m_{Ak.}$ Thus, each station k requires a station area $a(S_k)$ that is equal to the sum of areas required by the tasks assigned to station *k*.

This leads us to a new family of problems called TSALBP [1]. It may be stated as: given a set of *n* tasks with their temporal t_j and spatial a_j attributes ($1 \leq j \leq n$) and a precedence graph, each task must be assigned to a single station such that: (i) every precedence constraint is satisfied, (ii) no station workload time ($t(S_k)$) is greater than the cycle time (*c*), and (iii) no area required by any station ($a(S_k)$) is greater than the available area per station (*A*).

TSALBP presents eight variants depending on three optimization criteria: *m* (the number of stations), *c* (the cycle time) and *A* (the area of the stations). Within these variants there are four multiobjective problems depending on the set of criteria to be minimized (*m*, *c* and/or *A*). For more information about the problem we refer the interested reader to Chica et al. [5, 7].

3 Basics of the Nissan TSALBP Instance Generator Software (NTI-Gen)

The main goal of the NTIGen software is to create real-like TSALBP instances with different features to serve as a benchmark for any future research work. Although there are ALB instances available online and even a SALBP instance generator [9], there is not any existing source where TSALBP instances can be generated and referred. Also, as pointed out in the Introduction section, there is no instance containing production plans information.

Assembly lines in the automotive industry present a set of industrial features which condition the task and graph distribution of the problem instance. The user must be allowed to incorporate these industrial real-like features to the generated

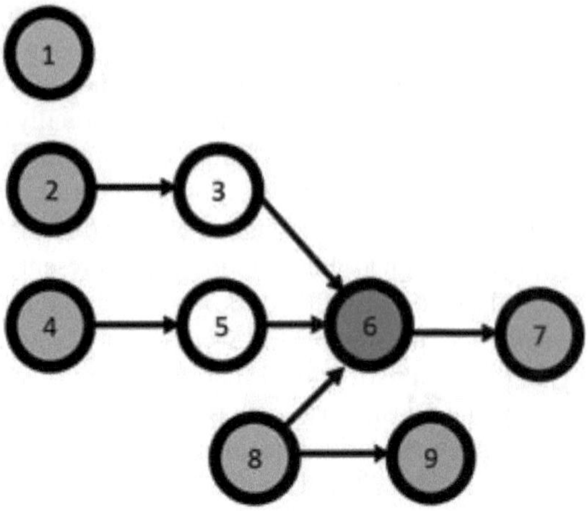

Fig. 1 A precedence graph with nine tasks showing examples of different kinds of tasks in an industrial context: chains of tasks, initial and final tasks, isolated tasks, and checkpoints

instances and these instances should be similar to the original Nissan instance context [7]. Concretely, the developed NTIGen software includes the following features, which are illustrated in Fig. 1:

- **Checkpoints**: They are assembly line points in which workers test the quality and completeness of a set of operations previously finished. If we consider these checkpoints as new tasks, the representation of a checkpoint in an assembly line graph is given by a task having a high number of preceding tasks (for instance, task 6 in Fig. 1).
- **Tasks without precedence**: In real industrial scenarios, such tasks are justified if there are operations unconditioned by other operations. They are commonly found in the engine and trim lines of the car manufacturing. In Fig. 1, tasks 1, 2, 4 and 8 have no precedence.
- **Final tasks**: Tasks in an assembly line which are associated to the most external and final operations of the product. They are represented as tasks with no successors in the precedence graph (tasks 1, 7 and 9 in Fig. 1).
- **Isolated tasks**: They can be performed at any part of the assembly of an item. An example of these kinds of tasks is this related with additional parts of a product which can be incorporated to the global product at any station. Task 1 in Fig. 1 is an isolated task as it has no precedence.
- **Operations aggregation**: This process comes up when some operations need the same tools or are done by the same worker. In this case, several tasks of the same stage are put together in just one task.
- **Operations breaking up**: If possible, it is used in the industrial context to detail the implementation of an operation in different operating tasks. It is useful for balancing an assembly line when the cycle time is reduced.
- **Chains of tasks**: They appear when there are strongly linked operations, normally in the same station or stage. A chain of tasks represents natural sequences of operations within the assembly process (see tasks 4, 5, 6 and 7 in Fig. 1)

4 Tuneable Parameters of NTIGenn

The features introduced in the previous sub-section can be parameterized by the NTIGen user to generate a customizable instance. NTIGen is also fed by a set of stages with some initial tasks. By default, these stages and tasks correspond to the original Nissan instance with 140 tasks and 21 workstations [7] although they can be modified by the user before launching the application. The user can set all the desired features by changing the parameters of an XML file. The most important input parameters are the following:

- **Number of tasks (*n*)**. This is an important parameter of the instance that enormously conditions its complexity. From the initial set of tasks, new operating tasks are generated by breaking up them until reaching the user needs. If we need fewer tasks than the original ones, they are merged at random. The new generated tasks are required to belong to the same or close stages than their original ones.
- **Processing times (*tj*)**. The processing time of each task t_j is randomly disrupted by a normal distribution within a user-defined interval. When creating or merging tasks, the processing times for the resulting tasks are reduced or duplicated, respectively. This is done to maintain the original situation of the Nissan instance.
- **Production plans**. The production plans are always set to the NSIO original plans. The processing times of the tasks for the different engine products are created by randomly modifying the original processing time t_j within the range $[0.9t_{j,}\ 1.1t_j]$.
- **Cycle time (*c*).** It is also disrupted independently from the processing times of the tasks. As done with t_j, the disruption is created within a user-defined interval. In our case, the new cycle time is set to a value within $[0.75c,\ 1.25c]$.
- **Required operation area (*aj*).** Task areas are specified by two-dimensional units, i.e. length (a_j) and width (b_j). The first dimension, a_j, is the truly useful variable for the TSALBP optimization. In the original instance, b_j is always set to one distance unit. To generate a new instance, the squared area of each task is always maintained by the generator but b_j is randomly changed to a set value. In our case, the set is given by $\{0.5,\ 0.75 \ldots\ 2.25\}$. This set of possible b_j values can be modified by the user of the NTIGen software. Therefore, the length of each task a_j, used for the optimization, is different for each generated TSALBP instance. As done with the processing times, a_j is reduced or duplicated when increasing or decreasing the number of tasks to try to maintain the original Nissan situation.

Apart from the operating tasks and their corresponding processing times and areas, NTIGen generates the precedence graph of the instance. These precedence relations are created between tasks of the same stage (generating chains) or different stages within a maximum window, set by the user, in order to link tasks which are industrially close. The minimum and maximum number of preceding tasks for a checkpoint in a problem instance can be set prior the instance generation. The same definition can be done for the number of initial, final, and isolated tasks.

NTIGen creates precedence relations until it reaches the required complexity of the graph which is another important feature of an ALB instance [3]. This complexity of the precedence graph is also a user parameter and it is measured by the order strength (*OS*) of the graph [8]. The *OS* is calculated from the graph in transitive closure. The transitive closure of a set of direct precedences *E* is given by $E^T = \{(i, j) | i \in V, j \in F_i^T\}$, with *V* being the set of nodes and F_i^T the set of indirect successors of the task *i*. The *OS* represents the number of ordering relations of the graph in a transitive closure with respect to all possible ordering relations: $OS = \frac{|E^T|}{\frac{n(n-1)}{2}}$.

The *OS* varies between *[0, 1]*. If *OS* is equal to 0 the instance has no precedence relations but if *OS* takes value 1, there is just one feasible sequence of tasks. The result after running the NTIGen software is a structured text file describing the generated instance with the list of tasks, their operating times and area, and their precedence relations. The precedence relations form the transitive reduction of the graph in order to minimize computational resources.

In addition, by changing the number of tasks, their processing time and area we can generate instances having different time variability (TV) and area variability (AV). Descriptors about the generated instance are listed after its creation to show the complexity of the graph, TV, AV, and the number of checkpoints, isolated, initial, and final tasks.

5 Examples of Some Generated TSALBP Instances

By using the NTIGen software, a set of eight new TSALBP real-like instances have been created (Table 1). The NTIGen software and this set of TSALBP instances are publicly available at http://www.prothius.com/TSALBP.

6 Concluding Remarks

The existing TSALBP formulation and previous ALB works do not cover an important real scenario where the same assembly line is devoted to produce mixed products and their demand is not fixed. Furthermore, the TSALBP instances of the literature were created by modifying ALB instances. The NTIGen software presented in this work allows researchers to create realistic TSALBP instances and production plans for future research. The generated TSALBP instances contain many real-like industrial features, e.g. checkpoints, isolated tasks, initial and final tasks, chains of tasks, or stages, which make the NTIGen software a practical tool for simulating the industrial conditions of an assembly line. Also, the NTIGen user can generate

Table 1 Main characteristics of the generated TSALBP instances

Features	P1	P2*	P3	P4	P5	P6	P7	P8
Random seed	24151	N/A	117017	21277	113683	56399	5869	73553
No. of tasks	100	140	190	220	280	320	376	420
Cycle time	199.97	180	207.07	222.42	221.62	169.552	186.65	137.751
OS	0.5	0.9	0.7	0.5	0.3	0.6	0.25	0.95
Precedences	156	293	314	304	407	435	548	608
Precs. window	5	N/A	5	1	2	1	3	2
TV	35.95	24	41.75	151.45	224.29	2742.28	901.34	1003.77
AV	500	513.86	266.67	300	400	200	300	133.33
Initial tasks	14	1	6	33	59	32	87	6
Final tasks	8	5	7	20	42	31	49	8
Isolated tasks	2	0	5	3	0	5	0	3
Checkpoints	3	N/A	0	6	7	1	12	0

* It corresponds to the original Nissan instance

instances with production plans, having different operation time for each task of the assembly line.

Acknowledgements This work has been supported by Ministerio de Economía y Competitividad under project SOCOVIFI2 (TIN2012-38525-C02-01 and TIN2012-38525-C02-02), and under PROTHIUS-III: DPI2010-16759, both including EDRF funding.

References

1. Bautista J, Pereira J (2007) Ant algorithms for a time and space constrained assembly line balancing problem. Eur J Oper Res 177:2016–2032
2. Baybars I (1986) A survey of exact algorithms for the simple assembly line balancing problem. Manage Sci 32:909–932
3. Bhattacharjee TK, Sahu S (1990) Complexity of single model assembly line balancing problems. Eng Costs Prod Econ 18:203–214
4. Boysen N, Fliedner M, Scholl A (2007) A classification of assembly line balancing problems. Eur J Oper Res 183:674–693
5. Chica M, Cordón O, Damas S, Bautista J (2010) Multiobjective, constructive heuristics for the 1/3 variant of the time and space assembly line balancing problem: ACO and random greedy search. Inf Sci 180:3465–3487
6. Chica M, Cordón O, Damas S (2011) An advanced multi-objective genetic algorithm design for the time and space assembly line balancing problem. Comput Ind Eng 61:103–117
7. Chica M, Cordón O, Damas S, Bautista J (2012) Multiobjective memetic algorithms for time and space assembly line balancing. Eng Appl Artif Intell 25:254–273
8. Dar-El EM (1975) Solving large single-model assembly line balancing problems—a comparative study. AIIE Trans 7:302–310
9. Otto A, Otto C, Scholl A (2011) SALBPgen—a systematic data generator for (simple) assembly line balancing. Jena Research Papers in Business and Economics 5
10. Scholl A, Becker C (2006) State-of-the-art exact and heuristic solution procedures for simple assembly line balancing. Eur J Oper Res 168:666–693

Appropriate Work Design in Lean Production Systems

Uwe Dombrowski, Eva-Maria Hellmich and Tim Mielke

Abstract The demographic change has a substantial impact on the age structure of manufacturing enterprises. The specific needs of older employees have to be considered thoroughly in the design of future work systems. Today, many enterprises organize their processes according to the principles of lean production systems. In order to achieve a sustainable implementation of age and aging appropriate work design, the existing lean production systems need an appropriate modification. The paper presents an analysis of today's work design concerning age and aging. Furthermore, it introduces four approaches for age and aging appropriate work in lean production systems.

Keywords Lean production system · Demographic change · Age and aging appropriate · Ergonomics · Occupational health and safety

1 Introduction

The demographic change is one of the key challenges that the European Union needs to overcome. Indicators for a demographic change are found in all EU-27 countries but some of them are affected more severely. For example, compared to the other EU-27 countries Germany shows one of the lowest fertility rates, one of the highest life expectancies and the oldest population which is already declining [8].

At the moment, about 50 million people in Germany are in an employable age. In 2060, it might be only 33 million. [2] A second important development in this context is the rising labor participation of older workers. In 2000, 38 % of employees

U. Dombrowski (✉) · E.-M. Hellmich · T. Mielke
Institute for Advanced Industrial Management, Technische Universität Braunschweig, Langer Kamp 19, 38106 Braunschweig, Germany
e-mail: u.dombrowski@tu-bs.de

E.-M. Hellmich
e-mail: e-m.hellmich@tu-bs.de

T. Mielke
e-mail: tim.mielke@tu-bs.de

C. Hernández et al. (eds.), *Managing Complexity*, Lecture Notes in Management and Industrial Engineering, DOI 10.1007/978-3-319-04705-8_15,

aged 55–64 were in gainful employment. Only ten years later, this share rose to 58 %. [1] Therefore, enterprises have to cope with these unprecedented challenges. In this context, health and especially occupational health and safety will gain in importance. Enterprises have to assure the achievement potential during the whole working life. In order to cope with the changing abilities of older workers, especially manufacturing enterprises have to improve the age and aging appropriate (A^3) work design.

Today's manufacturing industry designs its processes according to lean production systems (LPS), which represent state of the art manufacturing [14]. LPS are also called holistic production systems, which aim at the comprehensive and sustainable design of production [6]. However, practical experience shows that these systems focus on the improvement of quality, time and costs. The demographic change and the thereby rising importance of A^3 work design have not been regarded so far. For a sustainable consideration of the changes due to an older workforce, A^3 work design should be integrated in the widely spread lean production systems. Therefore, the principles, methods and tools of existing LPS have been analyzed to assess the actual significance of A^3 work design in LPS. Based on the results, four strategies were derived that show possibilities for further development of LPS towards A^3 work design.

2 Work Design in Lean Production Systems

In modern manufacturing enterprises, lean production systems specify the details of each and every work process. LPS claim to consider the three aspects technology, organization and people [13].

A lean production system (LPS) is "an enterprise-specific compilation of rules, standards, methods and tools, as well as the appropriate underlying philosophy and culture for the comprehensive and sustainable design of production. An LPS enables an enterprise to meet the requirements of today's business environment, taking into account technological, organizational, work-force-related and economic aspects" [4].

The superior goal of all LPS is the sustainable elimination of waste in all processes. [12, 14] In this context, waste is determined from a customer's point of view and includes all activities that do not add value to the product. As waste elimination is a basic approach in LPS, many descriptions exist. The most common are the following seven types of waste [10, 12, 14, 15]: Overproduction, Waiting, Transporting, Over-processing, Inventories, Unnecessary motion and Defects. Some authors name an eighth type, the waste of unused employee creativity [10].

Several methods and tools support the avoidance and elimination of waste. These methods and tools are embedded in a superior structure that links the enterprise's strategy to the principles, methods and tools of the LPS. Despite the enterprise-specific compilation of LPS, a general structure was identified, which is shown

Fig. 1 Organization and structure of a lean production system

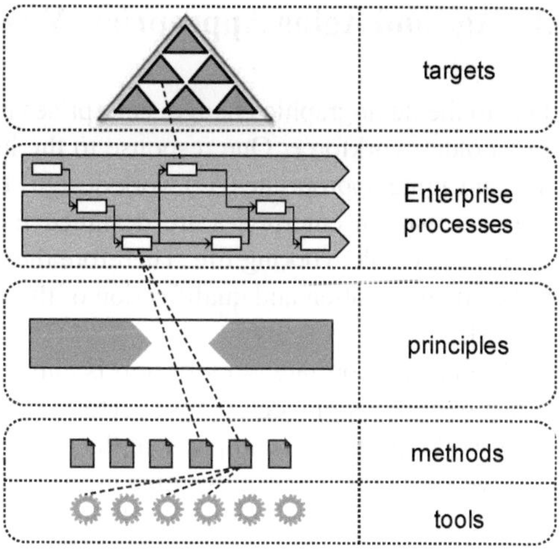

in Fig. 1. The fundamental elements of LPS are the enterprise's targets, processes, principles, methods and tools [14].

One of the main differences between LPS and traditional mass production systems is the improvement process. In contrast to mass production, LPS use the continuous improvement of all processes in small but frequent steps. This improvement needs various decentralized steps that contribute to the superior goal of zero waste [14].

The LPS principles are based on an enterprise-specific collection, which causes a variety of principles. Most of them can be traced back to the same eight basic principles that have been described in the LPS guideline of the German association of engineers [14]. These basic principles will be described in the following.

The **elimination of waste** is a fundamental principle that has already been mentioned above. Since waste is everything that does not contribute to customer value. The second principle is the **continuous improvement process** (CIP). Its aim is to question all current practices all the time and to improve them frequently. **Standardization** of processes is an important condition for the waste elimination and continuous improvement process. Standards help to sustain the improved state and show deviations from the desired process. The fourth principle, **zero defects**, contains methods and tools to prevent the appearance and identification of defects. The **flow principle** helps to avoid excess inventory, which results in shorter lead times. In ideal state, the lead time equals the processing time. The **pull principle** focuses on the material flow as well. According to this principle, every product has to be linked to customer demand. **Visual management** is used to illustrate the actual state and the current standards. Thereby, deviations from standards can be recognized at a glance. The principle of **employee orientation and management by objectives** includes methods and tools for leadership in LPS.

3 Age and Aging Appropriate Work

Due to the demographic change, enterprises have to adapt their future work design to an older workforce. One response to the increasing age of the workforce is the age and aging appropriate (A^3) work design. The A^3 work design should ensure that the processes in enterprises are designed for assuring the achievement potential during the whole working life. Therefore the A^3 work design contributes positively to health, motivation and qualification of the employees across their entire working life [3].

The age appropriate work design is aimed at adopting special measures for the group of older employees, whose performance has already changed in the course of their working lives. In comparison, the aging appropriate work design regards preventive measures. These preventive measures are supposed to maintain the achievement potential over the whole working live. Thereby false strains are avoided directly [9].

A lot of different measures can be attributed to the A^3 work design. In particular, these measures can be allcoated to six different aspects [3]: Ergonomic work design, Promotion of occupational health and safety, Job enrichment, Reduction of time pressure at work, Implementation of exculpatory working time models and Job rotation

4 Analysis of A^3 Work Design in Lean Production Systems

A recent study of the German Federal Ministry of Labour and Social Affairs shows that only 5.1 % of the examined enterprises design their processes under age and aging appropriate work aspects. Thereby enterprises have to focus on an A^3 work design to cope with changing conditions [1]. Based on these findings, the Institute for Advanced Industrial Management conducted a further analysis regarding A^3 work design. Since LPS represent state of the art manufacturing, it was investigated how A^3 work design is integrated in LPS.

The LPS were analyzed in terms of their direct impact on age and aging appropriate work design. Furthermore, it was of interest whether the different LPS could at least positively influence the A^3 work design if they do not have a direct impact.

The analysis was based on the LPS guideline of the German association of engineers (VDI 2870) and the LPS of 22 enterprises. The considered enterprises operate in ten different industry sectors. Thus, it was a heterogeneous group. Most frequently represented were the automotive original equipment manufacturers (OEM) (four enterprises) and suppliers (OES) (seven enterprises). Each of the 22 enterprises have a specific LPS considering their individual requirements. So, the LPS did not only vary in their number of levels of detail, design principles and methods. They also differed in terms of content.

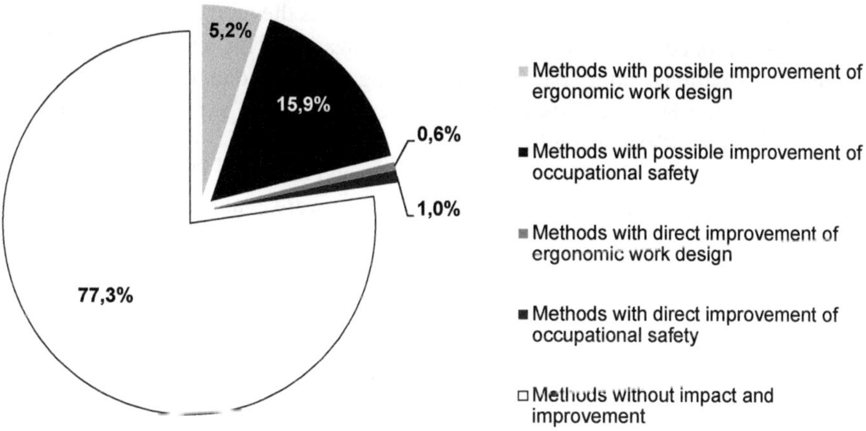

Fig. 2 A^3 Analysis of LPS

For comparability, all of the LPS examined were normalized to the structure of the VDI 2870. Based on this standardization, the targets, principles and methods of the various LPS were analyzed concerning their consideration of A^3 work design. As shown before, the A^3 work design can be divided into six aspects. In this analysis, only the aspects of ergonomic work design and promotion of occupational health and safety were examined. As regards the promotion of occupational health and safety, the focus was on the consideration of occupational safety.

At the target level, no direct consideration of ergonomic work design and occupational safety could be identified. The enterprise targets rather aim at involving the employees in the enterprise processes. They focus on increasing the employees' satisfaction, identification with the enterprise, motivation and longterm employment. These aspects, however, do not lead directly to an improvement of A^3 work design in terms of ergonomic work design and occupational safety. At least, they increase the motivation of employees and thus have a positive impact on the aging appropiate work design.

Furthermore, it was found that ergonomic work design is not considered in any of the analyzed LPS on the principle-level. However, occupational safety is fixed in five enterprises on this level.

In the next step, the LPS methods have been regarded. The analysis of the methods has shown that overall 21.1 % of the 805 considered methods could have a positive impact on ergonomic work design and occupational safety if they were applied with this aim. Therefore, they are improving the A^3 work design potentially or directly. As shown in Fig 2, 5.2 % of the methods may improve the ergonomic work design. An exemplary method is the job rotation. Job rotation means that employees change their jobs in regular intervals, which avoids a one-sided strain. Direct improvement of the ergonomic work design is included in only 0.6 % of the methods. One exemplary method for direct improvement in terms of the LPS is "ergonomic work analysis tools". The further analysis showed that 15.9 % of the methods could have a positive impact on occupational safety, such as the method 5S. Result of 5S is a well-organized workplace based on standardization. This leads to an improvement in occupational safety because of avoiding accidents due to misplaced items. Only 1 % of the 805 methods directly affect occupational safety. An exemplary method for this is "Visualized Safety".

Corresponding results have come out in a second analysis. Thereby, workshops and projects of a German automotive manufacturer were evaluated. It was checked whether aspects for improving the ergonomic work design were part of the projects and workshops or not. The analysis of a total of 52 workshops and nine projects showed that only 5.7 % of the workshops and 22.2 % of the projects have focused on an improvement in ergonomics.

In summary, it can be said that especially ergonomic work design is still not well enough considered in the examined LPS. In particular, it should be noted that no principle of the observed LPS directly improves the ergonomic work design. Also the consideration of ergonomic work design at the method level is very low. Only 0.6 % of the examined methods take aim at direct improvement of ergonomic work design. At the moment, occupational safety is considered in five out of 23 LPS at the principle level. Just 1 % of the methods have a direct impact on occupational safety. But the analysis has also shown, that many LPS methods could have a positive impact on A^3 work design.

5 Strategies for A^3 Work Design in LPS

The previous analysis of present LPS has shown that A^3 work design has not been considered consistently. As the results indicate, several methods already existing offer possibilities to improve A^3 work design. Many methods just have to be refocused on the specific topic and do not need much adaption. In summary, enterprises could use the potential of LPS for A^3 work design in order to cope with the new requirements due to the demographic change. Four strategies will be described in the following, which show four different ways to integrate A^3 work design into existing LPS. The strategies consider the causalities in the above described LPS structure and the findings from recent analyses of existing LPS. This is necessary to achieve the desired effects and to change the LPS sustainably. The four strategies are shown in Fig 3.

The **structural integration** describes an additional principle that regards A^3 work design. The principle has to refer to a strategic goal of the enterprise in order to ensure a consistent structure. This strategy allows clustering of particular methods and tools and assures their systematic use. If the LPS follows an A^3 principle, it will most likely be part of the visual LPS depiction. These depictions often serve as a logo or symbol of the LPS and are widely used in slideshows, brochures and other marketing material. Thereby, the structural integration supports the degree of awareness for A^3 work design. The most common depictions of LPS are a house, a circle and the enterprise's product [5].

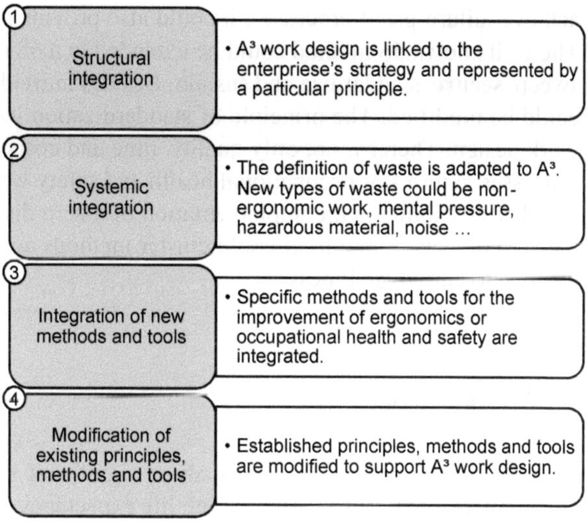

Fig. 3 Strategies for A^3 work design in LPS [7]

The second strategy is based on a **systemic integration** of A^3 work design. Therefore, the principle of waste elimination can be adapted. This principle represents the foundation for other principles and is a key element of every LPS. In general, this principle has the goal to eliminate activities that do not increase customer value [14]. Waste elimination could also contribute to eliminating activities that compromise the employee's health and safety. Waste would then be defined from the customer's and employee's point of view. Consequently, the description of waste should be improved by adding non-ergonomic work, mental pressure, hazardous material, noise or other unsafe working conditions.

The shop floor implementation of A^3 work design requires the **integration of A^3 methods and tools** in the existing LPS. This allows to continuously integrate A^3 work design into daily routines. The previously introduced analysis has shown a lack of methods that support A^3 work design in LPS. Especially methods and tools for the assessment and improvement of ergonomic work conditions should be integrated. Such methods are already widely known but are not part of the LPS and due to that, not part of work design. Many enterprises already use the ergonomic assembly worksheet (EAWS) [11]. If the EAWS would be integrated in LPS, a comprehensive application could be achieved. Another benefit would be the early ergonomic assessment during the design of the process. Besides EAWS, other methods have to be integrated. Especially the so called screening methods should be used. Their results are less detailed but easy to use and no special training is necessary.

The fourth strategy uses a **modification of existing principles, methods and tools**. Therefore, the systemic integration should have been applied. With the new understanding of waste, several basic LPS methods can contribute to A^3 work design. For example, PDCA, five whys or benchmarking are easily deployable on the improvement of ergonomic work conditions or occupational health and safety. Other methods might need a little adjustment. Poka yoke could not only be used to achieve failure-proof processes, it could also provide malposition-proof processes. The well known 5S method could be extended to a 6S method: sorting, set in order, sweep, **secure**, standardize and sustain. Besides individual methods, also principles could be modified. The principle of standardization would be very suitable for A^3 work design. Thereby, not only quality, time and costs would be regarded, the best processes in terms of occupational health and safety would be standardized as well. Another possibility is the implementation of a zero disease principle, derived from the zero defects principle. It could cluster methods and tools that reduce absenteeism due to employee illness.

6 Conclusions

Many industrial countries already show significant symptoms of a demographic change. The low fertility rate and high life expectancy result in a higher average age of workforces. Enterprises have to adapt their processes to the age specific requirements of their employees. The age and aging appropriate (A^3) work design combines approaches of occupational health and safety, ergonomics and age specific solutions like better lighting. The A^3 work design should be integrated in already existing and well established lean production systems (LPS) in order to achieve a sustainable application. An analysis has shown that A^3 work design is not suficiently regarded in presently existing LPS. Some LPS have implemented individual methods for ergonomics or safety but lack a comprehensive integration. Therefore, four strategies were introduced that show solutions to integrate A^3 work design in future LPS.

References

1. BMAS (2012) Fortschrittsreport “Altersgerechte Arbeitswelt”. Federal Ministry of Labour and Social Affairs, Berlin
2. BMI (2011) Demography report. Federal Ministry of the Interior, Berlin
3. Buck H, Kistler E, Mendius H-G (2002) Demographischer Wandel in der Arbeitswelt. Broschürenreihe, Demograpghie und Erwerbsarbeit, Stuttgart
4. Dombrowski U, Schmidt S (2008) Framework for the planning and control of lean production system implementation. Int J Agile Manuf
5. Dombrowski U, Palluck M, Schmidt S (2006) Strukturelle Analyse Ganzheitlicher Produktionssysteme. ZWF Z wirtsch Fabrikbetrieb 10:553–556
6. Dombrowski U, Mielke T, Schulze S (2010) Impact of lean production systems implementation on labor conditions. APMS, Cernobbio
7. Dombrowski U, Mielke T (2012) Entwicklungspfade zur Lösung des Demografieproblems in Deutschland. GITO-Verlag, Berlin
8. European Commission (2011) Demography report 2010: older, more numerous and diverse Europeans. Publications Office of the European Union, Luxembourg
9. Kistler E et al (2006) Altersgerechte Arbeitsbedingungen. BAuA, Dortmund

10. Liker JK (2004) The Toyota way. McGraw-Hill, New York
11. MTMaktuell (2011) Instruktoren aus fünf Ländern. EAWS. MTMaktuell 1:24–25
12. Ohno T (1988) Toyota production system. Productivity Press, New York
13. Spath D, Braun M (2003) Menschengerechte Arbeitsgestaltung als Führungsaufgabe. wt Werkstattstech online (1/2):75–79
14. VDI (2012) VDI 2870—Lean production systems. VDI, Beuth, Berlin
15. Womack JP, Jones DT (2003) Lean thinking. Free Press, New York

Two Simple Constructive algorithms for the Distributed Assembly Permutation Flowshop Scheduling Problem

Sara Hatami, Rubén Ruiz and Carlos Andrés Romano

Abstract Nowadays, it is necessary to improve the management of complex supply chains which are often composed of multi-plant facilities. This paper proposes a Distributed Assembly Permutation Flowshop Scheduling Problem (DAPFSP). This problem is a generalization of the Distributed Permutation Flowshop Scheduling Problem (DPFSP) presented by Naderi and Ruiz (Comput Oper Res, 37(4):754–768, 2010). The first stage of the DAPFSP is composed of f identical production factories. Each center is a flowshop that produces jobs that have to be assembled into final products in a second assembly stage. The objective is to minimize the makespan. Two simple constructive algorithms are proposed to solve the problem. Two complete sets of instances (small-large) are considered to evaluate performance of the proposed algorithms.

Keywords Distributed assembly flowshop · Permutation flowshop · constructive algorithms

1 Introduction

Assembly systems have been widely studied in the last decade given their practical interest and applications. An assembly flowshop is a hybrid production system where various production operations are independently and concurrently performed

S. Hatami (✉)
Departamento de Estadística e Investigación Operativa,
Facultad de Ciencias Matemáticas, Universitat de València, València, Spain
e-mail: hatami@alumni.uv.es

R. Ruiz
Grupo de Sistemas de Optimización Aplicada, Instituto Tecnológico de Informática,
Universitat Politècnica de València, Ciudad Politécnica de la Innovación,
Edifico 8G, Acc. B. Camino de Vera s/n, 46021 València, Spain
e-mail: rruiz@eio.upv.es

C. Andrés Romano
Research Group in Reengineering, Operations Management, Group Work and Logistics,
Excellence, Industriales, Universitat Politècnica de València Cno,
De Vera s/n, 46022 Valencia, Spain
e-mail: candres@omp.upv.es

C. Hernández et al. (eds.), *Managing Complexity*, Lecture Notes in Management and Industrial Engineering, DOI 10.1007/978-3-319-04705-8_16,

to make parts that are delivered to an assembly line [4]. Tozkapan et al. [8] considered a two-stage assembly scheduling problem by minimizing the total weighted flow time as an objective function. Al-Anzi and Allahverdi [1] addressed the model presented by [8] and minimized the total completion time of all the jobs by using metaheuristics to solve their model. This is just a small extract of the many existing papers in this regard.

From a manager's point of view, scheduling in these systems is more complicated than in single-factory settings. In single-factory problems, the only objective is to find a job schedule for a set of machines, while an important additional decision in the distributed problem is allocating jobs to suitable factories.

In this paper, flowshop scheduling is used as a production system for each factory or supplier in the distributed problem. The flowshop scheduling problem (FSP) is composed of a set of M of m machines where each job of a set N of n jobs must be processed in each machine. The number of operations per job is equal to the number of machines. The ith operation of each job is processed in machine i. Therefore, one job can start in machine i only after it has been completed in machine $i-1$, and if machine i is free. The processing times of each job in the machines are known in advance, non-negative and deterministic. In FSPs, a number of assumptions are made [2].

In the FSP, there are $n!$ possible job permutations for each machine. Therefore, the total number of solutions for a flowshop problem with m machines is $(n!)^m$. To simplify the problem, it is assumed that all machines have the same job permutation. With this simplifying assumption the FSP is referred to as Permutation Flowshop Scheduling Problem (PFSP) with $n!$ possible solutions. This problem is one of the most researched topics in the scheduling literature [6, 7, etc.]. The DPFSP can be viewed as a generalized version of the PFSP.

This paper studies the Distributed Assembly Permutation Flowshop Scheduling Problem (DAPFSP). It is a combination of the DPFSP and the Assembly Flowshop Scheduling Problem (AFSP), and consists of two stages: production and assembly. The first stage comprises of a set F of f factories or production centers where a set N of n jobs has to be scheduled. All factories are capable of processing all jobs and each factory is a PFSP with a set M of m machines. Factories are assumed to be identical. Processing times are denoted by $p_{ij},, j \in N$. The second stage is a single assembly factory with an assembly machine, M_A, which assembles jobs by using a defined assembly program to make a set T of t different final products. Each product has a defined assembly program. N_h and J_i are used, respectively, to represent product's h assembly program and the jobs that belong to the product's h assembly program, $N_h : \{J_j\}, J \in N_h$. Each product h has N_h jobs and job j is needed for the assembly of one product. Therefore, $\sum_{h=1}^{t} |N_h| = n$. Product h assembly can start only when all jobs that belong to N_h have been completed in the different factories. The considered objective is to minimize the makespan at the last assembly factory.

Despite the innumerable literature related to PFSP and AFSP, it seems that there are few studies about the DPFSP. Naderi and Ruiz [5] presented the DPFSP for the first time and developed six different MILPs , proposed two simple factory assignment rules and 14 heuristics based on dispatching rules, effective constructive heuristics and VND methods. To the best of our knowledge, no further literature exists on DAPFSP, so this is the first effort that considers the assembly flowshop problem in a distributed manufacturing setting.

The next section presents introduces two simple constructive algorithms, Sect. 3 describes a complete computational evaluation of the proposed algorithms. Finally, Sect. 4 offers conclusions, remarks and venues for future research.

2 Heuristic Methods

As mentioned in the paper of [5], the DPFSP is an NP-Complete problem (if $n > f$); accordingly, the DAPFSP with an additional assembly stage is certainly an NP-Complete problem (or rather, one should say that the associated decision problem is). Therefore, it is necessary to develop a heuristic approach to solve large-sized problems.

For the assignment of jobs to factories, the two rules (NR_1, NR_2), of [5] are used. Using these two factory allocation rules, two heuristics are presented to schedule jobs.

2.1 Heuristic 1

We first introduce some necessary notation. An example with $n=9$, $m=2$, $f=2$ and $t=3$, this is, 9 jobs, 2 factories with a flowshop of two machines each and three products to assemble, is employed to explain expressions and heuristics in some detail. The processing times of the 9 jobs on the first and second machines on factories are {1, 5, 7, 9, 9, 3, 8, 4, 2} and {3, 8, 5, 7, 3, 4, 1, 3, 5}, respectively. Assembly processing times of products on assembly machine are 6, 19 and 12 respectively. The products' assembly programs are: $N_1=\{3,4,6\}$, $N_2=\{1,2,8,9\}$ and $N_3=\{5,7\}$. π represents a product sequence, e.g., $\pi=\{1,3,2\}$ is a possible product sequence for the given example. As mentioned before, each product h is made up of $|N_h|$ jobs and π_h is the partial job sequence of product h, e.g., π_1:{6,4,3}, π_2:{1,9,8,2}, π_3:{7,5}. A complete job sequence, π_T, is constructed by putting together all partial job sequences, following the product sequence π, e.g., π_T:{6,4,3,7,5,1,9,8,2}.

The shortest processing time (SPT) is a well-known dispatching rule for the PFSP. Hence the SPT is used to determine the product sequence in the assembly machine.

Heuristic 1 begins by applying the SPT rule for the assembly operation times to obtain π, $\pi=\{1,3,2\}$. A heuristic which is based on [3] heuristic (FL) is applied on the jobs that belong to a given product.

The heuristic evaluates the completion times of the jobs that belong to product h, for example If, $h=1$. Set R_h is made by sorting jobs in ascending order of

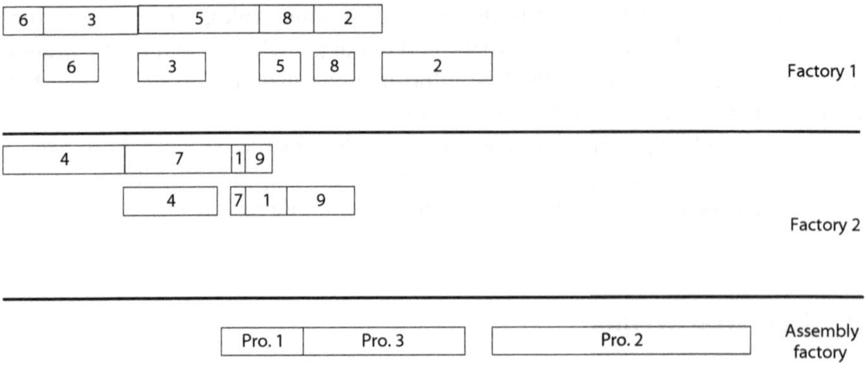

Fig. 1 Gantt chart of H_{11} for the example

completion times, $R_1=\{6,3,4\}$. Where completion times for set of jobs of the product 1, $N_1=\{3,4,6\}$ are $C_{23}=12$, $C_{24}=16$, $C_{26}=7$. The first two jobs of R_h are selected and inserted into S_h, $S_1=\{6,3\}$. All jobs' pairwise exchanges in S_h are checked and it is updated with the one that results in the best makespan, $C_{max(\{6,3\})}=15$ and $C_{max(\{3,6\})}=16$, S_1:$\{6,3\}$. The next step is removing the third job of R_h and inserting it in all possible positions of S_h , $C_{max(\{4,6,3\})}=25$, $C_{max(\{6,4,3\})}=24$ and $C_{max(\{6,4,3\})}=26$. The sequence with the best makespan will be selected, S_1 is updated to $\{6,4,3\}$. All possible sequences by carrying out pairwise exchanges between jobs are evaluated again, $C_{max(\{4,6,3\})}=25$, $C_{max(\{6,4,3\})}=24$, $C_{max(\{3,4,6\})}=27$. If a better makespan is obtained, then S_h is updated. The process continues until all jobs have been considered. S_h is the partial job sequence for product *h*, (π_h), $\pi_1=\{6,4,3\}$. By following the same method, the partial job sequences for the other products are: $\pi_2=\{1,9,8,2\}$ and $\pi_3=\{5,7\}$ with partial makespans of 20 and 18, respectively. π_T is constructed by putting together all π_h and jobs are assigned to factories from π_T by using NR_1 or NR_2, which respectively result in the H_{11} or H_{12} heuristics. Hence π_T is $\{6,4,3,5,7,1,9,8,2\}$. The final step is to assign jobs in π_T to factories by using NR_1/NR_2 to obtain the H_{11}/H_{12}. C_{max} of H_{11} and H_{12} are 55 and 53, respectively. The Gantt chart of the considered example after applying H_{11} is shown in Fig. 1.

2.2 *Heuristic 2*

The idea of the second heuristic is to give priority to products whose jobs are completed in the production stage sooner. This concept is noted as the earliest start time to assemble product *h*, E_h. The procedure that is used in H_{11} and H_{12} to find partial job sequences of products (π_h) also is used in heuristic 2. E_h, is calculated by using NR_1 or NR_2 to assign jobs in each partial job sequence to factories. For example, the earliest start times for assembling products by considering NR_2 are $E_1=15$, $E_2=15$, $E_3=12$. π is built by sorting E_h in ascending order.

3 Computational Evaluation

Two complete sets of instances have been generated to test the proposed heuristics. Four instance factors (*n.m, f,t)* are combined at the levels provided for small and large instances. In small instances, number of jobs (n) is tested at 5 levels, 8, 12, 16, 20 and 24, number of machines (m) has 4 levels, 2, 3, 4 and 5, both factors of number of factories (f) and number of products (t) have 3 levels, 2, 3 and 4. In the large instances, all factors have 3 levels and are; $n=\{100, 200, 500\}$, $m=\{5, 10\ 20\}$, $f=\{4, 6, 8\}$ and $t=\{30, 40, 50\}$.

Processing times in the production stage are fixed to $\cup\,[1,99]$ as it is usual in the scheduling literature. The assembly processing times depend on the number of jobs assigned to each product h as $U\left[1\times|N_h|, 99\times|N_h|\right]$. The total number of combinations in the small and large instances are $5\times4\times3^2=180$ and $3^4=81$, respectively. There are five replications per combination for small instances and ten replications for every large combination. Therefore, the total number of instances is 900 and 810, respectively. All instances are available at soa.iti.es.

3.1 *Heuristics Evaluation on Small Instances*

The four proposed methods $(H_{11}, H_{12}, H_{21}, H_{22})$ are tested. A MILP model is constructed for the small instances are solved with two commercial solver packages (CPLEX 12.3 and GUROBI 4.6.1). Serial (1 thread) and parallel (2 threads) and two time limits (900 and 3600 s) are tested with the solvers.

As the proposed heuristics are not expected to find an optimal solution, the Relative Percentage Deviation (RPD), is measured for comparisons. We measure RPD as follows: using the optimal solution or the best known solution, (OPT_{best}) and ALG_{SOL}, which reports the makespan obtained by a given algorithm for a given instance:

$$RPD = \frac{ALG_{SOL} - OPT_{best}}{OPT_{best}} \times 100$$

Table 1 provides the summarized results of the MILP and the average algorithm deviations from the best known solution for the small instances. They are grouped by n and f

MILP reports better results when compared to the proposed heuristics. CPU times to solve small instances with the proposed algorithms are negligible while most of the instances that are solved with the MILP. Therefore, the 3% average deviation of H_{22} needs to be contextualized.

In order to identify the best algorithm, a means plot and Tukey's Honest Significant Difference (HSD) intervals (99% confidence) for the four simple constructive heuristics is shown in Fig. 2. The second heuristic performs better in comparison with the other simple constructive heuristic and there is no significant difference between the rules used to assign jobs to factories.

Table 1 Relative Percentage Deviation (RPD) of MILP and proposed algorithms over the best known solution for the small instances

	Algorithms				
$f \times n$	MILP	H_{11}	H_{12}	H_{21}	H_{22}
2×8	0.00	14.62	13.61	6.91	5.99
2×12	0.01	13.70	12.78	5.74	5.17
2×16	0.42	12.52	11.40	5.77	5.10
2×20	1.26	9.92	9.28	4.25	3.48
2×24	2.70	7.75	7.38	4.07	3.81
3×8	0.00	11.35	9.96	4.57	3.15
3×12	0.00	9.96	9.13	3.03	2.55
3×16	0.06	10.10	9.16	3.77	3.14
3×20	0.35	9.86	8.93	2.72	2.19
3×24	1.18	7.65	6.37	3.00	2.40
4×8	0.00	9.03	8.01	2.16	1.25
4×12	0.00	5.63	4.53	1.82	1.38
4×16	0.04	7.21	6.34	2.86	2.27
4×20	0.23	6.80	6.00	2.96	2.61
4×24	0.44	5.13	4.42	2.00	1.59
Average	0.45	9.41	8.49	3.71	3.07

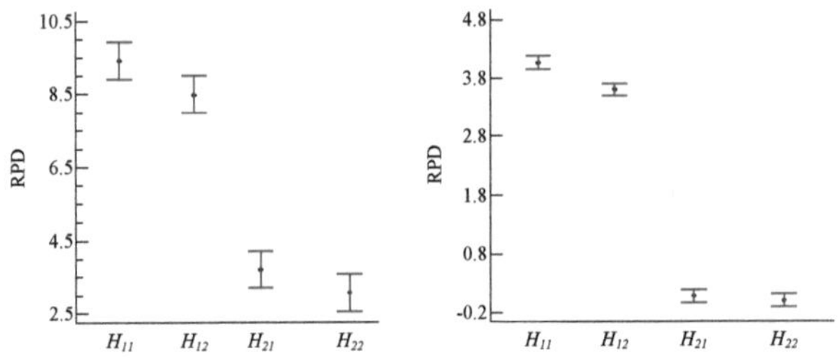

Fig. 2 Means plot and 99 % confidence level Tukey's HSD intervals of the relative percentage deviation for simple constructive heuristic methods for small instances on the left and for large instances on the right

3.2 Heuristics Evaluation on Large Instances

In this case, for calculating the RPD, only the best known solution is used as the MILP cannot be employed. A summarized result of the average RPD, considering number of factories, number of products and number of jobs, is shown in Table 2. Figure 2 shows a means plot (99 % confidence level Tukey's HSD intervals) of the proposed algorithms for large instances.

The second proposed algorithm performs better than the first one also for the large instances. NR_2 as a job assignment rule, reports better results on the first algorithm while job assignment rule on second algorithms does not have any significant effect. It is clear on Table 2, generally when the number of factories and jobs increases, finding a better solution becomes easier, while this trend has a reverse effect when the number of products increases. Proposed simple constructive algorithms use a very short time in order to solve problems (less than 0.01 s on average), therefore the details are not reported.

Table 2 Relative Percentage Deviation (RPD) for the proposed algorithms over the best known solution for the large instances

	Number of factories (f)			Number of products (t)			Number of jobs (n)			Average
Algorithms	4	6	8	30	40	50	100	200	500	
H_{11}	5.39	3.72	3.07	3.66	4.20	4.31	6.21	3.69	2.27	4.06
H_{12}	4.91	3.24	2.65	3.23	3.76	3.80	5.53	3.21	2.06	3.60
H_{21}	0.14	0.06	0.02	0.10	0.06	0.07	0.09	0.09	0.04	0.07
H_{22}	0.01	0.00	0.00	0.00	0.01	0.00	0.00	0.00	0.00	0.00

4 Conclusion and Future Research

To the best of our knowledge, this paper is the first attempt to generalize the Distributed Permutation Flowshop Scheduling Problem to the Distributed Assembly Permutation Flowshop Scheduling Problem, where there is more than one production center to process jobs and a single assembly center to make final products from produced jobs. Two constructive algorithms are proposed.

Computational evaluations were performed with two groups of small and large instances. Results show that in small instances MILP reported results perform better than the proposed algorithms. On the other side, the proposed methods consume very little CPU time in comparison with the MILP while they still produce reasonable solutions.

For future works, the setup time and distinct production factories can be considered in the presented model to make it more realistic. Applying metaheuristics like a Genetic Algorithm, Tabu Search, etc., may report better solutions if compared to our proposed simple heuristics.

References

1. Al-Anzi F, Allahverdi A (2006) A hybrid tabu search heuristic for the two-stage assembly scheduling problem. Int J of Oper Res 3(2):109–119
2. Baker KR (1974) Introduction to sequencing and scheduling. Wiley, New York
3. Framinan J, Leisten R (2003) An efficient constructive heuristic for flowtime minimisation in permutation flow shops. Omega Int J Manage Sci 31(4):311–317
4. Koulamas C, Kyparisis GJ (2001) The three stage assembly flowshop scheduling problem. Comput Oper Res 28(7):689–704
5. Naderi B, Ruiz R (2010) The distributed permutation flowshop scheduling problem. Comput Oper Res 37(4):754–768
6. Pan QK, Ruiz R (2012) Local search methods for the flowshop scheduling problem with flowtime minimization. Eur J Oper Res 222(1):31–43
7. Pinedo M (2012) Scheduling: theory, algorithms and systems, 4th edn. Springer, New York
8. Tozkapan A, Kirca O, Chung CS (2003) A branch and bound algorithm to minimize the total weighted flowtime for the two-stage assembly scheduling problem. Comput Oper Res 30(2):309–320

Applications of the Lagrangian Relaxation Method to Operations Scheduling

Juan José Lavios, José Alberto Arauzo, Ricardo del Olmo and Miguel Ángel Manzanedo

Abstract Lagrangian Relaxation is a combinatorial optimization method which is mainly used as decomposition method. A complex problem can be divided into smaller and easier problems. Lagrangian Relaxation method has been applied to solve scheduling problems in diverse manufacturing environments such as single machine, parallel machine, flow shop, job shop or even in complex real-world environments. We highlight the two key issues on the application of the method: the first one is the resolution of the dual problem and the second one is the choice which constraints should be relaxed. We present the main characteristics of these approaches and survey the existing works in this area.

Keywords Lagrangian relaxation · Scheduling · Combinatorial optimization · Integer programming

1 Introduction

The Lagrangian Relaxation method seeks the solutions of a complex optimization problem from the solutions of an easier problem, which is obtained by relaxing the complex constraints of the original problem. This method is mainly used as a decomposition method, where complex problems are divided into smaller and easier problems and has been applied to solve scheduling problems in diverse manufacturing environments including single machine, parallel machine, flow shop, job shop or

J. J. Lavios (✉)
Grupo INSISOC. EPS, Universidad de Burgos, Av. Cantabria s/n, 09006 Burgos, Spain
e-mail: jjlavios@ubu.es

J. A. Arauzo
Grupo INSISOC. EII, Universidad de Valladolid, Pso del Cauce 59, 47011 Valladolid, Spain
e-mail: arauzo@insisoc.org

R. del Olmo · M. Á. Manzanedo
Department of Civil Engineering, Higher Polytechnic School, University of Burgos, C/Villadiego, S/N, 09001 Burgos Spain
e-mail: rdelolmo@ubu.es

M. Á. Manzanedo
e-mail: mmanz@ubu.es

C. Hernández et al. (eds.), *Managing Complexity*, Lecture Notes in Management and Industrial Engineering, DOI 10.1007/978-3-319-04705-8_17,

even in complex real-world environments. The two key steps in Lagrangian Relaxation are (1) to decide which constraints are relaxed and (2) to choose which method should be used to solve the dual problem. These decisions determine the performance of the method.

The rest of the paper is organized as follows: in Sect. 2 we introduce the basic idea of Lagrangian Relaxation and discuss two of the main choices in the design of the method, which constraints to relax and which method apply to solve the dual problem; in Sect. 3, we review the main works that apply the Lagrangian Relaxation to the scheduling problem in a manufacturing environment; in Sect. 4, the conclusions are presented.

2 Implementation of the Lagrangian Relaxation in a Manufacturing Scheduling Problem

The basic idea of Lagrangian Relaxation is to remove some of the constraints and add them to the objective function. The relaxed constraints are added to the objective function multiplied by a variable (Lagrange multiplier) that penalize the violation of the constraints. The key is to relax the constraints such that make the problem difficult to solve.

The Lagrangian Relaxation method has been applied to different scheduling environments such as single machine, parallel machine, flow shop, job shop, real-world environments, projects and supply chain. Integer formulation can be used to define these problems. Total weighted tardiness or total weighted earliness and tardiness has been used as objective function. As additive functions, they can be easily decomposed into job dependent subfunctions.

In a manufacturing environment the main candidates to be relaxed are machine capacity constraints and task precedence constraints. The choice of which constraints are relaxed determinate the orientation of the problem decomposition. If capacity constraints are relaxed, the problem can be decomposed into job related subproblems, but if precedence constraints are relaxed, the problem can be decomposed into machine related subproblems. Wang et al. [33] and [6] indicates that the relaxation of precedence constraints causes important oscillations in the solution from iteration to iteration and prevent convergence of the algorithm. When the number of machines is large enough the relaxation of the capacity constraints leads to better lower bounds [3].

The Lagrangian Relaxation method transforms the solutions of the relaxed problem into feasible solutions of the original problem, trying to obtain near-optimal solutions. The optimal value of the relaxed problem is a lower bound of the optimal objective of the original problem. The subgradient method iteratively adjusts the Lagrangian multipliers to find the best lower bound of the optimal objective value of the original problem [11] and requires the optimization of all the subproblems. It has two main drawbacks: first, in some environments it may be difficult to obtain

Table 1 Problem environment, dual problem method and type of constraints relaxed

	Subgradient method		Surrogate gradient method	
Relaxed constraints	Capacity constraints	Precedence constraints	Capacity constraints	Precedence constraints
General scheduling problem	[23]			
Single machine	[15, 28, 32]	[17]		
Parallel machine	[10, 24, 30]			
Flow shop	[27]	[31]		
Job shop	[3, 5 ,6, 9, 16, 20–22, 33]	[13]	[18, 33, 35]	[4]
Complex environment	[1, 2, 26]			[7, 25, 29, 36]

the optimum of the relaxed problem; and second, the optimization of large size problems can be very time consuming.

To overcome these difficulties the surrogate subgradient method has been proposed. The advantage of this alternative is that the minimization of all the subproblems is not required to converge. Instead, only a near optimization is sufficient, so less computational effort is needed [34]. However, the updating direction is worse than in the subgradient method.

As is shown in Table 1 most researchers relax capacity constraints rather than precedence constraints and the surrogate subgradient method is preferred for complex environments or in large problems.

3 Applications of the Lagrangian Relaxation Method to Solve the Scheduling Problem in Manufacturing Environments

In this section, we review the main works that apply the Lagrangian Relaxation to the scheduling manufacturing problems; they are classified by the type of environment.

General Scheduling Problem Luh and Hoitomt [23] propose a generic methodology that uses Lagrangian Relaxation to solve a scheduling problem. It is applied to three different scheduling environments: individual operations to be scheduled in identical parallel machines; job orders consisting of multiple operations to be scheduled in identical parallel machines; and tasks related with general precedence constraints which have to be scheduled in different machines.

Single Machine Scheduling Problem Sun et al. [28] use the Lagrangian Relaxation algorithm to solve the problem of scheduling a single machine with sequence dependent setup time, which are modeled as capacity constraints.

Dewan and Joshi [8] decompose the problem into local job problems and the machine problem. The machine fixes the prices of time slots and jobs choose the time slots that optimize their costs. This arises in a dynamic environment where new job orders may enter at any time. Every time a job enter into the system the prices are recalculated, taking the last prices before the arrival of the new order as initial prices.

Jeong and Leon [15] apply a variation of the dual problem resolution to solve the problem of scheduling a shared resource of three production subsystems without using a central planning system. They search a fully distributed problem.

Tang et al. [32] deal with multiple immediate predecessors and successors. A forward and backwards dynamic programming algorithm is developed to solve the relaxed problem.

Parallel Machines Luh et al. [24] first applied the Lagrangian Relaxation method to schedule jobs on several identical machines. They define an iterative algorithm. A feasible program is obtained from the task sequence extracted from the relaxed problem.

Edis et al. [10] add machine eligibility restrictions. In these problems different machines can do the same job with a different performance. They develop a heuristic to build feasible programs from the relaxed problem.

Tang and Zhang [30] apply Lagrangian Relaxation method to rescheduling tasks on a parallel machine in a dynamic environment, where breakdown of machines is considered. The goal is to reschedule tasks with minimal modifications of the original problem, taking into account the efficiency of the new schedule.

Flow Shop Problem Lagrangian Relaxation is applied in [31] to the problem of scheduling a hybrid flow shop system. This system comprises various stages of production, each of which contains a number of machines working in parallel.

In Nishi et al. [27] the Lagrangian Relaxation algorithm is iteratively applied while new restrictions are progressively added. They make new proposals for improving the dynamic programming step.

Job Shop Problem Hoitomt et al. [13] relax the precedence and the capacity constraints. A list-scheduling algorithm is developed to generate a feasible schedule.

The main contribution of [5] is the use of dynamic programming techniques to solve the relaxed job shop problem at the job order level. This avoids additional relaxation of the precedence constraints.

Wang et al. [33] minimize earliness and tardiness of job orders. They compare several methods for solving the dual problem.

Chen et al. [6] propose to relax only the capacity constraints to decompose the problem into several subproblems associated to job orders. These problems are solved using dynamic programming technique. They apply it to real factory data.

Kaskavelis and Caramanis [18] and [35] apply the surrogate dual method to a separable problem (interleaved subgradient method).

The main contribution of [9] is the study of the relationship between the theory of auctions and Lagrangian Relaxation. The concept of multi-agent systems is used as a basis for the implementation of distributed systems in planning and production control. They define the problem as combinatorial auctions where the items sold are the time slots in which the planning horizon is divided. Liu et al. [20, 21] also define a reactive scheduling system where auctions are held in a machine whenever a time slot is free.

Chen and Luh [4] solve the job shop problem using the Lagrangian Relaxation method and relaxing the precedence constraints, rather than capacity constraints. The resulting subproblems are equivalent to solving a problem of scheduling a single machine or parallel machines.

Kutanoglu and Wu [19] combine the stochastic analysis with the dynamic adaptation of the system to improve the robustness of Lagrangian Relaxation method.

Baptiste et al. [3] compare the results obtained if either precedence constraints or capacity constraints are relaxed. They improve the solutions by local search.

Jeong and Yim [16] apply Lagrangian Relaxation to solve the job shop problem extended to the case of virtual enterprise with agents controlling one or more job orders and machines. Every subsystem calculates its own schedule. They use the protocol CICA, developed in [14], as a framework to solve the scheduling problem in a distributed manner using an agent-based environment.

Scheduling in Complex Environments Gou et al. [12] implement the problem of production scheduling in a real production plant using a multi-agent system. They organize the system as a hierarchy following a quasi-distributed holonic structure based on the relaxation of the capacity constraints and precedence of the job shop problem.

Zhang et al. [36] develop a macro-level scheduling method based on Lagrangian Relaxation method. Large problems with assemblies and disassemblies are studied. The structure of the resolution process is complex. Surrogate subgradient method is proposed to reduce the computation requirements.

Chen et al. [7] study a complex environment with multiple resources, setup times and transfer lots. A dynamic programming method solves the subproblems related to transfer lots easier than in the previous work.

Luh et al. [25] apply Lagrangian Relaxation to a problem of supply chain as an extension of a job shop problem. A variation of the Contract Net for is used as a communication protocol among agents.

Sun et al. [29] study complex process structures with coupling assemblies and disassemblies. They propose to relax only one of the precedence constraints to avoid oscillation problems. They add an auxiliary function penalty to improve the convergence of the method.

Arauzo [1] implements an auction mechanism for controlling flexible manufacturing systems. The analogy between auction and Lagrangian Relaxation process in the definition of the auction provides a robust mathematical method to the updating prices step.

Araúzo et al. [2] propose a multi-agent system for project portfolio management. Projects negotiate the allocation of shared resources by an auction. The analogy between the Lagrangian Relaxation method and auctions is used as a modeling tool. This approach allows managing the company's current projects and provides decision criteria for the acceptance or rejection of new projects.

4 Conclusions

We reviewed the studies that have applied the Lagrangian Relaxation method to the resolution of scheduling operations in recent years. The scope has been diverse and covers most types of problems. The application of the method has been performed with two types of relaxation of constraints: capacity or precedence constraints. The choice of one or another implies respectively an approach towards either job or machine decomposition of the problem. The dual problem has also been solved with two different methods: the subgradient method and the surrogate gradient method. The objectives of these studies aim to improve the speed of resolution and the stability of the method solving these problems.

Acknowledgements The authors acknowledge support from the Spanish Ministry of Science and Innovation Project CSD2010-00034 (SimulPast CONSOLIDER-INGENIO 2010) and by the Regional Government of Castile and Leon (Spain) Project VA056A12-2.

References

1. Arauzo JA (2007) Control distribuido de sistemas de fabricación flexibles: un enfoque basado en agentes. Thesis. University of Valladolid
2. Araúzo JA, Pavón J, Lopez-Paredes A, Pajares J (2009) Agent based Modeling and Simulation of Multi-project Scheduling. In MALLOW, The Multi-Agent Logics, Languages, and Organisations Federated Workshops
3. Baptiste P, Flamini M, Sourd F (2008) Lagrangian bounds for just-in-time job-shop scheduling. Compu Oper Res 35(3):906–915
4. Chen H, Luh PB (2003) An alternative framework to Lagrangian relaxation approach for job shop scheduling. Eur J Oper Res 149(3):499–512
5. Chen H, Chu C, Proth J-M (1995) More efficient Lagrangian relaxation approach to job-shop scheduling problems. In: Proceedings of the IEEE International Conference on Robotics and Automation, pp 496–501
6. Chen H, Chu C, Proth J-M (1998) An improvement of the Lagrangean relaxation approach for job shop scheduling: a dynamic programming method. IEEE T Robotic Autom 14(5):786–795
7. Chen D, Luh PB, Thakur LS, Moreno Jr, J (2003) Optimization-based manufacturing scheduling with multiple resources, setup requirements, and transfer lots. IIE Trans 35(10):973–985
8. Dewan P, Joshi S (2000) Dynamic single-machine scheduling under distributed decision-making. Int J Prod Res 38(16):3759–3777
9. Dewan P, Joshi S (2002) Auction-based distributed scheduling in a dynamic job shop environment. Int J Prod Res 40(5):1173–1191
10. Edis EB, Araz C, Ozkarahan I (2008) Lagrangian-based solution approaches for a resource-constrained parallel machine scheduling problem with machine eligibility restrictions. In Proceedings of the 21st international conference on Industrial, Engineering and Other Applications of Applied Intelligent Systems: New Frontiers in Applied Artificial Intelligence. Springer, Wrocław, pp 337–346
11. Fisher ML (2004) The Lagrangian relaxation method for solving integer programming problems. Manage Sci 50(Suppl 12):1861–1871
12. Gou L, Luh PB, Kyoya Y (1998) Holonic manufacturing scheduling: architecture, cooperation mechanism, and implementation. Comput Ind 37:213–231

13. Hoitomt DJ, Luh PB, Pattipati KR (1993) A practical approach to job-shop scheduling problems. IEEE T Robotic Autom 9(1):1–13
14. Jeong I-J, Leon VJ (2002) Decision-making and cooperative interaction via coupling agents in organizationally distributed systems. IIE Trans 34(9):789–802
15. Jeong I-J, Leon VJ (2005) A single-machine distributed scheduling methodology using cooperative interaction via coupling agents. IIE Trans 37(2):137–152
16. Jeong I-J, Yim S-B (2009) A job shop distributed scheduling based on Lagrangian relaxation to minimise total completion time. Int J Prod Res 47(24):6783
17. Jiang S, Tang L (2008) Lagrangian relaxation algorithm for a single machine scheduling with release dates. In: Second International Symposium on Intelligent Information Technology Application, pp 811–815
18. Kaskavelis CA, Caramanis MC (1998) Efficient Lagrangian relaxation algorithms for industry size job-shop scheduling problems. IIE Trans 30(11):1085–1097
19. Kutanoglu E, Wu SD (2004) Improving scheduling robustness via preprocessing and dynamic adaptation. IIE Trans 36(11):1107
20. Liu N, Abdelrahman MA, Ramaswamy S (2004) A multi-agent model for reactive job shop scheduling. In: Proceedings of the Thirty-Sixth Southeastern Symposium on System Theory, pp 241–245
21. Liu N, Abdelrahman MA, Ramaswamy S (2007) A complete multiagent framework for robust and adaptable dynamic job shop scheduling. IEEE TSyst Man Cy C 37(5):904–916
22. Luh PB, Feng W (2003) From manufacturing scheduling to supply chain coordination: the control of complexity and uncertainty. J Syst Sci Syst Eng 12(3):279–297
23. Luh PB, Hoitomt DJ (1993) Scheduling of manufacturing systems using the Lagrangian relaxation technique. IEEE T Automat Contr 38(7):1066–1079
24. Luh PB, Omt D, Max E, Pattipati KR (1990) Schedule generation and reconfiguration for parallel machines. IEEE T Robotic Autom 6(6):687–696
25. Luh PB, Ni M, Chen H, Thakur LS (2003) Price-based approach for activity coordination in a supply network. IEEE T Robotic Autom 19(2):335–346
26. Masin M, Pasaogullari MO, Joshi S (2007) Dynamic scheduling of production-assembly networks in a distributed environment. IIE Trans 39(4):395–409
27. Nishi T, Hiranaka Y, Inuiguchi M (2007) A successive Lagrangian relaxation method for solving flow shop scheduling problems with total weighted tardiness. In: IEEE International Conference on Automation Science and Engineering, pp 875–880
28. Sun X, Noble JS, Klein CM (1999) Single-machine scheduling with sequence dependent setup to minimize total weighted squared tardiness. IIE Trans 31(2):113–124
29. Sun T, Luh PB, Min L (2006) Lagrangian relaxation for complex job shop scheduling. In: Proceedings of the IEEE International Conference on Robotics and Automation, pp 1432–1437
30. Tang L, Zhang Y (2009) Parallel machine scheduling under the disruption of machine breakdown. Ind Eng Chem Res 48(14):6660–6667
31. Tang L, Xuan H, Liu J (2006) A new Lagrangian relaxation algorithm for hybrid flowshop scheduling to minimize total weighted completion time. Comput Oper Res 33(11):3344–3359
32. Tang L, Xuan H, Liu J (2007) Hybrid backward and forward dynamic programming based Lagrangian relaxation for single machine scheduling. Comput Oper Res 34(9):2625–2636
33. Wang J, Luh PB, Zhao X, Wang J (1997) An optimization-based algorithm for job shop scheduling. Sădhana 22(part 2):241–256
34. Zhao X, Luh P, Wang J (1997) The surrogate gradient algorithm for Lagrangian relaxation method. In: Proceedings of the 36th IEEE Conference on Decision and Control 1:310, 305
35. Zhao X. Luh PB, Wang J (1999) Surrogate gradient algorithm for Lagrangian relaxation. J Optimiz Theory App 100(3):699–712
36. Zhang Y, Luh PB, Narimatsu K et al (2001) A macro-level scheduling method using Lagrangian relaxation. IEEE T Robotic Autom 17(1):70–79

Toward Various Exact Modeling the Job Shop Scheduling Problem for Minimizing Total Weighted Tardiness

Mohammad Namakshenas and Rashed Sahraeian

Abstract In this paper, two different mixed integer programming (MIP) and one constraint programming (CP) models are formulated for classical job shop problem with the aim at minimization of the total weighted tardiness as objective function. The proposed models are solved and compared with well-known benchmarks in the job shop literature, using IBM ILog Cplex software. Examination and comparison of these exact models suggest that one formulation performs much more efficiently than others, namely CP model, in triple criteria: First, number of generated variables; Second, solution time and Third, complexity scale.

Keywords Job shop · Total weighted tardiness · Mixed integer programming · Constraint programming

1 Introduction

The classical job shop problems are the most prevalent issues in scheduling processes in factories in which high number of products should be produced each with custom orders. The usage of job shop algorithms is not restricted to production and manufacturing environment, but they can be implemented in services and other applications, too. With the rapid development in computational technologies, mixed integer programming (MIP) and other high-tech modeling concepts for solving scheduling problems are significantly receiving attention from researchers. Although, there is no efficient solution methodology due to the NP-hard nature of these problems, mathematical programming is the first step to develop an effective heuristic or algorithm.

Contrary to job shops with makespan objective, literature on solution procedures to the total weighted tardiness job shop scheduling problem (TWT-JS) is very limited. Given the only branch-and-bound algorithm for this problem proposed by [5] the remaining approaches mainly based on local search [3] and shifting bottleneck

M. Namakshenas (✉) · R. Sahraeian
Department of Industrial Engineering, College of Engineering, Shahed University, Tehran, Iran
e-mail: m.namakshenas@shahed.ac.ir

R. Sahraeian
e-mail: sahraeian@shahed.ac.ir

C. Hernández et al. (eds.), *Managing Complexity*, Lecture Notes in Management and Industrial Engineering, DOI 10.1007/978-3-319-04705-8_18,

methods [4]. In the widely used three-field notation of Graham, TWT-JS is written as:

$J_m \,||\, \sum w_j T_j$, where $T_j = max\ (0,\ c_j - d_j)$. The assumptions made for the classic job-shop problem are summarized here. The processing times are known, fixed, and independent of the sequence. All jobs are ready for processing at time zero. No preemption is allowed, i.e. once an operation has started it must be completed before another operation may be started on that machine. Also, recirculation is prohibited, i.e. all jobs should meet machines only one time and each job should meet all machines. Only one job may be processed on a machine at any instant of time. No restrictions are placed on the routings of jobs. By this supposition, the problem can be declared as: *Given n jobs to be processed on m machines in a job shop with no restrictions on the routing constraints of jobs, determine the optimal job sequence on each machine in order to minimize the total weighted tardiness.*

2 Mixed Integer Formulation

In this paper, we use the following notations for MIP formulation procedure:

Parameters & Sets:

n number of jobs (N: jobs set);
m number of machines (M: machines set);
p_{jh} processing time of job j on machine h;
w_j weight of job j;
d_j due date of job j;
r_{jlh} 1, if the lth operation of job j requires machine h; $o.w.0$;

Decision variables:

c_j completion time of job j;
s_{jh} starting time of job j on machine h;
Z_j $w_j.\max\ (0,\ c_j - d_j)$;
E_{kh} $w_k.max\ (0,\ c_k - d_k)$ on machine h;
x_{jkh} 1, if job j scheduled in position k on machine h; $o.w.0$;
y_{ijh} 1, if job j follows job i on machine h (not necessarily immediately); $o.w.0$;

2.1 *Disjunctive Approach*

Disjunction based (DJ) formulation is closely intertwined with the disjunctive graph representation of the job shop. However, the relationship between disjunctions is implicit, which allows us to define the starting times s_{jh} with combination of a key binary variable, say y_{ijh}. Also, It suffices to work with these variables on the basis of $i<j$.

$$\text{Objective: Min} \sum_{j \in N} Z_j \tag{1}$$

Constraints:

$$\sum_{h \in M} w_j r_{jmh}(s_{jh} + p_{jh}) - w_j d_j \le Z_j, \quad \forall j \in N, \tag{2}$$

$$s_{ih} - s_{jh} \ge p_{jh} - My_{ijh}, \quad \forall i, j \in N, \forall h \in M, \text{with } i < j, \tag{3}$$

$$s_{jh} - s_{ih} \ge p_{ih} - M(1 - y_{ijh}), \quad \forall i, j \in N, \forall h \subset M, \text{ with } i < j, \tag{4}$$

$$\sum_{h \in M} r_{jlh}(s_{jh} + p_{jh}) \le \sum_{h \in M} r_{j,l+1,h} s_{jh}, \quad \forall j \in N, \forall l \in M - \{n\}, \tag{5}$$

$$Z_j, s_{jh} \in \text{nonnegative integers}, \quad \forall i, j \in N, \forall h \in M.$$
$$y_{ijh} \in \text{binary},$$

Constraints set (2) is used for linearization of the max function in TWT-JS and based on operational variable, namely r_{jlh}. Constraints set (3) ensures that if the starting time of job j precedes i on machine h, then there must be delay with duration of processing job j on machine h and it must relax the constraints set (4) and vice versa. Also, Constraints set (5) imposes operational precedence of a job on specified machine. To be precise, it declares that starting time of operation $l+1$ should occur after the completion of operation l for job j.

2.2 *Hybrid Approach*

An alternative formulation can be constructed by deploying both the sequence-position variables and the precedence variables, namely hybrid formulation (HB). In order to establish a consistent model using two sets of variables, we define a variable v_{kh} for declaration the starting time of the scheduled job in the kth position for processing on machine h.

$$\text{Objective: Min} \sum\nolimits_{k \in N} \sum\nolimits_{h \in M} E_{kh} \tag{6}$$

Constraints:

$$E_{kh} \ge \sum_{j \in N} \sum_{t = 1..k} x_{jth} p_{jh} w_j - \sum_{j \in N} x_{jkh} d_j w_j, \quad \forall k \in N, \forall h \in M, \tag{7}$$

$$\sum_{j \in N} x_{jkh} = 1, \quad \forall k \in N, \forall h \in M, \tag{8}$$

$$\sum_{k \in N} x_{jkh} = 1, \quad \forall j \in N, \forall h \in M, \tag{9}$$

$$v_{kh} - v_{k-1,h} \geq \sum_{j \in N} x_{j,k-1,h} p_{jh}, \quad \forall k \in N - \{1\}, \forall h \in M, \tag{10}$$

$$M\left(2 - \sum_{h \in M} r_{jlh} x_{jth} - \sum_{h \in M} r_{j,l+1,h} x_{jkh}\right) + \sum_{h \in M} r_{j,l+1,h} v_{kh}$$

$$-\sum_{h \in M} r_{jlh} v_{th} \geq \sum_{h \in M} r_{jlh} p_{jh}, \quad \forall j,k,t \in N, \forall l \in M - \{m\}, \forall j,k \in N, \forall h \in M. \tag{11}$$

$Z_j, v_{kh} \in$ nonnegative integers,

$x_{jkh} \in$ binary,

In hybrid formulation, the first set of constraints (7) is used to linearized the expression $w_j.max\ (0, c_j - d_j)$ and this allows to calculate the objective function based on the position of jobs. The second (8) and third (9) set of constraints impose position sequence for jobs, the fourth set (10) indicate that starting time of consecutive position of jobs must have delay with duration of processing time of the previously positioned job, and eventually last set (11) impose operational constraints of each job.

3 Constraint Programming Formulation

Constraint programming (CP) has been proven very efficient for solving scheduling problems [2]. This approach is highly applicable in two cases: (1) A non-convex solution space that comprised myriad of locally optimal solutions, (2) Multiple disjunctions or globally-defined constraints, which results in poor information retrieved by a linear relaxation of the problem.

A pure disjunctive programming with interval variables s_{ij} (starting time of job i on machine j in a directed graph) is represented as follows:

$$\text{Objective: Min} \sum_{j \in N} Z_j \tag{12}$$

Constraints:

$$w_j (s_{ij} + p_{ij} - d_j) \leq Z_j, \quad \forall i \in N, \forall j \in M, \tag{13}$$

$$s_{ik} - s_{ij} \geq p_{ij}, \quad \forall i \in N, \forall (j,k) \in M, \text{ with } < i, j > \to < i, k >, \tag{14}$$

Table 1 Syntax definition for declared OPL local or global constraints

Syntax	Definition
Cum	Accumulation over interval variables
Precedes	Declares precedence constraint on interval variables
Requires	Declares resource requirement
Wait	Declares precedence constraint on the assignment of resource variables

$$(s_{ih} - s_{jh}) \vee (s_{jh} - s_{ih}) \geq p_{(jh) \vee (ih)}, \quad \forall (i, j) \in N, \forall h \in M, \tag{15}$$

$$s_{ij}, Z_j \geq 0, \quad \forall i \in N, \forall j \in M. \tag{16}$$

In this formulation, the first set of constraints (13)is considered for linearization of the objective function. The second set (14) ensure that operation (i, k) cannot start before the completion of operation (i, j). The third set of constraints (15) observe ordering among operations of different jobs that have to be assigned on the same resource.

The CP model, in contrast to IP model, is highly contingent upon the CP package used for modeling the problem because of the discrepancies in functional structures of various modeling languages [1]. In this study, ILOG's OPL Studio 11.1 is used as the modeling language. Also, Table 1 presents the syntax definition of the constraints used in OPL model.

A basic OPL modeling framework involves a set of intervals (jobs) that need to be assigned on a set of resources (e.g. machines, operators, etc.). An interval variable is a decision variable with three components, i.e. a start and end time and a duration, logically linked together. For interval variables, the search methodology does not enumerate the values in the variables domain and the search space is usually independent of the problem domain size. The search space is estimated as the number of possible orderings of the *n* interval variables and the complexity of problem is estimated as *n.log2 (n)*. Then, the OPL formulation can be written as follows:

$$\text{Objective}: \text{Min } \boldsymbol{cum}_{i \in N} \{w[i].\ \boldsymbol{max}(C[i] - d[i], 0)\} \tag{17}$$

Constraints:

$$C[i] \quad \textbf{assigns} \quad OPR[i, nb\_resource].end, \quad \forall i \in N, \tag{18}$$

$$OPR\,[i, nb\_resource] \quad \textbf{precedes} \quad C[i], \quad \forall i \in N, \tag{19}$$

$$OPR\,[i, j] \quad \textbf{precedes} \quad OPR[i, j+1], \quad \forall i \in N, \forall j \in M - \{n\}, \tag{20}$$

$$OPR\,[i,j] \;\; \textbf{requires} \;\; MAC[OPR[i,j],j], \quad \forall i \in N, \forall j \in M, \tag{21}$$

$$\begin{aligned} &MAC\,[OPR[i,j],j] \;\; \textbf{waits} \\ &MAC\,[OPR[i,j],j+1], \quad \forall i \in N, \forall j \in M-\{n\}, \end{aligned} \tag{22}$$

In this OPL formulation, we deployed two key variables, namely *OPR* as interval variable and *MAC* as resource variable or machines. The objective allows the solver to accumulate over sequences of interval variables and to select the minimum value corresponding to that combination. The first set of constraints (19) declares that jobs on their last operations cannot override *C*[*i*] interval. This auxiliary variable helps the modeler to impose a global deadline constraint for the entire planning horizon. The constraints set (20) adjusts the precedence conception, and (21 indicates all interval variables should be dedicated to *unary* resource variable (*MAC*). Note that there is no *alternative* resource for interval variables, therefore they must be dedicated to only one source. The last set of constraints (22) suggests that resource assignment should not be overlapped.

4 Test Problems and Experiments

In order to evaluate the capabilities of proposed models, it should be tested on difficult problem instances. Also, our attempt should be focused on conditions under which the solution procedure is most severely challenged. Moreover, the ideal condition for conducting such experiment could exist in usage of data which are representative of real scheduling problems, but providing such pure data is almost impossible. We obtained our job shop problem benchmarks among well-known test problems in the literature, e.g. ABZ05, ABZ06, MT10, etc. This benchmarks initially intended for makespan case are revised by adding a due date and a weight for each of jobs. The latter are constructed following scheme of [4]: They suggested that 20 % of the customers are very important, 60 % of them are of average importance, and the remaining 20 % are of less importance. Therefore, we assume that w_j=(4,4,2,2,2,2,2,2,1,1) for a typical 10×10 (job×machine) instance. According to following equation, the due date of job j is set to be equal to the floor of the sum of the processing times of its operations multiplied by a due date tightness factor β. Hence, $d_i = \left\lfloor \beta.\sum_{k=1}^{m} p_{ik} \right\rfloor$, $\beta = 1.3, 1.5, 1.6.$

4.1 Computational Results and Analysis

Table 2 displays the size of the MIP and CP formulations for all three alternatives. As the table indicates, the CP formulation contains the smallest number of constraints

Table 2 Model comparison based on the number of generated variables and constraints

	Variables							
	Binary		Interval	Non-negative		Constraints		
Size	DJ	HB	CP	DJ	HB	DJ	HB	CP
4×4	24	64	20	21	40	64	254	16
6×6	90	216	36	43	89	216	1230	36
8×8	224	512	72	73	133	512	3832	64
10×10	450	1000	110	11	192	1000	9390	100
12×12	792	1728	156	157	291	1728	19572	144
15×15	1575	3375	240	241	453	3375	48135	225

and variables in comparison with the other formulations. The hybrid formulation (HB) has the highest number of variables. It can be immediately inferred that the last set of constraints possess high portion of constraints in this model, but it is obviously not a tight enough set to take advantage of this property. The disjunction formulation (DJ) has a moderate number of constraints by a wide margin, which seems to account for its efficiency to solve larger problems. In spite of the fact that the HB formulation is conceived to be very tight, the size of the formulation finally contributes to its longer computation times. Table 2 signals the fact that number of binary variables in DJ and HB formulation almost tend to grow exponentially by the rise of problem size. Moreover, solving the CP model is roughly analogous to solving a pure integer model, because solution procedures for facing with CP models only have been built and designed on the integer base.

Our experiments involved solving each test problem using ILog CPLEX 11.1 on a 2.66GHZ Intel Core 2 Quad Processor (4 MB Cache) with 6 GB of memory. We restricted the solver with a 3600-s time limit in order to terminate a specific run if the optimal solution had not been verified in that period of time. The runs information are summarized in Table 3. For each formulation, the table displays the number of problems solved within the time limit for the full set of 30 problems, the average time (in second), and the maximum time. We also kept track of the maximum and average gap for the cases in which an optimal solution was not found by the solver and for better addressing the convergence of the model. (A gap is the difference between the solution returned by truncated run and the optimum, computed as a ratio to the optimum.) The results indicate that the HB formulation is very weak, it is also inefficient compared to other formulations. The CP formulation solves most of the benchmarks. Also, we found that DJ formulation converges as fast as CP to the optimum in some cases, but it takes so much time to prone and explore the branches and prove optimality. As the table displays, tightness factor has dramatic impact on average and maximum time to solve a problem, because it restricts the search space and triggers fast convergence. There remains no skepticism about the performance of CP: it outperforms other formulations in all measures.

Table 3 Summary of computational results

Tightness factor (β)	Model	Problem solved	Average time (sec)	Maximum time (sec)	Average gap (%)	Maximum gap (%)
1.3	DJ	27	985.13	3600+	11.11	30.00
1.3	HB	11	2549.67	3600+	46.25	100.00
1.3	CP	30	748.02	1985.22	0.00	0.00
1.5	DJ	21	2914.65	3600+	73.12	97.01
1.5	HB	5	3600+	3600+	100.00	100.00
1.5	CP	22	2771.36	3600+	25.97	43.12
1.6	DJ	17	3600+	3600+	100.00	100.00
1.6	HB	2	3600+	3600+	100.00	100.00
1.6	CP	23	2789.27	3600+	65.74	100.00

5 Conclusions

That a scheduling problem can be formulated through mathematical programming does not insinuate that there is an available satisfactory standard solution procedure. TWT-JS is a very hard problem either based on enumeration or on heuristics.

We conducted some computational experiments based on two integer model and one OPL model, namely CP, using state-of-the-art software package, CPLEX11.1. We tested the efficiency of our proposed formulations by well-known job shop 10×10 benchmarks each with different computational complexity. The CP formulation solves nearly all difficult problems in a reasonable amount of time, because search space is usually independent of the problem domain size. Also, we found that formulation based on disjunction concept (DJ) provides a highly consistent perspective. It produces least number of variables and constraints in comparison with HB formulation and yields a tight model. When it fails to prove optimality after an hour of computation time, it preserves optimality gap.

For scheduling researchers, the implication is that standard optimization approaches are able to solve moderate-sized scheduling problems in a reasonable amount of time. Specialized algorithms would seem to be preferable only for larger problems in size. Furthermore, the attention paid to CP formulation may be justified if they provide insight into combinatorial perspective, but CP model are desirable when mathematical models come to fail in description of highly complex search spaces.

References

1. Edis EB, Ozkarahan I (2011) A combined integer/constraint programming approach to a resource-constrained parallel machine scheduling problem with machine eligibility restrictions. Eng Optimiz 43(2):135–157
2. IBM Ilog CP (2009) Detailed scheduling in IBM ILOG CPLEX optimization studio with IBM ILOG CPLEX CP Optimizer. http://cpoptimizer.ilog.com. Accessed 16 sept 2012

3. Mati Y, Dauzère-Pérès S, Lahlou C (2011) A general approach for optimizing regular criteria in the job-shop scheduling problem. Eur J Oper Res 212:33–42
4. Pinedo M, Singer M (1999) A shifting bottleneck heuristic for minimizing the total weighted tardiness in a job shop. Nav Res Log 46:1–17
5. Singer M, Pinedo M (1997) A computational study of branch and bound techniques for minimizing the total weighted tardiness in job shops. IIE Trans 30(2):109–118

Set-up Continuity in Tactical Planning of Semi-Continuous Industrial Processes

David Pérez Perales, Maria del Mar Eva Alemany Díaz, Francisco-Cruz Lario Esteban and V. S. Fuertes-Miquel

Abstract In most of production planning models developed in a hierarchical context at the tactical level setup changes are not explicitly considered. Its consideration includes decisions about the allocation and lot sizing of production, known as CLSLP problem. However the CLSLP problem does not account for set-up continuity, specially relevant in contexts with lengthy set-ups and where product families minimum run length are almost are similar to planning periods. In this work, a MILP model which accounts for this set-up continuity inclusion is modelled, solved and validated over a simplified real-case example.

Keywords Set up continuity · Semicontinuous processes · Tactical planning

1 Introduction

In the majority of the production planning models developed in a hierarchical context at the tactical level, the capacities at each stage are aggregated and setup changes are not explicitly considered. However, if at this level the setup times involve an important consumption capacity and have been completely ignored, this may lead to an overestimation of the real capacity availability which, in turn, may lead to unfeasible events during the subsequent disaggregation of tactical plans. Considerable savings may be also be achieved through optimum lot-sizing decisions.

D. P. Perales (✉) · M. del Mar Eva Alemany Díaz · F.-C. Lario Esteban
Centro de Investigación en Gestión e Ingeniería de Producción, Universitat Politècnica de València, Camino de Vera s/n, 46022 Valencia, Spain
e-mail: dapepe@omp.upv.es

M. del Mar Eva Alemany Díaz
e-mail: mareva@omp.upv.es

F.-C. Lario Esteban
e-mail: fclario@omp.upv.es

V. S. Fuertes-Miquel
Dpto. de Ingeniería Hidráulica y Medio Ambiente, Universitat Politècnica de València, Camino de Vera s/n, 46022 Valencia, Spain
e-mail: vfuertes@upv.es

C. Hernández et al. (eds.), *Managing Complexity*, Lecture Notes in Management and Industrial Engineering, DOI 10.1007/978-3-319-04705-8_19,

However, accounting for setup times at the tactical level would mean simultaneously including decisions about the allocation and lot sizing of production. This problem is known as the capacitated lot-sizing and loading problem (CLSLP) [3]. Nevertheless, CLSPP does not consider the set-up continuity over discrete periods of time, that is, it assumes that if a product family is manufactured in two periods of time a double set-up should be considered, leading to an underestimation of the real capacity availability. It becomes necessary to model the set-up continuity in these cases so that only one set-up is considered, and therefore saving one. This is particularly important in industrial sectors such as ceramic [2], food [4], textile, etc., since they cope with very lengthy setup times in their manufacturing semicontinuous processes and at the same time their product families minimum run length are almost or equal to the planning period.

In this article, a manner to model the former set-up continuity is proposed. It is applied and validated in a one-stage production process of the ceramic sector, so it may be adapted to a larger model for a specific situation. The rest of the paper is arranged as follows. Section 2 describes the problem being studied. In Sect. 3, a deterministic MILP model to solve the problem is presented. Section 4 reports a numerical example to validate the model. Section 5 offers some conclusions.

2 Problem Description

Production in ceramic SCs usually includes several stages such as presses-glazing lines, kilns and sorting-packing. In this work, only the first one is characterised, although it may also be extrapolated to the second one, with similar characteristics. This presses-glazing stage is made up of one or several production lines in parallel with a limited capacity. Different product families can be processed by each production line. Changeovers from one product family to the next incur setup time and costs. Given the important setup times, when a certain product family is manufactured on a specific line, it should be produced in an equal or greater amount than the minimum lot size. At the tactical level, an Aggregate Plan (AP) is defined for product families, while at the operational level, the Master Plan (MP) is defined for finished goods. Tactical planning must account for two aspects, setup times and its continuity over consecutive planning periods, because the set-up are lengthy and the product families run length (3 weeks) are similar to the planning periods (1 month). These aspects are crucial to get accurate capacity availability estimation, which will constraint the MP.

3 Problem Modeling

A deterministic MILP model has been developed to solve the ceramic tactical planning problem. This model has been simplified since the main objective is formulate and validate the set-up continuity constraints. A supply chain-based extended

Table 1 Indices

f	Product families (F) (f=1...F)
l	Production lines (L) (l=1...L)
t	Periods of time (PT) (t=1...T)

Table 2 Parameters

dmd_{ft}	Demand of F f in PT t
$costinv_f$	Inventory cost of a F in a PT
$costsetupl_{fl}$	Setup cost of F f on L l
$tfabl_{fl}$	Time to process a F f on L l
$tsetupl_{fl}$	Setup time for F f on L l
$lminf_{fl}$	Minimum lot size of F f on L l
$capfabl_{lt}$	Production capacity available (time) of L l during PT t
$inv0_f$	Inventory of F f at the start of the first PT
M1, M2	Very large integres
nfamilias	Number of F
$betal0_{fl}$	The L l is prepared to manufacture the F f at the start of the first PT

version of this model may be found in Alemany et al. [1], but without accounting for the set-up continuity. The objective is to minimize the total cost(set-up and inventory) over the time periods of the planning horizon. Decisions will have to simultaneously deal with not only the allocation of product families to production lines with a limited capacity, but also with the determination of lot sizing. Another decisions regard to set-up continuity modelling. For example those which allow to know the first and the last product family processed on a production line in a planning period, so that one changeover can be saved if the last one processed in t and the first one in t + 1 are the same. Or those which allow to process the minimum lot size between two consecutive periods with no change over. All of them are later explained. The indices, parameters, and decision variables are described in Tables 1–3, respectively.

$$Min \sum_t \sum_l \sum_f \cos tsetupl_{fl} * ZL_{flt} + \sum_t \sum_f \cos tinv_f * INV_{ft} \quad (1)$$

subject to:

$$INV_{ft} = inv\,0_f + \sum_l PFL_{flt} - dmd_{ft}, \quad \forall \text{ f, t } = 1 \quad (2)$$

$$INV_{ft} = INV_{ft-1} + \sum_l PFL_{flt} - dmd_{ft}, \quad \forall \text{ f, t} > 1 \quad (3)$$

$$\sum_f tfabl_{fl} * PFL_{flt} + \sum_f tsetupl_{fl} * ZL_{flt} \le capfabl_{lt}, \quad \forall \text{ l, t 1} \quad (4)$$

Table 3 Decision variables

INV_{ft}	Inventory of F f in PT t
PFL_{flt}	Amount of F f manufactured on L l in PT t
YL_{flt}	Binary variable with a value of 1 if F f is manufactured on L l in PT t, and with a value of 0 otherwise
XL_{flt}	Binary variable with a value of 1 if L l is ready to manufacture the F f in PT t, and with a value of 0 otherwise
ZL_{flt}	Binary variable with a value of 1 if L l if a setup takes place of F f on L l in PT t, and with a value of 0 otherwise
WL_{lt}	Binary variable with a value of 1 if more than one F f is manufactured on L l in PT t, and with a value of 0 otherwise
$ALFAL_{flt}$	Binary variable with a value of 1 if L l is prepared to manufacture the F f at the start of PT t, and with a value of 0 otherwise
$BETAL_{flt}$	Binary variable with a value of 1 if L l is prepared to manufacture the F f at the end of PT t, and with a value of 0 otherwise

$$PFL_{flt} \le M1 * XL_{flt}, \quad \forall \text{ f, l, t} \tag{5}$$

$$PFL_{flt} \le M2 * YL_{flt}, \quad \forall \text{ f, l, t} \tag{6}$$

$$l\min f_{fl} * (ZL_{flt} + ZL_{flt+1} - YL_{flt+1}) \le PFL_{flt}, \quad \forall \text{ l, f, t} \tag{7}$$

$$l\min f_{fl} * (ZL_{flt} + ZL_{flt+1} + YL_{flt} + YL_{flt+1} - 2) \le PFL_{flt} + PFL_{flt+1}, \quad \forall \text{ l, f, t} \tag{8}$$

$$YL_{flt} \le PFL_{flt}, \quad \forall \text{ f, l, t} \tag{9}$$

$$YL_{flt} \le XL_{flt}, \quad \forall \text{ f, l, t} \tag{10}$$

$$ZL_{flt} \le YL_{flt}, \quad \forall \text{ f, l, t} \tag{11}$$

$$ALFAL_{flt} - betal\,0_{fl} \le \sum_{f} ZL_{flt}, \quad \forall\ f,\ l,\ t = 1 \tag{12}$$

$$ALFAL_{flt} - BETAL_{flt-1} \le \sum_{f} ZL_{flt}, \quad \forall \text{ f, l, t>1} \tag{13}$$

$$BETAL_{flt} - ALFAL_{flt} \le \left(\sum_{f} XL_{flt} \right) - 1, \quad \forall \text{ f, l, t} \tag{14}$$

$$\sum_{f} ALFAL_{flt} = 1, \quad \forall \text{ l, t} \tag{15}$$

$$\sum_{f} BETAL_{flt} = 1, \quad \forall \text{ l, t} \tag{16}$$

$$ALFAL_{flt} \leq XL_{flt}, \quad \forall \text{ f, l, t} \tag{17}$$

$$BETAL_{flt} \leq XL_{flt}, \quad \forall \text{ f, l, t} \tag{18}$$

$$3 * XL_{flt} - \sum_{f} XL_{flt} \leq ALFAL_{flt} + BETAL_{flt}, \quad \forall \text{ f, l, t} \tag{19}$$

$$2 * XL_{flt} - ALFAL_{flt} + betal0_{fl} \leq 2 * ZL_{flt}, \quad \forall \text{ f, l, t} \tag{20}$$

$$2 * XL_{flt} - ALFAL_{flt} + BETAL_{flt-1} \leq 2 * ZL_{flt}, \quad \forall \text{ f, l, t} \tag{21}$$

$$\sum_{f} ZL_{flt} \leq nfamilias * (3 - ALFAL_{flt} - BETAL_{flt} - betal0_{fl}), \quad \forall \text{ f, l, t=1} \tag{22}$$

$$\sum_{f} ZL_{flt} \leq nfamilias * (3 - ALFAL_{flt} - BETAL_{flt} - BETAL_{flt-1}), \quad \forall \text{ f, l, t>1} \tag{23}$$

$$2 - \sum_{f} YL_{flt} \leq 2 * (1 - WL_{lt}), \quad \forall \text{ l, t} \tag{24}$$

$$\left(\sum_{f} YL_{flt} \right) - 1 \leq nfamilias * WL_{lt}, \quad \forall \text{ l, t} \tag{25}$$

$$ALFAL_{flt} + BETAL_{flt} \leq (2 - WL_{lt}), \quad \forall \text{ f, l, t} \tag{26}$$

The objective function (1) expresses the minimization of the setup costs of the FPs on the Ls and the inventory costs of the Fs at the end of the manufacturing process.

Constraints (2) and (3) are the inventory balance equations of in-process and finished Fs, respectively. Constraint (4) ensures that the capacity required for the setup of Fs and the manufacturing of the lots assigned to each L do not exceed the capacity available on each L in each PT. Constraint (5) indicates that a F can only be manufactured on a L in a PT if the L has previously be prepared to manufacture the F in such a PT. Constraint (6) indicates that a F can only be manufactured on a L in a PT if it has previously been decided to manufacture the F on the L in such a

PT. Constraint (7) guarantees that should a certain amount of a F be manufactured on a L, it is equal to or above the minimum lot size established for the F on that line. Constraint (8) allows not to manufacture the minimum lot size established for a F on a L in a PT, if either the F was the last one manufactured in the previous PT and the first one manufactured in the next PT, or the F is the only one manufactured during two consecutive PTs. However, it guarantees in both cases that the total amount of F manufactured will be superior to its minimum lot size. Constraint (9) establishes that if there is no amount of F manufactured on a L in a PT then it is not allowed to manufacture the F on the L in such a PT. Constraint (10) establishes that if a F is manufactured on a L in a PT, then the L has been previously prepared to manufacture the L in such a PT. Constraint (11 establishes that if a F is not manufactured on a L in a PT, then there is no setup on the L in such a PT. Constraints (12) and (13 ensure that if a L "status" at the start of a PT is different from the "status" of the L at the end of the previous PT, then at least one setup has to be made on the L in such a PT. Constraint (14) indicates that if a L does not change its "status" during a PT, then the L is already prepared (either at the start or the end of such a PT) to manufacture the same F. Constraints (15) and (16) guarantee that a L can be only prepared to manufacture just one F, in the start and in the end of a PT, respectively. Constraints (17) and (18) ensures that if a L is not prepared to manufacture a F in a PT, then that F cannot be either the first or the last, respectively, for which the L was prepared in such a PT. Constraint (19) indicates that if a L is only prepared to manufacture just one F in a PT, then the L should be prepared either at the start or the end of such a PT to manufacture the F. Constraints (20) and (21) indicate that it is only possible to save a single changeover on a L in a PT if the L is prepared at the start of the current PT to manufacture the same F for which it was prepared at the end of the previous PT. Constraints (22) and (23) indicate that if the "status" of a L at the start and the end of a current PT is equal to the "status" at the end of the previous PT, then just one or no F is manufactured. Constraint (24) assures that if one or no F is manufactured on a L in a PT, then WL=0, although the contrary case does not imply WL=1. For this it is implemented constraint (25). Constraint (26) guarantees that if more than one F is manufactured on a L in a PT, no one of them can be the first and the last at the same time in such a PT. Therefore, only in the case in which one or no F is manufactured on a L in a PT is possible that ALFAL=1 and BETAL=1 for that F.

4 Validation

An example to validate the model is described. Data of product families demand and production lines capacity in each PT and specific data of product families on production lines (inventory cost, minimum lot size, etc) are respectively shown in Figs. 1–2.

Just a few representative values of the decision variables that lead to the optimum solution and help to validate the set-up continuity are shown in Table 4.

F	dmdft					
	t1	t2	t3	t4	t5	t6
F1	100	125	135	140	150	130
F2	125	110	135	150	125	115
F3	140	125	110	130	115	125
F4	100	125	135	140	150	130
F5	125	110	135	150	125	115
F6	140	125	110	130	115	125

L	capfablt					
	t1	t2	t3	t4	t5	t6
L1	50	70	70	50	70	70
L2	70	50	70	50	50	70
L3	50	50	50	70	70	50

Fig. 1 Data of product families (*F*) demand and production lines (*L*) capacity in each TP (*t*)

F	L	costinvf	inv0f	tsetuplf	costsetuplf	tfablf	lminff	beta0lf
F1	L1	0,1	50	2	35	0,1	160	0
F2		0,15	50	2,5	30	0,25	180	1
F3		0,2	50	3	40	0,2	175	0
F4		0,15	50	3.5	45	0,2	160	0
F5		0,25	50	2,5	30	0,1	180	0
F6		0,1	50	3	45	0,15	170	0
F1	L2	0,1	50	2	35	0,1	160	0
F2		0,15	50	2,5	30	0,25	180	0
F3		0,2	50	3	40	0,2	175	1
F4		0,15	50	3.5	45	0,2	160	0
F5		0,25	50	2,5	30	0,1	180	0
F6		0,1	50	3	45	0,15	170	0
F1	L3	0,1	50	2	35	0,1	160	1
F2		0,15	50	2,5	30	0,25	180	0
F3		0,2	50	3	40	0,2	175	0
F4		0,15	50	3.5	45	0,2	160	0
F5		0,25	50	2,5	30	0,1	180	0
F6		0,1	50	3	45	0,15	170	0

Fig. 2 Data of product families (*F*) on production lines (*L*)

These results confirm that the described constraints are valid to model the setup continuity and that the minimum lot size can be splitted between two consecutive periods of time in case a F is the last to be manufactured on a L in a PT t and the first to be manufactured on the same L in PT t+1.

A representative example may be seen in Table 4, for example for F5, which is manufactured on L1 in PTs t=1 and t=2.

Table 4 Amount (m^2) of product families (F) manufactured on production line L1 in PT t

			t1	t2	t3	t4	t5	t6
L1	PFL	F1			70	165		
		F2	75					
		F3		125	110	130	140	
		F4						
		F5	75	145				
		F6			240		115	125
	XL=YL	F1			1	1		
		F2	1					
		F3		1	1	1	1	
		F4						
		F5	1	1				
		F6			1		1	1
	Betal0=F2	ALFAL	F2	F5	F3	F1	F3	F6
		BETAL	F5	F3	F1	F3	F6	F6
	ZL	F1			1			
		F2						
		F3		1		1		
		F4						
		F5	1					
		F6			1		1	
		WL	1	1	1	1	1	

5 Conclusions

This work presents a deterministic MILP model to solve the tactical planning problem for the production in the ceramic sector, although in may be extrapolated to another semi-continuous production sectors.

Its main contributions are on one hand that the accounting for setup times at the tactical level which implies including decisions about the allocation and lot sizing of production (CLSLP problem). Not many works accounts for it in tactical planning. On the other hand the consideration of set-up continuity constraints, especially important in contexts with lengthy set-ups and where product families minimum run length are almost or equal to the planning period.

Both contributions help to achieve a more accurate capacity availability estimation in the tactical level so it may lead to feasible and more efficient events during the subsequent disaggregation into operational plans.

Since the main objective of this work is this set-up continuity validation, the model has just been applied in a simplified real one-stage ceramic production process and only some of the results are shown. These results show how this model may be adapted to a larger model for a specific situation.

Acknowledgements This research has been carried out in the framework of the project funded by the Spanish Ministry of Economy and Competitiveness (Ref. DPI2011-23597) and the Polytechnic University of Valencia (Ref. PAID-06-11/1840) entitled "Methods and models for operations planning and order management in supply chains characterized by uncertainty in production due to the lack of product uniformity" (PLANGES-FHP).

References

1. Alemany MM, Alarcón F, Lario FC, Boj JJ (2009) Planificación agregada en cadenas de suministro del sector cerámico. 3rd International Conference on Industrial Engineering and Industrial Management XIII Congreso de Ingeniería de Organización Barcelona-Terrassa, 2–4 septiembre 2009
2. Alemany MM, Boj JJ, Mula J, Lario FC (2011) Mathematical programming model for centralised master planning in ceramic tile supply chains. Int J Prod Res 48:5053–5074
3. Özdamar L, Birbil SI (1998) Hybrid Heuristics for the capacitated lot sizing and loading problem with setup times and overtime decisions. Eur J Oper Res 110:525–547
4. Romsdal A, Thomassen MK, Dreyer HC, Strandhagen JO (2011) Fresh foodsupply chains; characteristics and supply chain requirements. 18th international annual EurOMA conference. Cambridge, UK, Cambridge University

Estimating Costs in the EOQ Formula

Pilar I. Vidal-Carreras, Jose P. Garcia-Sabater, Maria Valero-Herrero and Cristina Santandreu-Mascarell

Abstract The EOQ formula (Harris, Fact Mag Manage 10(2):135-6-152, 1913) provides a balance between setup costs and holding costs in the system. This formula has been widely developed in the literature. However in the industrial reality, it is often difficult to know the exact value of these setup and holding costs. In this paper, we develop a formula to estimate lot size from the values known in the company. It is verified that the behavior of these formulas meets expectations.

Keywords EOQ · Inventory management · Setup cost · Holding cost

1 Introduction

Defining an inventory management system in a company involves setting up the technique employed (reorder point management, provisioning theoretical, etc.) and determining its control parameters. Essentially, the control parameters required for managing inventory levels are minimum and maximum level of the stock of each product that the company wishes to maintain in the system.

P. I. Vidal-Carreras (✉) · J. P. Garcia-Sabater
Grupo ROGLE. Dpto. de Organización de Empresas, Universidad Politécnica de Valencia, Camino de Vera S/N, 46022, Valencia, Spain
e-mail: pivicar@omp.upv.es

J. P. Garcia-Sabater
e-mail: jpgarcia@omp.upv.es

M. Valero-Herrero
Dpto. de Organización de Empresas, Universidad Politécnica de Valencia, Camino de Vera S/N, 46022 Valencia, Spain
e-mail: mavaher@upv.es

C. Santandreu-Mascarell
IGIC. Dpto. de Organización de Empresas, Universidad Politécnica de Valencia, Camino de Vera S/N, 46022 Valencia, Spain
e-mail: crisanma@omp.upv.es

C. Hernández et al. (eds.), *Managing Complexity*, Lecture Notes in Management and Industrial Engineering, DOI 10.1007/978-3-319-04705-8_20,

The theoretical resolution of this problem is based on the classic EOQ formula, Economic Order Quantity [4]. This EOQ formula provides a balance between the setup costs and holding costs in the system. The EOQ model involves having to know the value of the setup costs and holding costs of products. Yet apart from the consideration that all costs are costs of opportunity, there are various reasons why these costs are generally unknown to companies. Despite this, it is still necessary to establish inventory levels, not so much to optimize the system, but to keep it under control.

The EOQ model has been investigated in depth in the literature. One of its variants that this paper is interested in is "joint pricing and inventory optimization problem". One of the earliest papers on pricing and inventory is that by [14], who proposed a link between pricing and inventory control. Lee [7] presented a geometric programming approach to determine a profit-maximizing price and order quantity for a retailer. Most recently, [10] developed an EPQ inventory model that determines production lot size, marketing expenditure and a product's selling price.

As price is an obvious strategy to influence demand, researching inventory models with price-dependent demand have attracted much attention. You and Chen [15] developed an EOQ model of seasonal goods with spot and forward purchase demands. Recently, Mo et al. [9] proposed an EOQ model with stock and price sensitive demand.

Several researchers have studied the effect of delayed payments on the EOQ. Goyal [3] was the first to develop a model for delaying payment to the supplier, which makes all the usual assumptions of the classic EOQ model, save when payment is due. More recently, [5] assumed that the supplier would offer the retailer a partial delay in payments when the order quantity is smaller than a predetermined quantity. Jaggi et al. [6] proposed a model in which demand is linked to the credit period offered by the retailer to customers. Taleizadeh et al. [12] offered an EOQ problem under partial delayed payment.

However for situations in which a company does not have a possible cost estimate of EOQ formulas, some papers assume that setup costs are proportional to the time setup, and that holding costs are proportional to the cost unit. Yet they do not indicate what this proportionality is based on [1, 2, 13]. Furthermore, the number of characteristics and variables addressed by models and algorithms has grown in parallel with the development of computational capacity signal equipment. Nevertheless, one feature observed, which is similar to what other authors have reported [8, 11] is that most companies are still using Excel spreadsheets for planning, scheduling and controlling their operations. Thus despite the costs associated with their manual management, these tools appear to be "more effective" in daily management since most companies prefer them.

Thus the present paper presents an alternative resolution to the problem of estimating costs for inventory management in the industrial reality. For this purpose, the problem of analyzing information that is typically available in a company is proposed in Sect. 2. Next some EOQ formulas that have been adapted to that information are explained in Sect. 3. Then these formulas are integrated into a simple Excel tool and their behavior is checked in Sect. 4. The conclusions and future works are presented in Sect. 5.

Table 1 Notation

i	Item index i = 1…n
Q_i	Lot size of item i (units)
d_i	Demand of item i (units/unit of time)
ca_i	Holding cost of item i (monetary units)
cs_i	Setup cost of item i (monetary units)
ts_i	Setup time of item i (time units)
η_i	Density of value of item i (monetary units/packaging units)
pe_i	Pallet equivalent to item i (monetary units/pallet)

2 Statement of the Problem and Notation

Having to know setup costs and holding costs presents difficulties in current inventory management systems. While the setup time is either available in companies or is easy to measure, establishing *a priori* setup costs is not. This is because setup costs are actually an opportunity cost of use of installed capacity. Similarly, holding costs are an opportunity cost of the storage capacity available in the company.

In order to be able to clearly analyze the problem, taking the case of a machine of a goods manufacturer is considered appropriate. The machine required for each production run change needs preparation, which implies time and resources. For this change (setup), workers are available to prepare the machine for production. However, the company does not know the cost associated with this setup. So it can approximately make these setup costs proportional to the setup time [1]. The notation used in the sequent sections is shown above in Table 1.

It is not possible to define the holding costs of each product *a priori*. Companies usually have limited storage capacity and a amount of cash limit that they wish to maintain. Each product has specific characteristics relating to both its economic value and the volume it occupies. If the company's financial constraint is important, it can be assumed that the costs associated with storage are proportional to the amount of money in stock or, likewise, to the unit cost of the products. If space limitation in the company is important, one can assume that storage costs are proportional to the number of locations used. In some case, this value may reduce the number of pallets in stock. In both situations, and as mentioned earlier, the exact value of these storage costs in the company are not known. Generally therefore, the different departments involved in the company establish a maximum storage limit, according to which inventory policies are usually defined.

3 A Method to Calculate Costs

According to the classical EOQ formula [4], and as shown in (1), calculating the economic lot to manufacture product i, by considering its demand d_i, requires having to consider a series of costs. On the one hand, a setup cost, cs_i,, has been considered in relation to the change in the production run; that is, the process of preparing machinery to produce product i. On the other hand, there is a holding cost of manufactured product i, ca_i:

$$Q_i = \sqrt{\frac{2d_i cs_i}{ca_i}} \tag{1}$$

Yet as previously discussed, the value of the cost of this setup and holding cost required in the formula is not always known in the industrial reality (1). For the setup cost, it is assumed that setup time ts_i is known. So it is appropriate to assume that setup cost cs_i, is proportional to setup time ts_i in accordance with a proportionality constant that we call cs, whose units are (monetary units/ units of time):

$$cs_i \ \alpha \ ts_i \Rightarrow cs_i = cs * ts_i \tag{2}$$

The company should set the constant cs value according to the value of the time spent by operators on the setup. If a company employs highly automated processes in which workers only participate directly in the process for a specific fraction of time, this constant cs may have a very low value because idle manpower exists for a given time period. Conversely, if a company demands high manpower requirement or if this is scarce, the cs value should be higher. This constant cs is independent of products and depends only on the company's characteristics.

Holding cost, ca_i, which appears in Formula (1), is not exact information. However, and as mentioned before, it can be assumed that this value is proportional to the unit cost of the product, to the volume that the product occupies in the warehouse, or to a combination of both variables.

Firstly, the holding cost may be proportional to cv_i according to a proportionality constant that we call ca_1:

$$ca_i \ \alpha \ cu_i \Rightarrow ca_i = ca_1 * cu_i \tag{3}$$

The company should set the value of constant ca_1 according to its financial conditions. For example, if the company has a mortgage loan at a high interest rate, the rate constant should take a high value, and something similarly happens if goods are perishable stocks. The constant value of ca_1 usually differs in the life cycles of companies.

Should the cost be proportional to the storage space that the product takes up, there is a proportionality constant named ca_2. The volume of the product is defined with the equivalent pallet variable, pe_i. This is appropriate for assuming that the cost of storage ca_i is proportional to pe_iaccording to proportionality constant ca_2 whose units are (monetary units/ pallet).

$$ca_i \ \alpha \ pe_i \Rightarrow ca_i = ca_2 * pe_i \tag{4}$$

Besides, the holding cost should be considered inversely proportional to the density value (5). The density value is called η_i, and is defined as the monetary units on each equivalent pallet (5). It is easy to obtain a variable if the company knows the units on each pallet and their cost.

$$ca_i \, {}^{1}\!/\!_{\alpha} \; \eta_i \Rightarrow ca_i = ca_2/_{\eta_i} \tag{5}$$

$$\eta_i = \frac{cu_i}{pe_i} \tag{6}$$

Thus by combining (4) and (6), the following formula for situations in which the cost is proportional to the storage space occupied by the products is obtained (7):

$$ca_i = \beta ca_2 \frac{pe_i}{\eta_i} \tag{7}$$

In order to include the consideration that the cost is proportional to the storage unit cost of products (3) and to the space that they occupy (7), the following formula (8) is defined:

$$ca_i = \alpha ca_1 cu_i + \beta ca_2 \frac{pe_i}{\eta_i} \tag{8}$$

With this formula, the various combinations of storage costs in the company are represented. Variable α represents the weight of the company's financial condition, and the weight of β represents the volume restrictions in the warehouse. Since it is a combination of both, the sum of variables α and β follows:

$$\alpha + \beta \leq 2 \tag{9}$$

α and β confer the system flexibility in specific situations without having to rethink the values of ca_1 and ca_2. If the setup costs value (2) and the holding costs value (8) are included in the lot size formula, (1) then the result is (10):

$$Q_i = \sqrt{cs}\sqrt{\frac{2d_i ts_i}{\alpha ca_1 cu_i + \beta ca_2 \frac{pe_i}{\eta_i}}} \tag{10}$$

This formula allows the company to obtain the lot size of a product from its known variables, such as product demand, unit cost, and its equivalent pallet setup time required to produce it. The company should set the value of constants cs, ca_1, ca_2, α and β cs, which are common to all the products and allow adjustments to be made in different circumstances. For example, if the company warehouse is full, the storage

Table 2 Numerical example

Data								Result
alpha	1	ca1 (€/€day)	0,00042					
beta	1	ca2(€/palletdía)	0,2					
		cs (€/hour)	4					
Item	Demand (uds/day)	Unit cost (€)	Setup Time (hour)	Units per pallet	Equivalent Pallet	Value of Density		Lot Size
i	di	cui	tsi	udi/palet	pei	ni		Qi
1	100	5	3	100	0,010	500		1072,28
2	150	4	1	150	0,007	600		847,96
3	200	3	1	200	0,005	600		1130,62
4	250	2	1,5	250	0,004	500		1895,55

costs associated with volume are expected to increase, which implies constant αca_2. If the warehouse is empty, the financial criteria would overlap and should increase the value of variable βca_2.

Then, the lot size value of specific circumstances is reevaluated:

If the user considers relevant for the company only the financial limitations, and space constraints are not assumed, then $\beta=0$ so it would not be necessary to specify the value of constant ca_2. So the lot size formula would be as so:

$$Q_i = \sqrt{cs}\sqrt{\frac{2d_i ts_i}{\alpha ca_1 cu_i}} \Rightarrow Q_i = \frac{\sqrt{cs}}{\sqrt{\alpha ca_1}}\sqrt{\frac{2d_i ts_i}{cu_i}} \tag{11}$$

If the company's space limitations are the relevant issue there are no significant financial constraints, it is assumed that $\alpha=0$, and it would not be necessary to specify the value of constant ca_1. So the lot size formula would be as so (12):

$$Q_i = \sqrt{cs}\sqrt{\frac{2d_i ts_i}{\beta ca_2 \frac{pe_i}{\eta_i}}} \Rightarrow Q_i = \frac{\sqrt{cs}}{\sqrt{\beta ca_2}}\sqrt{\frac{2d_i ts_i}{\frac{pe_i}{\eta_i}}} \tag{12}$$

4 Analysis of the Behavior of Formulae

The formulas presented here are provided on an Excel spreadsheet. Thus lots sizes are obtained by introducing known values, as shown in Table 2.

The behavior of formula 10 is analyzed and considered to be complete for an item, particularly item 1 in Table 2. Figure 1 shows the evolution of the lot size of this item by varying different costs. Figure 1a depicts lot size growth by increasing variable *cs* as this is an increase in the costs associated with the setup; for example, manpower restrictions. Figure 1b shows that a decrease in lot size is able to increase

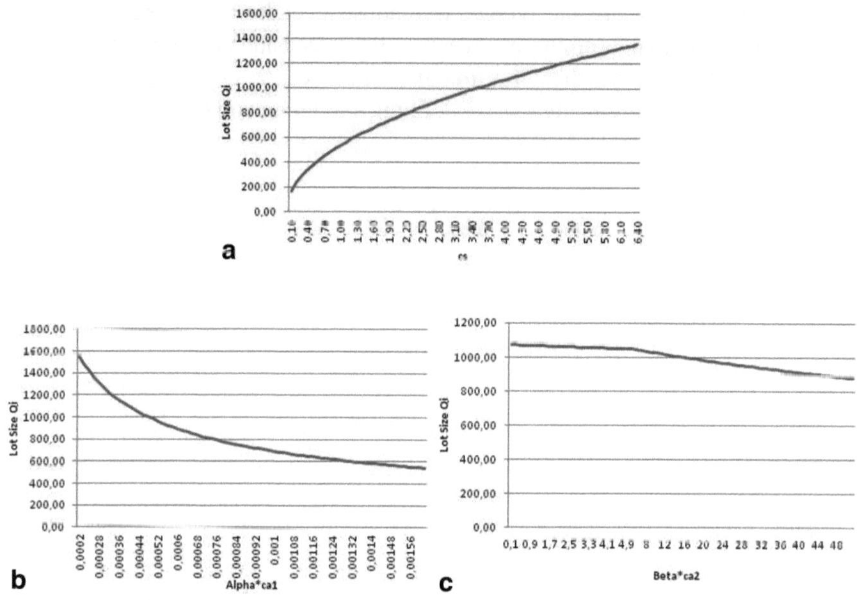

Fig. 1 Evolution of lot size according to **a** *cs*. **b** αca_1 and **c** βca_2

variable αca1. This decrease is suitable for variable αca1 which affects part of the holding cost. Figure 1c shows a much slower decrease in the lot size of the variable by increasing βca_2 It should be taken into account that the holding cost in the formula is associated with volume restrictions, and its relationship with the density value is also contemplated, thus it does not contribute that much in the formula. This is why it is necessary to significantly increase the value of βca_2 in order to observe the expected decrease in lot size.

Should formula 10 not consider the density value, but only the space occupied by products through variable pe_i, given the behavior of the formula in relation to βca_2 minor variations will be seen, which are similar to those shown in Fig 1c.

5 Conclusions

In this work, the classical EOQ formula of [4] is adapted to the industrial reality, in which it is often complex to include the values of the setup and holding costs. This work proposes simple formulas to obtain setup and holding costs based on the data known by the company. These costs can be easily interpreted by the person employing the tool. The behavior of this approach has been analyzed and it is considered reasonable and useful. Future work will consist in integrating these formulas into an inventory tool that can contemplate more considerations that companies may have, such as: available capacity, the average warehouse level desired, average inventory level, the maximum and minimum inventory limits considered for some products, etc. This tool should allow the definition of the final lot sizes by considering all the constraints and, from these, the lot size defining maximum and minimum levels per product.

Acknowledgments The work described in this paper has been supported by Project "CORSARI MAGIC DPI2010-18243" from the Spanish Ministry of Science and Innovation.

References

1. Bomberger EE (1966) A dynamic programming approach to a lot size scheduling problem. Manage Sci 12(11):778
2. Brander P, Segerstedt A (2009) Economic lot scheduling problems incorporating a cost of using the production facility. Int J Prod Res 47(13):3611–3624
3. Goyal SK (1985) Economic order quantity under conditions of permissible delay in payments. J Oper Res Soc 44:785–795
4. Harris FW (1913) How many parts to make an once. Fact Mag Manage 10(2):135-6-152
5. Huang YF (2007) Economic order quantity under conditionally permissible delay in payments. Eur J Oper Res 176(2):911–924
6. Jaggi CK, Goyal SK, Goel SK (2008) Retailer's optimal replenishment decisions with credit-linked demand under permissible delay in payments. Eur J Oper Res 190(1):130–135
7. Lee WJ (1993) Determining order quantity and selling price by geometric programming: optimal solution, bounds, and sensitivity. Decision Sci 24(1):76–87
8. Meyer B (2004) Value-adding logistics for a world assembly line. Bonifatius Verlag, Paderborn
9. Mo J, Mi F, Zhou F, Pan H (2009) A note on an EOQ model with stock and price sensitive demand. Math Comput Model 49(9):2029–2036
10. Sadjadi SJ, Oroujee M, Aryanezhad MB (2005) Optimal production and marketing planning. Comput Optim Appl 30(2):195–203
11. Shirodkar S, Kempf K (2006) Supply chain collaboration through shared capacity models. Interfaces 36(5):420–432
12. Taleizadeh AA, Pentico DW, Saeed Jabalameli M, Aryanezhad M (2013) An EOQ model with partial delayed payment and partial backordering. Omega 41(2):354–368
13. Vidal-Carreras PI, Garcia-Sabater JP, Coronado-Hernandez JR (2012) Economic lot scheduling with deliberated and controlled coproduction. Eur J Oper Res 219(2):396–404
14. Whitin TM (1955) Inventory control and price theory. Manage Sci 2(1):61–68
15. You PS, Chen TC (2007) Dynamic pricing of seasonal goods with spot and forward purchase demands. Comput Math Appl 54(4):490–498

Part III
Logistics and Supply Chain

Order Promising Process for Supply Chains with Lack of Homogeneity in the Product

Maria del Mar Eva Alemany Díaz, Andrés Boza, Angel Ortiz and Raúl Poler

Abstract Traditionally, during the Order Promising Process (OPP), the homogeneity of different available units of the same finished good to be committed to customers has been assumed. However, this assumption is not valid for manufacturing contexts with Lack of Homogeneity in the Product (LHP). In this paper, special LHP-dimensions that affect the OPP are outlined. Based on them, specific LHP availability and allocation rules are defined.

Keywords Lack of homogeneity in the product · Order promising process

1 Introduction

The Order Promising Process (OPP) refers to the set of business activities that are triggered to provide a response to customer order requests. During the OPP, when a new customer order request arrives, it is necessary to compute whether there are enough uncommitted real or planned finished goods (FGs), materials and/or resources available to fulfill the new order on time. Traditionally, the homogeneity of different available units of the same FG to be promised to customers has been assumed. This homogeneity characteristic has allowed the accumulation of uncommitted FG availabilities from different resources and time periods to satisfy the same customer order.

However, this homogeneity assumption is not valid for manufacturing contexts with Lack of Homogeneity in the Product (LHP). LHP contexts are characterized by

---

M. del Mar Eva Alemany Díaz (✉) · A. Boza · A. Ortiz · R. Poler
Centro de Investigación de Gestión e Ingeniería de la Producción (CIGIP),
Universitat Politècnica de València, Camino de Vera s/n, 46022 Valencia, Spain
e-mail: mareva@omp.upv.es

A. Boza
e-mail: aboza@omp.upv.es

A. Ortiz
e-mail: aortiz@cigip.upv.es

R. Poler
e-mail: rpoler@cigip.upv.es

C. Hernández et al. (eds.), *Managing Complexity*, Lecture Notes in Management and Industrial Engineering, DOI 10.1007/978-3-319-04705-8_21,

the existence of units of the same FG that differ in some characteristics (subtypes) that are relevant for the customer. LHP appears in certain industries like ceramics, textile, wood, marble, horticulture, tanned hides and leather goods [1]. These firms are obliged to classify lots of the same FG into different subsets of homogeneous quantities (subtypes).

LHP has a direct impact on the OPP in several ways. On the one hand, classifying one same FG into several subtypes increases the number of references and the information volume to be processed, thus complicating system management. On the other hand, the alternatives for allocating real or planned uncommitted subtypes quantities to customer orders substantially increases and the homogeneity requirement complicates the search for a feasible and optimal solution. This is due to the fact that different subtypes of the same FG cannot be used to promise a specific customer order. Therefore, the typical way of calculating the accumulated ATP will not be valid. Not accomplishing this homogeneity requirement can lead to returns, product image and company deterioration, which may involve less customer satisfaction and even loss of customers. Furthermore, the real homogeneous quantities available of the same FG to be promised to customers are not known until their production has finished. This aspect proves to be a problem when customers' orders have to be promised, reserved and served from the homogeneous units available derived from the planned production. Therefore, it is necessary to anticipate as much as possible the future availability of homogeneous quantities (subtypes) in order to serve customers with the required quantities and homogeneity level required on time.

For this reason, the main purpose of this paper is to explicitly highlight the inherent characteristics of LHP manufacturing systems that impact on the OPP by the definition of the so called LHP-dimensions (Sect. 2). Based on these dimensions, it will be possible to decide the most suitable way of considering the different LHP availability levels during the OPP in order to make reliable commitments with customers (Sect. 3). Finally, the impact of allocation rules for uncommitted homogeneous quantities to customers is analyzed (Sect. 4).

2 LHP-Dimensions for OPP

Different dimensions should be analyzed in order to determine the impact of LHP specific situation to the OPP. With the aim of better understanding the concepts, the OPP in ceramic sector [1] is taken as an example.

LHP Customer Order Characteristics Dimension. It is important to keep in mind that LHP becomes a managerial problem due to the homogeneity requirements of customers. As in most companies, it will be essential to know from the customer order: the **requested products** (one, several or a product-pack), the **unit measure** for each product (that can be dependent on the customer class: units, pallets or trucks), the **quantity** and the **due date**. But LHP introduces a new customized aspect in order proposals: the **homogeneity type required** by the customer among the ordered products. The customer may require uniformity between components of a product (pearls on a necklace) or between units of the same product (ceramic tiles) or between different products of a product-pack (chairs and a dining table). In addition, the customer **can specify the value of the homogeneity attributes required** or, in case there are subtypes of the same item with different value, the **maximum price** willing to pay. The **homogeneity tolerance** makes reference to the customer sensibility in the permissible range of variation of LHP attributes to consider that two units of a subtype are homogeneous. This order characteristic can be or not explicitly defined by customers. For instance, in the ceramic sector, customers specify the quality of the product (value) though for each quality, several subtypes exist (the same quality tile with different tones and gages), they do not specify other homogeneity attributes (like the tone and gage). Note that the **order size** becomes a very relevant order proposal aspect for LHP environments because the larger the orders size, the more difficult will be to meet the uniformity requirement among all their units.

LHP origin Dimension: Raw Materials & Transformation Activities. LHP origin provides us with information about the inclusion or not of LHP characteristics in the availability levels. LHP can be originated by the heterogeneous characteristics of **raw materials** and/or components, for instance when they are directly originate from nature. LHP can also appear due to the own **transformation activities (LHP-activities)** or **environmental characteristics (LHP-factors)** that introduce certain variability in the processed items, even when the input material is homogeneous. With the aim of being more exhaustive in the determination of the LHP origin, we extend the LHP definition introducing the terms Lack of Homogeneity in Raw Materials (**LHRM**), Lack of Homogeneity in Intermediate Products (**LHIP**) and Lack of Homogeneity in Finished Goods (**LHFG**).

For instance, in the ceramic sector LHP is originated for the different characteristics of the raw materials (clays) and components (glazes) but also for the production lines and kilns (LHP-activities) due to temperature and humidity (LHP-factors). Therefore, in ceramic sector, LHRM, LHIP and LHFG exist.

Subtypes Dimension: LHP Items & Classification Stages. In order to ensure that customers are served with the required homogeneity, LHP supply chains are obliged to introduce one or several classification (sorting) stages along their productive process. Figure 1 shows the relationship among customer order characteristics and the definition of subtypes. The number and location of classification stages can be influenced by the **homogeneity type** required by customer (i.e. among components, among units of the same FG or among different FGs). For each sorted item, the **classification criteria** and the **values** they can take (discrete or continuous) should be identified. This provides us with the **number of subtypes** of the same LHP-item. The **classification criteria** maintains a closely relationship with the **homogeneity attributes** in FG required by the customer. At the same time, the possible **values** allowed of each **attribute** depend on how much exigent customers are (**homogeneity tolerance**). Finally, it is important to know not only the existing **subtypes** but also their **appearance** in the processed quantities by the SC transformation activities. This aspect provide us with information about if different subtypes can be

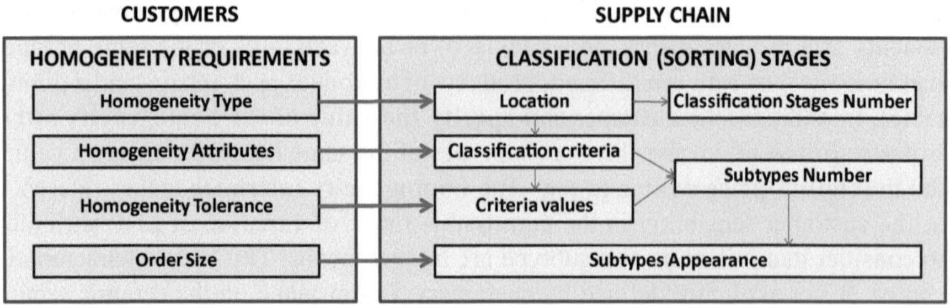

Fig. 1 Relationship between homogeneity requirements and classification stages

found either in the **same lot** or **among lots** of the same resource or among different resources at the same or different point of **time**. Furthermore, it will be interesting to found if the relation between the input quantity of a LHP-item and the **subtype output quantities** are constant or variable. The **subtype appearance** jointly with the **order size** will condition not only the units to be produced in planned lots (master plan) but also the anticipation of future homogeneous quantities available (LHP availability Levels) used during the OPP.

In ceramic sector, the homogeneity type is defined among tiles of the same model. Subtypes are classified only once, at the end of the manufacturing process, based on their surface defects, color and thickness. This classification criteria provide subtypes of the same FG that differ in quality (first, second, third and waste), tone and gage. Subtypes appear among lots processed in different resources and, even, in the same resource at different points of time due to environmental conditions.

3 Characterizing LHP Availability Levels

The above LHP dimensions present a direct impact on the way of representing and calculating the different LHP uncommitted availability levels to answer customer requests. In this section traditional availability levels are first described in order to better understand the differences from LHP availability levels.

Traditional Availability Levels During the OPP, when a new customer order request arrives, it is usually to check whether there are enough uncommitted quantities of different availability levels for the due date requested. Three usual availability levels are defined: ATP (Available-To-Promise), CTP (Capable-To-Promise) and MATP (Material-Availability-To-Promise) (Fig. 2).

Available-To-Promise (ATP) includes the calculation of the uncommitted quantity, either real or planned, of the items stocked at the customer order decoupling point, which is based on the master production schedule. In principal, ATP can be represented on any stage of the supply chain, e. g. finished goods, components, or raw materials. The decision where to represent ATP best for a certain business is strongly linked with the location of the customer order decoupling point [2]:

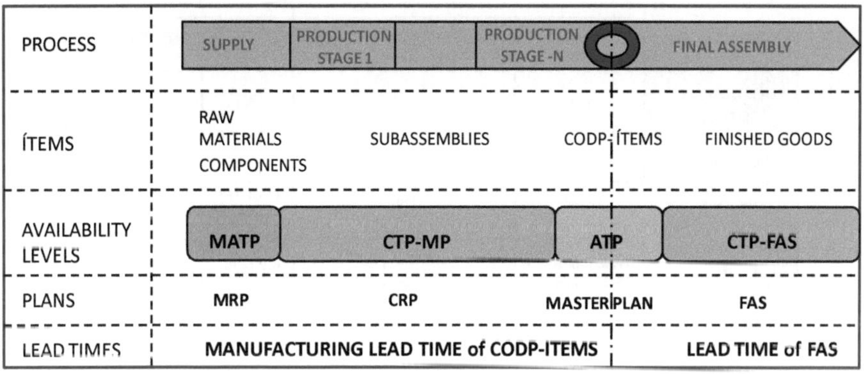

Fig. 2 Availability levels for traditional OPP

finished goods for MTS, main component and subassemblies for ATO and raw materials for MTO.

If there is not enough ATP to commit an order, the calculation of the **Capable-To-Promise** (CTP) quantities can be required. CTP represents the uncommitted available capacity, either real or planned, of those productive resources involved in the product fulfillment of a customer order. CTP can be checked either to produce additional quantities of CODP-items or to ensure that there is enough resources capacity to assembly CODP-items for obtaining finished goods. In the reviewed literature these two concepts are not clearly differentiate. For this reason, it is our proposal to define **two kinds of CTP**: **CTP-MP** and **CTP-FAS**. CTP-MP represents the uncommitted capacity of productive resources upstream the CODP to produce additional quantities of CODP-items. Consuming CTP-MP implies modify the Master Plan. CTP-FAS represents the uncommitted capacity of productive resources downstream the CODP to carry out the Final Assembly Schedule. Finally, there are authors [3] that distinguish between uncommitted capacity (CTP) and **Material-Available-To-Promise** (**MATP**). In this case MATP represents the not yet assigned net capacities of materials.

LHP Availability Levels. Taking as the starting point the traditional availability levels defined to promise orders, the impact of LHP in their definition and calculation is analyzed. The question is to determine in **which LHP situations make sense to compute LHP availability levels** with the aim of more precisely anticipating the future uncommitted homogeneous quantities. When modeling **future planned LHP availabilities** it is possible to **define** them in terms of **specific subtypes** or simply try to estimate the **future subsets of homogeneous quantities** without anticipating the specific subtype. As it will be shown, it strongly depends on the customer order characteristics. The decision about defining availability levels taking into account the LHP aspect is influenced not only by the **CODP location**, as usual, but **also** by its **relative position as regards the classification stages**. Figure 3 shows by column when makes sense to express the uncommitted availability levels in LHP terms. Upstream the CODP it is possible to consider **MATP-LHP**

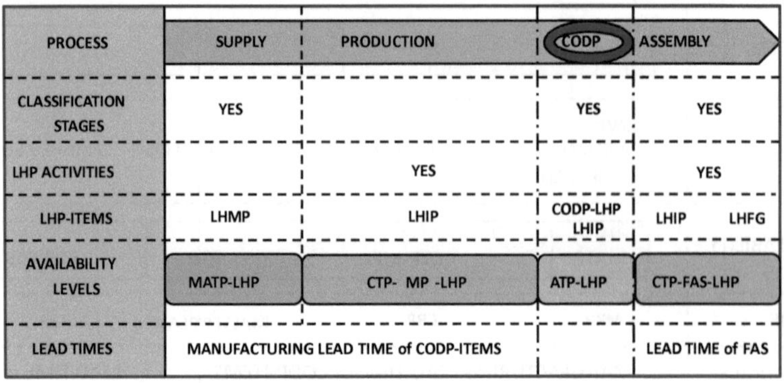

Fig. 3 LHP availability levels for OPP

if LHRM and classification stages upstream the CODP just at the supply activity exist. The **ATP-LHP** consideration is recommendable when LHRM or LHIP and classification stages exist upstream the CODP. Furthermore, the strong relationship existing between the Master Plan and the ATP should be taken into account when modeling the ATP. If the Master Plan is expressed in terms of specific subtypes or homogeneous quantities, the ATP-LHP should be expressed in concordance. However, in certain circumstances it is possible to disaggregate the Master Plan expressed in CODP-items to CODP-LHP items if it is known the appearance of subtypes in lots and the splitting of homogeneous quantities.

Modeling MATP-LHP or ATP-LHP at the level of subtype will be necessary when: (1) customer not only requires homogeneity but also defines the required subtype in the order, (2) different MATP-LHP or ATP-LHP subtypes once classified do not present the same value (qualities) (3) later transformation activities can be dependent on the subtype and (4) it is known the way MATP-LHP or ATP-LHP subtypes affect the ATP-LHP and LHP finished goods, respectively.

Modeling CTP-MP-LHP and CTP-FAS-LHP makes sense when there are LHP activities upstream and/or downstream the CODP, respectively, and it is known the LHP-factors and their impact on homogeneity. Along these lines, it is very important to keep in mind not only the existing **subtypes** but also their **appearance** in the processed quantities by the SC transformation activities. This aspect provide us information about if different subtypes can be found either in the **same lot** or **among lots** of the same resource or among different **resources** at the same or different point of **time** and helps us to determine the level of detail when calculating the uncommitted capacity (per resource, group of resources, etc.) and its utilization to produce additional uncommitted availability of items.

When using the above CTP-LHP levels to plan new production quantities, it will be necessary to consider the constant or variable relation between the input quantity of a LHP-item and the **subtype output quantities**.

Ceramic sector mainly follows an MTS strategy with one classification stage at the end of the process. Usually, customers only specify the FG quality within their orders and they do not specify the subtype. The ATP-LHP is obtained for the first quality quantities of the Master Plan and different lots produced by different machine or by the same machine in different periods of time are considered non homogeneous. In other words, ATP-LHP is not expressed in terms of subtypes but only homogeneous subsets are anticipated based on the same lot, resource and time period, not being possible to accumulate them to serve a customer order.

4 Allocation Rules for ATP-LHP

When discrete ATP-LHP is defined in terms of subtypes, different ATP-LHP subtypes cannot be mixed to serve the same customer order. When discrete ATP-LHP is only defined by subsets of homogeneous quantities without differentiating the subtype, the only constraint consists of not allowing accumulate discrete non-homogeneous ATP-LHP quantities. Therefore, as it has been described, the ATP-LHP existence restricts the possible ways of accumulating ATP-LHP. For instance, in ceramic sector due to LHP it will impossible to accumulate ATPs from different production lines or different time periods for the same FG. For this case, the choice of reserving ATP from a specific homogeneous subset of a FG affects subsequent promises, being necessary to define **allocation rules**.

As an illustrative example, let us assume for specific time period t for given FG that there are two homogeneous ATPs, one with a value of atp1 = 800 units and the other with a value of atp2 = 320 units. Let us also assume that the arrival of three orders (o1, o2, o3) takes place one after another with due date ddo1 = ddo2 = ddo3 = t, and that the first includes a requested quantity of 250 units (qo1), the second has a requested quantity of 600 units (qo2) and the third a requested quantity of 70 units (qo3), for all of which a real-time response must be given. In Fig. 4 two possible alternatives to assign ATP to the incoming orders are shown.

When promising the first order, it is possible to reserve the ATP of both the homogeneous quantities atp1 and atp2. Solution A assigns 250 units of atp1 meanwhile

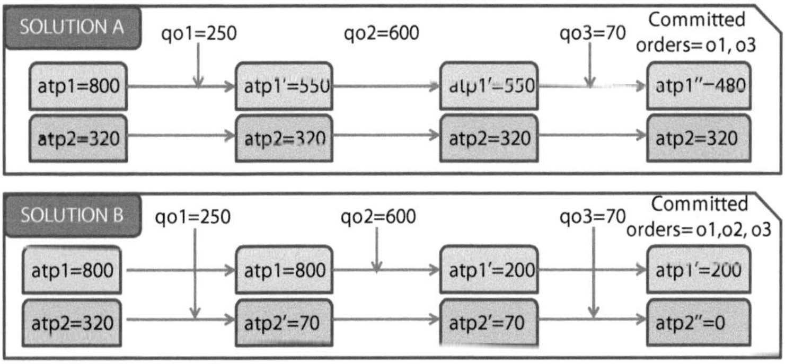

Fig. 4 Impact of allocation rules of ATP-LHP on committed orders

solution B assign 250 units of atp2 for the first order, being the updated atp equal to atp1'=550 and atp2'=70, respectively. Solution A cannot commit the second order, meanwhile solution B assign ATP from atp1, being the updated atp1'=200. Finally, the third order can be committed in two cases. From the examples above, we may deduce that when there are several alternatives to assign homogeneous ATPs of different subtypes to an order, the policy formulated to select one of these alternatives affects future commitments and, therefore, system performance. Therefore, for maximizing customer service level and supply chain performance, properly ATP-LHP allocation rules should be defined.

Acknowledgements This research has been carried out in the framework of the project funded by the Spanish Ministry of Economy and Competitiveness (Ref. DPI2011-23597) and the Polytechnic University of Valencia (Ref. PAID-06-11/1840) entitled "Methods and models for operations planning and order management in supply chains characterized by uncertainty in production due to the lack of product uniformity" (PLANGES-FHP).

References

1. Alemany MME, Lario F-C, Ortiz A, Gómez F (2013) Available-to-promise modeling for multi-plant manufacturing characterized by lack of homogeneity in the product: an illustration of a ceramic case. Appl Math Model 37:3380–3398
2. Kilger C, Meyr H (2008) Demand fulfilment and ATP. In: Stadtler H, Kilger C (eds) Supply chain management and advanced planning. Concepts, models, software, and case studies, 4th edn. Springer, Berlin, pp 191–198
3. Meyr H (2004) Supply chain planning in the German automotive industry. OR Spectr 26(4):447–470

Modeling the Master Plan for Supply Chains with Lack of Homogeneity in the Products

Maria del Mar Eva Alemany Díaz, Llanos Cuenca, Angel Ortiz and David Pérez Perales

Abstract Master Planning of Supply Chains (SCs) with Lack of Homogeneity in the Products (LHP) strongly differs from other SCs. Although LHP affects SCs of different sectors, an absence of a common research body exists. In this paper, the characterization of LHP dimensions for Master Planning and their modeling is described. To validate the proposal an application to the fruit SC is presented.

Keywords Master planning · Supply chains · Lack of homogeneity in the product · Model

1 Introduction

Supply chains (SCs) master planning is a complicated task due to the existence of a huge number of decisions, constraints, objectives (sometimes conflictive), possible alternatives to be evaluated and the presence of uncertainties. For the case of SCs with Lack of Homogeneity in the Product (LHP), this planning task becomes even more complex as it can be described below. One of the main LHP consequences is the existence of units of the same finished good that differ in some characteristics originating the existence of subtypes. LHP becomes a managerial problem when finished goods (FGs) are not homogeneous and customers require homogeneity in the requested quantities. Though LHP is present in SCs of a variety of sectors (ceramic, tanned hides, leather goods…) and some models have been developed [5] there is not a common body of research that allows study the relevant characteristics

M. del Mar Eva Alemany Díaz (✉) · L. Cuenca · A. Ortiz · D. Pérez
Centro de Investigación de Gestión e Ingeniería de la Producción (CIGIP),
Universitat Politècnica de València, Camino de Vera s/n, 46022 Valencia, Spain
e-mail: mareva@omp.upv.es

L. Cuenca
e-mail: llcuenca@cigip.upv.es

A. Ortiz
e-mail: aortiz@cigip.upv.es

D. P. Perales
e-mail: dapepe@cigip.upv.es

C. Hernández et al. (eds.), *Managing Complexity*, Lecture Notes in Management and Industrial Engineering, DOI 10.1007/978-3-319-04705-8_22,

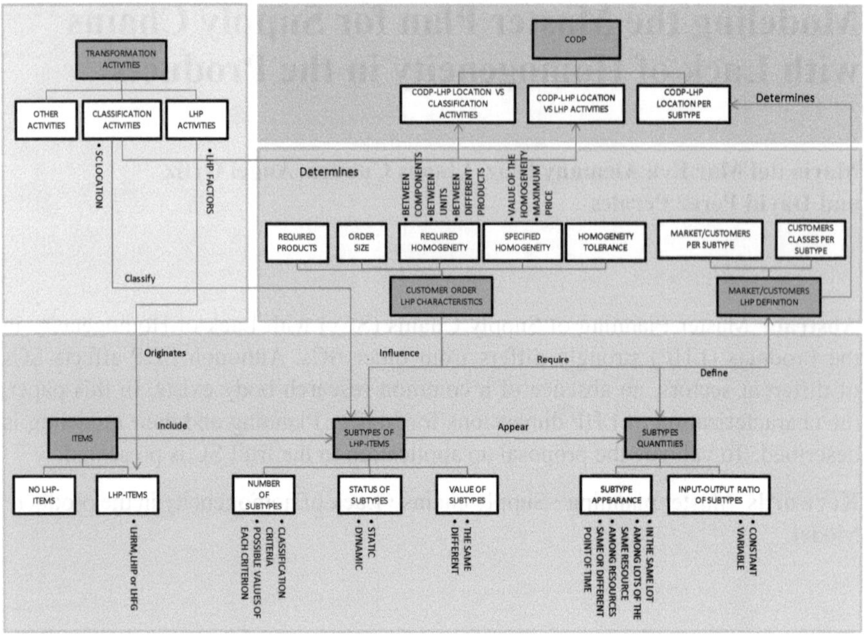

Fig. 1 Master plan LHP-dimensions and their relationship

to take into account for properly define the SC Master Planning. In this paper the LHP dimensions relevant for master planning (Sect. 2) and their modeling (Sect. 3) are described. To validate the proposed framework and to clarify concepts, an application to the fruit supply chain is reported (Sect. 4).

2 LHP Dimensions for Supply Chain Master Planning

In this section, the different dimensions characterizing LHP that are relevant for the Master Planning of Supply Chains modeling point of view are described. The objective is to provide a guideline about what LHP characteristics are merely of being considered during the Master Planning and then, how to model them. To achieve this objective four **dimensions** are identified: (1) Transformation Activities, (2) Items, (3) Customers/Market and (4) the Customer Order Decoupling Point (CODP). In Fig. 1, the different **elements** composing each dimension, their attributes as well as the relationship among them are shown.

For LHP contexts, two types of **transformation activities** additional to the typical ones are crucial: the **classification activities** and the **LHP activities**. After classification activities different subsets of homogeneous items are obtained. In fact, the key LHP element will be the location of classification stages along the SC as well as items classified in each one of them. Furthermore, with the aim of anticipating the homogeneous quantities available in production plants it will be necessary to identify the transformation activities that introduce heterogeneity in the process (**LHP-activities**) and the variables that cause it (**LHP-factors**). It will be important to define the relationship between the heterogeneity origin and the productive resources (on the same machine, between machines, in different periods of time). This helps us to know the degree of detail for modeling resources in the master planning (see Sect. 3). As regards **Items**, two kinds can be distinguished: **LHP-items** and **NO-LHP-items**. LHP-items are those non homogeneous items that affect the lack of homogeneity in finished goods. We distinguished between the Lack of Homogeneity in Raw Materials (**LHRM**), Lack of Homogeneity in Intermediate Products (**LHIP**) and Lack of Homogeneity in Finished Goods (**LHFG**), respectively. Only **LHP items** are classified in to different **subtypes** based on certain classification criteria. A subtype of an item presents a specific value of the aspects of the classification criteria. Certain attributes are inherent to the subtypes like its number, status and value. The **number of subtypes** will depend on the classification criteria for sorting LHP items and their possible values. If the value of all the attributes defining a subtype remains the same along time, the **status** of the subtype will be static, otherwise it will be considered as dynamic. Furthermore, different subtypes of the same LHP-item can have the same or different **economic value**. Usually, different economic values imply the existence of several **qualities** and the appearance of **undesirable stocks** for subtypes with low economic value. Finally, it is important to know not only the existing **subtypes** but also their appearance in the **processed quantities** by the SC transformation activities. The **subtype appearance** provide us with information about if different subtypes can be found either in the same lot or among lots of the same resource or among different resources at the same or different points of time. Furthermore, it will be interesting to found if the relationship between the **input** quantity of a LHP-item and the **output quantities** of each subtype are constant or variable.

As regards the **customers**, different SC Markets and Customers can exist depending on the subtypes they ordered Because LHP management problem arises from the homogeneity requirement imposed by customers, it is crucial to identify the customizable parameters of order proposals affecting LHP. As in most companies, it will be essential to know from the customer order, the **requested products** (one, several or a product-pack), the **unit measure** for each product (that can be dependent on the customer class: units, pallets or trucks), the **quantity** and the **due date**. But LHP introduces a new customized aspect in order proposals: the **homogeneity type required** by the customer among the ordered products. The customer may require uniformity between **components of a product** (pearls on a necklace) or between **units of the same product** (ceramic tiles) or between **different products of a product-pack** (chairs and a dining table). In addition, the customer **can specify the value of the homogeneity attributes required** or, in case there are subtypes of the same item with different value, the **maximum price** willing to pay. The **homogeneity tolerance** makes reference to the customer sensibility in the permissible range of variation of LHP attributes to consider that two units of a subtype can be considered homogeneous. Note that the **order size** becomes a very relevant factor for LHP because the larger the orders size, the more difficult will be to meet the uniformity requirement among all their units.

The SC strategy for accomplishing with customer requirements will be defined by the **Customer Order Decoupling Point** location. For LHP SC, different CODP locations can exist depending on the subtypes. Furthermore, for modeling the master plan it will be essential to know the relative position between the CODP and the classification activities and the CODP and the LHP location activities.

Figure 1 shows the main **relationships** among the different **dimensions** and their **elements**. As it can be observed, LHP activities originate LHP-items that are sorted into different subtypes by classification activities. Subtypes of LHP items appear in processed quantities. Customer order LHP characteristics influence the subtype definition. Indeed, subtypes classification criteria will be conditioned by the homogeneity characteristics required by customers in products. The customer homogeneity tolerance is determinant for defining the number of subtypes, i.e. lower the homogeneity tolerance, higher the number of subtypes to be defined. When defining the quantities of subtypes to be obtained the required quantities of subtypes become crucial. Finally, for LHP SC management, different CODP can be defined depending on the subtype. Furthermore, for LHP manufacturing environments it is crucial not only the CODP location but also its relative position as regards the classification stages and LHP-Activities. This aspect represents the starting point to analyze LHP implications in the master plan that, in turn, will constitute the basis for choosing the LHP characteristics relevant to be modeled.

3 Modeling LHP for SC Master Planning

The objective is to derive a master plan (MP) that anticipate the expected homogeneous quantities of each subtype (**LHP Master Plan**) and made properly decisions in order to accomplish with the customer requirements in terms of homogeneity, quantity and due dates. For LHP MP modeling purposes we distinguish **five SC planning elements**: Items, Transformation Activities (resources), LHP-factors, Time and Market/Customers. The Time element (not previously considered in the LHP dimensions) is included for introducing the own meaning of planning.

Two questions are of relevance when modeling the Master Plan: (1) what SC elements of each dimension should be considered and (2) what is the level of detail for representing them. To answer the first question, the CODP position in the SC is crucial. As in the traditional SC Master Planning, only those SC transformation activities upstream the CODP and their processed items are object of our analysis and, therefore, are susceptible of being modeled. Additionally the LHP-factors affecting the SC physical system upstream the CODP could be also taken into account in order to anticipate the different subtypes quantities and the subtype evolution in case dynamic state of subtypes.

As regards the second question, the level of detail for representing the different SC elements in the Master Planning will be conditioned to a certain extent by the

LHP characteristics chosen to be modeled. In the following paragraphs we provide with a guideline about what LHP characteristics are susceptible to be modeled in the Master Plan for the different elements described above. The existence or not of Classification and LHP Activities upstream the COPD provides us with the key for answering the question about the level of detail for representing the different SC elements (items, transformation activities, LHP-factors, Time and Market/Customers) in the LHP Master Plan.

- **ITEMS: When to model the Master Plan in terms of subtypes?** Two kinds of items are of relevance for Master Planning purposes: the items stored at the CODP (CODP-items) and the other items belonging to the BOM of CODP-items (i.e. items supplied or processed upstream the CODP). If there are classification activities upstream the CODP, the Master Plan should be expressed in terms of **CODP-LHP items** and other **LHP items upstream the CODP**. This is equivalent to deal with subtypes of items in the Master Plan instead of the typical items. The number of subtypes for each item will be known for the LHP dimension. To deal with subtypes in the Master Plan increases the number of references and, therefore, increases the Master Plan complexity. Therefore, to consider subtypes in the MP makes sense if one or more of the following situations occurs:
  - CODP-LHP item forecasts anticipate some LHP customer order characteristics like:
    - Forecasts are expressed in terms of CODP-LHP items. This will be the maximum degree of detail.
    - Forecasts are expressed in terms of customer classes which will be defined based on the **order size, type of homogeneity and/or value of the homogeneity required.**
  - To deal with LHP items helps better defining the lot sizes in order to serve an integer number of the different customer orders classes with homogeneous quantities. In this case, the **sizing of the planned subtype quantities will allow** a better balance between supply and demand, increasing customer service level and SC costs.
  - There are data, objectives or constraints in the MP that strongly depend on the subtypes. For instance different economic values for subtypes could imply to consider different costs of supplied material or different revenues for each subtype.
- **TRANSFORMATION ACTIVITIES: What Transformation Activities include in the LHP Master Plan?** Analogously to items, only those transformation activities **upstream the CODP-LHP** are candidate to be modelled. Traditionally the bottleneck resources are considered in the MP. For LHP contexts, additional considerations are made. It would be interesting to model **the LHP Activities and Classification Stages of this SC part** if the impact of LHP activities on the LHP items upstream **the CODP is known:** for instance, if there is some information about the appearance of subtypes in planned production quantities in order to anticipate them. It will be sense to model the divergent flow of items at classification activities when it is possible to estimate the input-output

ratio of subtypes (i.e. subtype quantities in a lot or between lots in terms of percentages are known).

- **LHP-FACTORS: When to consider LHP factors in the Master Plan?** If there is a known relationship about how certain factors, for instance, temperature or humidity, affects the LHP appearance (subtypes), either in the static or dynamic state, the Master Plan should be taken these LHP factors into account. For this situation some classification criteria could be expressed by a function of the LHP factors for anticipating the appearance of the different subtypes.
- **TIME: Which is the time unit to be used in the Master Plan?** The time unit used in the master plan, i.e, the planning period length is influenced by the periodicity of decisions considered. But when the subtype state is dynamic (perishability, obsolescence,…) and it is known the magnitude of the time period that influence its state (LHP-time unit), the Master Plan can be expressed in this LHP time unit depending on the level of detail to be achieved. If the LHP time unit is higher than the Master Plan horizon, it will not make sense to model the dynamic state of the subtype.
- **CUSTOMER/MARKET ORDER: When to consider customer/market classes based on the subtypes in the Master Plan?** If the location of the CODP depends on the subtype, it will be necessary to express the Master Plan in terms of subtypes. Traditionally, the forecast demand of CODP-items has been expressed in aggregate terms. However, to better match subtypes supply with demand it will be interesting to model different customer classes based on the subtype, order size and required homogeneity type. This is recommendable if we are able to forecast the demand of customers in terms of subtypes. To model different customer classes make sense when the profit heterogeneity of demand depends on the subtype revenue (different values of subtypes), costs (different value of LHP items of subtypes) and customer strategic importance.

4 Fruit Supply Chain Application and Conclusions

In order to better understand the described concepts and to validate them, an application to the fruit supply chain is presented.

LHP Dimensions for Fruit Supply Chain: The basic transformation activities in fruit supply chains are [7]: (a) growing and harvesting, (b) processing, (c) washing, sorting and grading, (d) packaging and labelling, (e) storage and distribution and (f) retailing. There are **several classification (sorting and grading) activities** located in different points along the productive process with the aim of eliminate **waste** and classify fruits into **several qualities** based on different attributes. Blanco et al. [3] distinguishes the following classification activities: (1) a pre-classification stage where non-tradable fruit (waste) is separated for juice production, (2) a quality classification stage where tradable fruit is separated in several categories, and some waste is also produced at this stage, (3) the different qualities are further classified by size or weight (depending on the available technology) in several types at the gauge classification stage where some waste is also produced. There are **several LHP-transformation activities** along the Fruit SC that impact on the LHFG, especially on the decay and therefore, on the fruit quality: harvesting, fruit treatment (washing and classification) and handling, storage and transport. **LHP-factors** can be dependent on the resource (landscape), the time (harvesting period) and the environmental conditions (temperature and humidity). The **LHP origin** in Fruit SC is mainly due to the lack of homogeneity in the raw materials (LHRM), i.e. the fruit obtained directly from the nature. This LHRM jointly with LHP-Activities originates the existence of LHIP and LHFG. FGs in the fruit sector mainly differ in variety, quality, packaged and labeling.

The existing subtypes of LHP-items depend on **the classification stage** and the **classification attributes**. There are several classification attributes in the fruit sector [1, 3, 4] that in this work we have categorized into: inherent (fruit variety), sanitary (fungus presence), aesthetic (color, shape, ripeness, imperfections, ...) and functional (shelf life, juice quantity, ...) issues. The **number of subtypes** depends on the possible combinations of the above attributes that originates different qualities for a specific fruit variety. The worst values of these attributes define the non-tradable fruit (**byproducts**: separated for other purposes) and the **waste fruit**. The **status** of the fruit is **dynamic because quality depends on the time (perishable goods)**. Amorim et al. [2] distinct two perishable goods classes: those with **fixed and random lifetime.** Obviously the **value of subtypes (qualities)** is different and dynamic. As regards **LHP subtype quantities, subtypes appears** among different producers (fruit from different landscapes), different production lots (fruit of collected from the same landscape at different time), and among units of the same production lot. The input/output relation is variable but can be estimated, for instance, by fixed percentages [3]. In relation to the **Market/customer LHP definition**, the high uncertainty in the quantities and qualities obtained of each fruit variety lead to the **market diversification** (local, national and international) [6]. There is also a great importance of the spot market to face with the supply uncertainty. Each market presents **order characteristics** that are different in terms of quantities, subtypes, packaging, labeling, lead times, distribution modes and sales price. Finally, **multiple CODP-locations** are possible for Fruit SCs. Van der Vorst [6] and [7] indicates that there are several elements that influence the CODP positioning upstream (perishability, decay, divergent product flows, high demand uncertainty and high degree of customization). Therefore, in the most general case, classification and LHP-activities exist upstream the CODP.

- **Modeling LHP for Fruit Supply Chain Master Planning:** There are several reasons to define the Fruit SC Master Plan in terms of CODP-LHP items (subtypes) [3]: (1) independently of the CODP location, there will be always CODP-LHP items because LHP in Fruit SCs is always originated from the raw material, (2) several classification activities usually exist upstream the CODP, (3) customer requires different fruit qualities and each quality has different value (cost and selling price). Because different classification stages exist upstream the CODP it is recommendable to consider them for properly modeling the subtypes flow by taking into account the different balance equations at different classification stages. Different approaches exist to model the perishability aspect and their associated LHP-factors. In case perishability exists two cases could be formulated: (1) when the shelf-life is fixed beforehand, and (2) when the shelf-life is indirectly a decision variable influenced by the environmental setting. For this last case manufacturers can estimate the shelf-life of products throughout the supply chain based on external factors using the knowledge of predictive microbiology (see [2]). Different customer classes should be considered based on the required quantities and subtypes, and therefore, there is recommendable to express demand forecasts based on customer classes.

This paper puts forward the relevant characteristics that it is necessary to take into account for properly define the Master Planning in SCs with LHP.

Acknowledgements This research has been carried out in the framework of the project funded by the Spanish Ministry of Economy and Competitiveness (Ref. DPI2011-23597) and the Polytechnic University of Valencia (Ref. PAID-06-11/1840) entitled "Methods and models for operations planning and order management in supply chains characterized by uncertainty in production due to the lack of product uniformity" (PLANGES-FHP).

References

1. Ahumada O, Villalobos JR (2011) Operational model for planning the harvest and distribution of perishable agricultural products. Int J Product Econ 133:677–687
2. Amorim P, Günther H, Almada-Lobo B (2012) Multi-objective integrated production and distribution planning of perishable products. Int J Product Econ. doi:10.1016/j.ijpe.2012.03.005
3. Blanco AM, Masini G, Petracci N, Bandoni JA (2005) Operations management of a packaging plant in the fruit industry. J Food Eng 70:299–307
4. Mowat A, Collins R (2000) Consumer behaviour and fruit quality: supply chain management in an emerging industry. Suppl Chain Manage: An Int J 5(1):45–54
5. Mundi I, Alemany MME, Boza A, Poler R (2013) A model-driven decision support system for the master planning of ceramic supply chains with non uniformity of finished goods. Stud Info Control J. (In press)
6. Van der Vorst JGAJ (2000) Effective food supply chains: generating, modeling and evaluating supply chain scenarios. Doctoral thesis. Wageningen University
7. Verdouw CN, Beulens AJM, Trienekens JH, Wolferta J (2010) Process modelling in demand-driven supply chains: a reference model for the fruit industry. Comput Electron Agri 73:174–187

Packaging Logistics. A Case Study in Dairy Sector

Jesús García Arca, J. Carlos Prado Prado and A. Trinidad González-Portela Garrido

Abstract No many companies give importance to the impact of packaging design for achieving logistic efficiency and sustainability in the supply chain. This paper sets out to illustrate how the adoption of "Packaging Logistics" approach makes it possible to obtain competitive advantages throughout the supply chain. To illustrate this statement, in this paper, not only the conceptual field of this concept is developed, but also its application, analysing a case study in dairy sector.

Keywords Packaging · Logistics · Supply chain · Strategy

1 Introduction

Nowadays, companies must deal with the challenges, not only in constant innovation in terms of new products and processes, shorter life cycles or increased commercial range, but also in terms of the demand for ever lower prices, with increasingly improved quality standards and service. Thus, current markets could be characterized as turbulent, volatile and global, so many organizations are searching for a more efficient management of their supply chains as a source of competitive advantages [6].

In this context, in recent years there has been a double phenomenon of strong impact on supply chains' efficiency: first, globalization of supply chains and, secondly, the constant increased costs of raw materials, particularly, the oil. The combination of these two phenomena is not a trivial issue because, strategically, underscores the urgency of action in pursuit of maximum performance in logistics activities undertaken across the supply chain (transport, handling, storage, production,…), eliminating "waste" or activities that do not add value to the market (in line with the Kaizen approach, "Just in Time" (JIT) or "Lean Manufacturing") but also developing and implementing innovations in processes and products.

J. García Arca (✉) · J. Carlos Prado Prado · A. T. González-Portela Garrido
Grupo de Ingeniería de Organización (GIO), Escola de Enxeñería Industrial,
Campus Lagoas-Marcosende, C/Maxwell, 36310, Vigo, Spain
e-mail: jgarca@uvigo.es

C. Hernández et al. (eds.), *Managing Complexity*, Lecture Notes in Management and Industrial Engineering, DOI 10.1007/978-3-319-04705-8_23,

Likewise, the growing sensitivity in society as regards a responsible management should imply that the supply chain management should be enlarged to take in the concept of sustainability and its three axes associated: environmental, economic and social [5].

Beyond the individual framework of a company, this sustainability concept should be extended to the other companies in the supply chain, whereby all their organizations should take an active part in implementing logistic processes that could be considered as sustainable [1, 4, 5]. In this context, sustainability and efficiency should be considered as complementary [10].

2 Packaging Logistics

In the conceptual framework commented in the previous heading, packaging is one of the key elements that makes it possible to provide support for the combined action of efficiency and sustainability strategies. Thus, beyond the traditional (but nonetheless important) view of packaging as a means of protecting products, over the last few years, new design requirements have been added for packaging: on the one hand, to improve the differentiation capacity of the product, and on the other, to improve the efficiency of the product at logistic level.

Furthermore, this contribution of packaging to efficiency in logistics should be considered not only in terms of its direct view (in the processes of supplying, packing, handling, storing and transport), but also reversely (re-use, recycling and/or recovery waste from packaging) [2]. All this has, in practice, meant the development of specific legislations (e.g., European Directive 94/62/EC; 1994).

In this context, authors such as Saghir [12], García-Arca and Prado-Prado [8] or Bramklev [3] identify in packaging three main functions: the commercial function, the logistics function and the environmental function. Also, in order to put these functions into practice, it is essential to consider the packaging as a system comprising three levels [12]: the primary packaging (also known as the "consumer packaging"), the secondary packaging (a group of several primary packages; known as "transport packaging") and the tertiary packaging (involving several primary or secondary packages grouped together on a pallet).

When contemplating packaging as a whole, the natural interaction among different levels would become manifest, depicting the dependence among them. In fact, the adaptation of a set level of packaging should not be contemplated if the integration of the set of all the levels of grouped form is not also considered.

Nowadays, the choice of a packaging is usually subject only to costs' considerations. However, packaging design affects costs both directly (costs of purchasing and waste management) and indirectly (packing, handling, storage and transport). It is precisely this indirect way that makes difficult an adequate understanding of the repercussions of certain decisions in packaging design [8].

With this broader view of packaging, the integration of logistics and the packaging design has been conceptualized in the term "packaging logistics", particularly emphasizing the operational and organizational repercussions [8, 9]. Shagir considers "packaging logistics" as "the process of planning, implementing and controlling the coordinated packaging system of preparing goods for safe, efficient and effective handling, storage, retailing, consumption and recovery, reuse or disposal and related information combined with maximizing consumer value, sales and hence profit".

As a result of the packaging logistics implementation, it is possible to deal with the search for packaging able to meet the needs of the companies based on the possibilities associated with the combinations in the packaging structure (primary, secondary and tertiary) and with the four main decisions to be taken in design: selection of the materials, dimensions, groupings (the number of packs/package) and "graphic artwork" (or the aesthetic design of the packaging).

After the conceptual development of "Packaging Logistics", the main objective of this paper is to illustrate the potential of its applying in the dairy supply chain. For theory testing, the authors have adopted the "action research" approach, directly participating in the "packaging logistics" implementation in a dairy company, leading and coordinating the project. Thanks to this involvement, the researchers have the opportunity to witness the process, not only as mere observers, but also as real "agents of change" in intervention and knowhow compiling processes [11]. Action research can be seen as a variant of case research [13], but whereas a case researcher is an independent observer, an action researcher "... is a participant in the implementation, but simultaneously wants to evaluate a certain intervention technique..." [7].

The analyzed company, based in Galicia (Northwest Spain), is one of the most important manufacturers in dairy Spanish market (among the 12 main manufacturers), with an annual turnover of over 100 million € and in its factory provides jobs to over 250 employees. Currently markets various dairy products such as milk, liquid yoghurts, cream and butter, milkshakes and cheeses. To make the study, we have focused on milk briks (the company packs more than 100 million L/year). Particularly, we have studied the 1 L milk brik with cap (primary packaging), grouped in packs of 6 briks (secondary packaging) and palletized in EUR pallet (tertiary packaging). The analysis was complemented by a literature review and a field study of dairy products (based on briks) in three supermarkets chains in Galicia (Northwest Spain).

3 Packaging Logistics in Action

The Brik was developed by Ruben Rausing in 1951 in Lund (Sweden). It can be made for up to six different layers and for guidance, a brik pack would comprise 75 % cardboard, 20 % plastic (Polyethylene) and 5 % of aluminum. Despite its usefulness to preserve perishable liquid foods (including milk) without refrigeration and preservatives and its good logistical efficiency (volumetric occupation), this packaging is still blames environmental misbehavior. However, this difficulty of recycling has improved as technology evolves in separation of layers.

The logistics of milk briks has no special requirements of conservation (temperature) as it happens with other milky products like cheese, yoghurt and cream (with

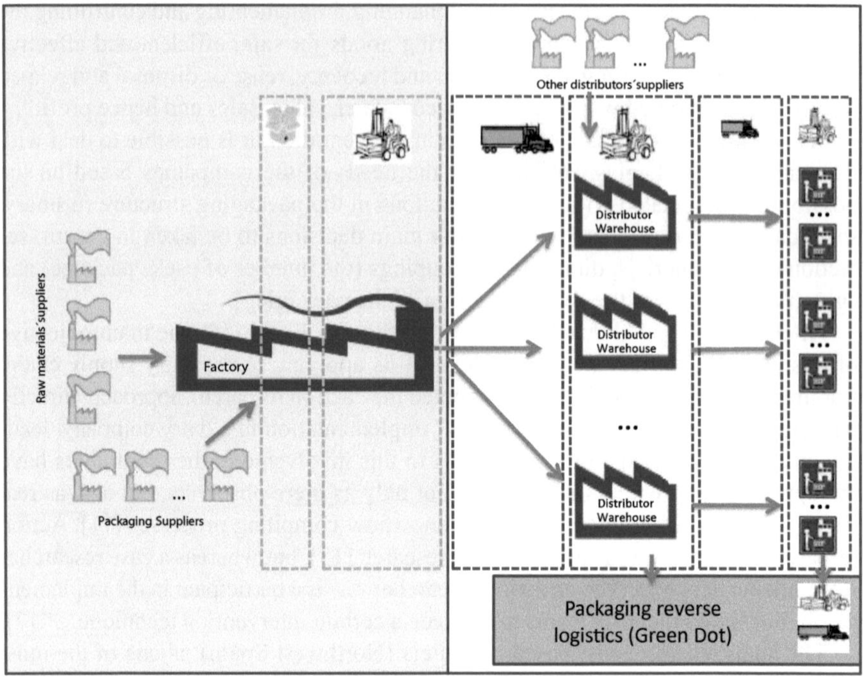

Fig. 1 Supply chain in the dairy company

a specific supply chain). In this context, the supply chain of the dairy company selected can be represented in Fig. 1. In this figure, we have paid special attention to the processes from packaging purchases, packing and physical distribution to reverse logistics. Table 1 shows the costs associated with the studied part of the chain and the level of packaging affected (Brik, pack and/or pallet).

Furthermore, in the analysis 4 milk brik formats were selected (see Fig. 2, A, B, C and D) as well as 5 of the most widely used formats of packs (see Table 2, A.1, A.2, B.1, C.1 and D.1). Among all these combinations the analyzed company used brik A and pack A.1.

Regarding the process of packaging purchases, the final price depends on the type of brick format, material and weight (A, B, C and D), as well as purchase volume (economies of scale). As a simplification for the analysis, it was considered that for the same volume of purchase, the final cost of each brik depends on its individual weight (see Table 3), although, also would be affected by the number of layers and the type of material of each layer.

Also, the cost of pack (secondary packaging) was determined by the type of materials, its weight and the number of briks in each pack. Furthermore, the packing process is highly automated, although their flexibility and adaptation to different formats of briks and packs is low (high impact of setup). This aspect limits, in general, the coexistence of various formats of briks and packs in the same manufacturing line so that, in practice, these lines are specialized.

Table 1 Costs affected by packaging design decisions

Processes	Units of cost	Impact on supply chain
Packaging purchases	Brik, pack (purchases)	Manufacturer
Packing	Brik, pack (setups, productivity)	Manufacturer factory
Handling, storage and picking	Pallet, pack (picking)	Manufacturer's warehouse
Transport (mainly full truck)	Pallet	Manufacturer and distributors
Handling, storage and picking	Pallet, pack (picking)	Distributors' warehouses
Transport (mainly combined truck)	Pallet	Distributor and supermarkets
Handling	Pallet, pack	Supermarkets
Reverse logistics	Brik and pack waste (Green Dot)	Manufacturer

Fig. 2 Different types of briks (from left to right: *A*, *B*, *C* and *D*)

With regard to the physical distribution (handling, storage and transport), the efficiency of palletizing is conditioned by the type of brik but also by the part of the supply chain that focuses on the analysis (see Table 2). In this sense, the milk briks pallet has a high density and high consumption. A priori, this product could be distributed efficiently, optimizing the activities of handling, storage and transport, looking for a larger number of liters per pallet, within the constraints of strength of brik and pack. In this regard, the maximum number of layers per pallet is conditioned not only by the type of brik, but also by the location of the cap (other formats without cap can withstand more layers).

However, the type of transport between manufacturer and distributors' warehouses is "full load truck" (maximum load limit of 33 EUR pallets and 24.4 t). Traditionally, manufacturers have not paid much interest in improving the volumetric efficiency of pallet (although there are significant differences as shown in Table 2), since the weight determines the maximum number of pallets per truck. In fact, in the company analyzed, the number of pallets per truck does not exceed 30. All this significantly affects, not only to the efficiency of handling and storage in the warehouses of the manufacturer, distributors and supermarkets, but also in transport between distributors' warehouses and supermarkets.

Table 2 Logistic analysis of alternatives (briks and packs)

Type and Brik dimensions (mm; W*L*H)	Type of pack and dimensions (mm; L*W*H)	Palletization	Pallet height (m, incl. pallet)	Pallet weight (kg; incl. pallet)	Transport efficiency
Type A (60*90*195)	Type A.1 Cardboard box (wrap-around) (282*127*216)	720 briks 24 packs/layer; 5 layers/pallet	1.33	807	*30 pallets/truck 21,600 briks/truck (initial solution)*
Type A (60*90*195)	Type A.2 Cardboard tray and plastic cover (282*128*210)	720 briks 24 packs/layer; 5 layers/pallet	1.3	800	*30 pallets/truck 21,600 briks/truck (no improvement)*
Type B (65*70*252)	Type B.1 Plastic cover (219*130*265)	768 briks (+6.66 %) 32 packs/layer; 4 layers/pallet	1.31	848	*28 pallets/truck 21,504 briks/truck (−0.44 %)*
Type C (71*75*204)	Type C.1 Plastic cover and cardboard sheet (227*150*205)	864 briks (+20 %) 24 packs/layer; 6 layers/pallet	1.4	951	*25 pallets/truck 21,600 briks/truck (no improvement)*
Type D (62*70*239)	Type D.1 Cardboard tray and plastic cover (228*128*245)	816 briks (+13.3 %) 34 packs/layer; 4 layers/pallet	1.23	903	*27 pallets/truck 22,032 briks/trailer (+2 %)*

Table 3 Costs of Green Dot of each brik and pack

Brik model	Brik weight (without cap) (g)	Brik Green Dot (million €/year)	Pack model	Type of pack	Pack weight (g)	Pack Green Dot (million €/year)	Total Green Dot (million €/year)
A	38	1.227	A.1	Cardboard box (wrap-around)	87	0.098	*1.325*
A	38	1.227	A.2	Cardboard tray (C) and plastic cover (P)	44 (C) 12 (P)	0.149	*1.376 (+3.44 %)*
B	39	1.26	B.1	Plastic cover	15	0.118	*1.378 (+3.90 %)*
C	36	1.162	C.1	Plastic cover (P) and carboard sheet (C)	14 (P) 22 (C)	0.135	*1.297 (−2.11 %)*
D	36	1.162	D.1	Cardboard tray (C) and plastic cover (P)	51 (C) 13 (P)	0.160	*1.322 (−0.23 %)*

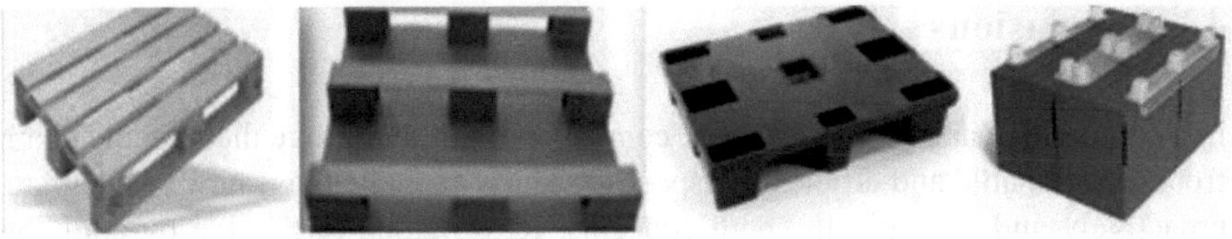

Fig. 3 Alternatives for configuration of palletized loads (From *left* to *right*: wood pallet, cardboard pallet, plastic pallet and loading ledge)

In particular, in the latter transport, the type of truck changes, not only in the capacity (typically with less capacity vehicles), but also in the configuration of goods, due to milk pallets are combined with other pallets of products with lower densities (mono-reference and/or multi-reference pallets). By combining these different types of products, generally, the average weight in each pallet on this new truck is reduced, enabling a priori a better pallet volume. This brings additional advantages in supermarkets that have opted to present directly at the point of sale the milk pallet (minor handling and better occupancy in supermarkets).

Even, company can adopt alternatives that reduce the weight of traditional wooden EUR pallet (25 kg per pallet) to gain useful load capacity on trucks. Among these alternatives are (see Fig. 3): the plastic pallet (6 kg.), the cardboard pallet (12 kg.) or the "loading ledge" (1.5 kg. approximately). In an initial full load truck (30 pallets), the analyzed manufacturer could earn up to 800 kg (option with loading ledge), equivalent to an additional pallet on each truck (3.33 %). In the rest of the options, also an additional pallet per truck is load except in the option D. However, any of these alternatives would require a change in the "pool system" (exchange of items). Besides, the "loading ledge", moreover, requires changes in the palletizing system at the manufacturer's premises. Therefore, it has not been considered all these options in the final analysis.

Finally, at the level of reverse logistics, the Green Dot cost also depends on the selection of the brik and the pack. Table 3 summarizes the total costs of Green Dot per each alternative (Ecoembes fees in 2013 are: 0.323 €/kg. Brik; 0.068 €/kg. cardboard; 0.472 €/kg. plastic).

As a final decision, it was decided to choose the most efficient brik format in logistics (format D), since is the alternative with major level of total savings (see Tables 2 and 3). This alternative involves no substantial changes in the system of packing briks and packs and implies savings in handling and storage of over 16,000 pallets a year (a reduction of 11.7 %). Likewise, an annual reduction in the number of full trucks to 92 trucks (2 % reduction) is achieved. Only in transport between the manufacturer and distributors, this change should involve savings of 60,000 €/year (total savings of at least 35,000 €/year). Additional savings could be achieved thanks to the reduction of handling and storage in manufacturers, distributors and supermarkets, but also in the costs of transport between the distributors' warehouses and supermarkets.

4 Conclusions

In a competitive and global scenario, companies should improve their supply chain from a sustainable and efficient perspective. The real challenge is how to integrate, proactively and strategically, both concepts. Redesigning packaging by applying the "Packaging Logistics" concept is an example of this integration as it was illustrated in the dairy company. In this case, the supply chain as a whole has succeeded in making substantial savings at logistics and environmental level.

References

1. Andersen M, Skjoett-Larsen T (2009) Corporate social responsibility in global supply chains. Supply Chain Manage: Int J 14(2):77–87
2. Azzi A, Battini D, Persona A, Sgarbossa F (2012) Packaging design: general framework and research agenda. Packag Technol Sci 25:435–456
3. Bramklev C (2009) On a proposal for a generic package development process. Packag Technol Sci 22:171–186
4. Carter CR, Rogers DS (2008) A framework of sustainable supply chain management: moving toward new theory. Int J Phys Distrib Logist Manage 38(5):360–387
5. Ciliberti F, Portrandolfo P, Scozzi B (2008) Logistics social responsibility: standard adoption and practices in Italian companies. Int J Prod Econ 113:88–106
6. Christopher M (2005) Logistics and supply chain management: strategies for reducing cost and improving service, 3ª edn. Financial Times, London
7. Coughlan P, Coghlan D (2002) Action research for operations management. Int J Op Prod Manage 22(2):220–240
8. García-Arca J, Prado-Prado JC (2008) Packaging design model from a supply chain approach. Supply Chain Manage: Int J 13(5):375–380
9. Hellström D, Saghir M (2006) Packaging and logistics interactions in retail supply chain. Packag Technol Sci 20:197–216
10. Mejías-Sacaluga A, García-Arca J, Prado-Prado JC, Fernández-González AF (2011) A model for the corporate social responsibility adoption in the supply chain management. Dir Organ 45:20–31
11. Prado-Prado JC (2000) El proceso de mejora continua en la empresa. Pirámide, Madrid
12. Saghir M (2002) Packaging logistics evaluation in the Swedish supply chain. Lund University, Lund
13. Yin RK (2002) Applications of case study research. Sage, Thousand Oaks

Best Practices in Sustainable Supply Chain Management: A Literature Review

Ana María Mejías Sacaluga and Juan E. Pardo Froján

Abstract On the basis of a content analysis, this paper explores the evolution of best practices from the traditional approach to cost efficiency in supply chain management towards the current context of sustainability. In this sense, the paper shows a comprehensive review of the best practices that supply managers will need to engage in to create a sustainable supply chain. Our analysis suggests that the practices that lead to a more sustainable supply chain management are in equal parts best practices in traditional supply chain management or slight modifications of existing practices, and innovative practices.

Keywords Best practices · Literature review · Logistics · Sustainable supply chain management (SSCM)

1 Introduction

As the new economic order unfolded, people recognized that profits and profitability were only one element in the long-term success of companies and economies. Other important factors are the future of people and the future of planet Earth [7]. These legitimate new concerns are captured in measures as the triple bottom line. According to Carter and Rogers [3], the triple bottom line suggests that at the intersection of social, environmental and economic performance, there are activities that organizations can engage in which not only positively affect the natural environment and society, but which also result in long-term economic benefits and competitive advantage for the firm.

These new trends of thinking underwent a great development in recent times. It was in the first decade of 2000 when concepts of Corporate Social Responsibility

A. M. Mejías Sacaluga (✉) · J. E. Pardo Froján
Grupo OSIG, Dpto. de Organización de Empresas y Marketing,
Escuela de Ingeniería Industrial, Universidad de Vigo, C/Maxwell 9,
Campus Lagoas-Marcosende, 36310 Vigo, Spain
e-mail: mejias@uvigo.es

J. E. Pardo Froján
e-mail: jpardo@uvigo.es

C. Hernández et al. (eds.), *Managing Complexity*, Lecture Notes in Management and Industrial Engineering, DOI 10.1007/978-3-319-04705-8_24,

(CSR), Green or Sustainability and Supply Chain Management (SCM) began to be jointly dealt with in literature. In this paper, the terms sustainability (or "green") are used as synonyms for corporate social responsibility [13]. According to Seuring and Müller [11], we define sustainable supply chain management (SSCM) as "the management of material, information and capital flows as well as cooperation among companies along the supply chain while taking goals from all three dimensions of sustainable development (economic, environmental and social) into account which are derived from customer and stakeholder requirements".

However, although the literature of SSCM has been evolving rapidly, most of the companies are still searching for the best way to implement sustainability principles into their supply chain. Likewise, Andersen and Skjoett-Larsen [1] point out that despite many multinational corporations' efforts to implement social and environmental issues in their supply chains, a gap exists between the theoretical application of sustainability in SCM in theory and the implementation in practice.

For other authors [5], the literature on SSCM is still limited, and focuses on the study of the concrete fields (for example, there is very limited research that explores the social dimension of sustainability), and it is scant in literature reviews.

In this context, and taking into account that the question for companies has become not whether to commit to a strong environmental, health and safety record, but how to do so in the most cost-effective manner [7], the aim of this paper is to explore the evolution of best practices from the traditional approach to efficiency in the SCM to the current context of sustainability. This paper conducts a comprehensive literature review from a holistic perspective, integrating the triple bottom line with all the logistics activities along the SC: purchasing, production, warehouse and transportation, and reverse logistics.

2 Methodology

A literature review is a systematic, explicit, and reproducible design for identifying, evaluating, and interpreting the existing body of recorded documents [4]. Literature reviews usually have two objectives: first, they summarize existing research by identifying patterns, themes and issues; second, they may constitute an initial step in the theory development process [8].

From a methodological point of view, literature reviews can be comprehended as content analysis. Content analysis is a method for the objective, systematic, quantitative and reliable study of published information, i.e. a suitable method for comprehensive literature reviews [9]. A process model for a content analysis is described in four steps [11]:

1. Material collection: the material to be collected is defined and delimitated.
2. Descriptive analysis: formal aspects of the material are assessed.
3. Category selection: structural dimensions and related analytic categories are selected, which are to be applied to the collected material.

4. Material evaluation: the material is analyzed according to the structural dimensions. This should allow identification of relevant issues and interpretation of results.

According to Spens and Kovács [12], content analysis can be used as an instrument for determining key ideas and themes in publications but also, such as in our case, for measuring comparative positions and trends in reporting.

2.1 Sample and Delimitations of the Research

The first step in content analysis is to determine the documents to be analyzed and the units of analysis. As this literature review aims to study the evolution of best practices in logistics activities that lead to a more SSCM, we selected logistics-related journals for the sample. The logistics journals selected are perceived as of the highest quality and represent the state of the art of logistics research (in alphabetical order): *European Journal of Purchasing & Logistics Management, Harvard Business Review, International Journal of Integrated Supply Management, International Journal of Logistics Management, International Journal of Operations & Production Management, International Journal of Physical Distribution & Logistics Management, International Journal of Production Economics, International Journal of Production Research, Journal of Business Logistics, Journal of Cleaner Production, Journal of Operations Management, Production and Operations Management, Purchasing and Supply Management, Supply Chain Management: an International Journal, Transportation & Distribution* and other journals.

The time period of the literature review was defined between 1990 and 2011. After a first quick content check, identified articles were included or excluded from the analysis. This first phase involved using the preliminary keyword "best practices" to guide the research by identifying the papers that explicitly included that keyword in this title, abstract and/or full text. The papers that did not meet the practical approach in conducting the literature review were excluded (i.e., mathematical models of SSCM, theoretical approaches related to different organizational theories, technical issues—minimizing CO_2 emissions, …—or case studies focused in specific fields—energy, oil and gas industry…).

Taken the stated delimitations into account, a total of 105 papers were identified. The list of revised papers is available on request.

2.2 Categorization in the Coding Scheme

Content analysis builds on a coding scheme that is developed on the basis of a theoretical framework. In order to derive patterns in the presentation and reporting of information, content analysis involves the codifying of information into pre-defined categories [6].

The second phase of the research involved the application of the keywords listed below and different combination of them:

- Cost (reduction)
- Efficiency
- Environment (al)
- Green
- Lean (operations, manufacturing)
- Logistics (management, network)
- Manufacturing
- Operations management
- Performance
- Production (management)
- Purchasing (management)
- Reverse logistics (or supply chain)
- Social responsibility
- Supplier
- Sustainability (or sustainable)
- Supply chain management
- Transportation
- Vendor
- Warehouse (management)

3 Frequency Analysis

As we stated before, 105 papers were identified in the literature review in the research period (1990–2011). While 1991 is the first year where papers about traditional best practices (economic efficiency approach) in Logistics/SCM were found, the greatest number of publications on these practices is found in the time period between 1995 and 2001. Instead, the greatest number of publications on sustainable best practices is found from 2006 onwards. The distribution of the papers in the research period is shown in Figs. 1 and 2.

4 Best Practices: from Cost Efficiency to Sustainability

As a result of the content analysis, the best practices identified were classified into two broad categories: traditional (economic efficiency point of view) and sustainable. Then, for each category, the practices were sorted in four subcategories based on their application: throughout the supply chain, in purchasing management, in production management, in warehouse and transportation (W&T) management and in reverse logistics management.

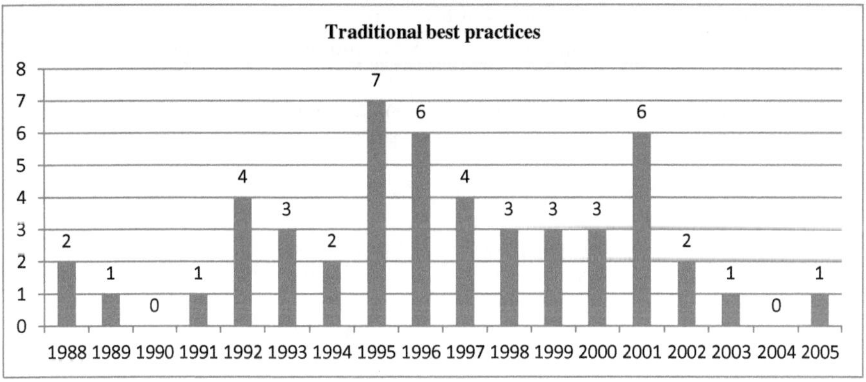

Fig. 1 Distribution of papers related to traditional best practices across the research period (Note: three relevant papers were founded in 1988 and 1989 were added to the research period)

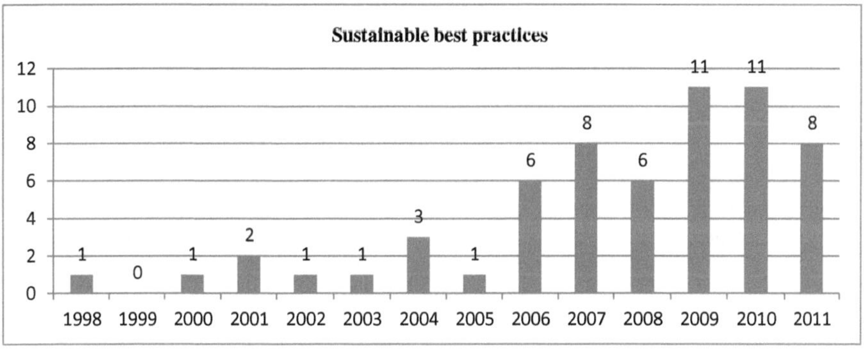

Fig. 2 Distribution of papers related to sustainable best practices across the research period

We show a synthesis of our findings due to reduction reasons. First, we list the traditional and sustainable best practices with general application throughout the entire supply chain. Then, we show in Table 1 the evolution of the best practices for each logistics area, identifying: traditional practices without and with continuity in the context of sustainability (TP-Discontinuity, TP-Continuity), sustainable practices which are modifications of existing practices (SP-Modification) and innovative sustainable practices (SP-Innovation).

This is the list of traditional and sustainable best practices with general application:

- Traditional practices:
  - Quality and environmental management systems implementation (under the international standards ISO 9000, ISO 14000 and Eco-Management and Audit Scheme of the European Union).
  - Coordination between buying and supplying organizations.
  - ICT implementation.

Table 1 Best practices: from efficiency to sustainability

TP-discontinuity	TP-continuity	SP-modification	SP-innovation
Purchasing			
Supplier requirements to reduce cost annually	Supplier selection	Long-term relationships with suppliers	Local suppliers development
Supplier requirements of warehouses	Supplier certification	Codes of conduct	Reward systems linked to sustainability
	Collaborating with suppliers Supplier development	Traceability and collaboration with suppliers Ensuring supplier continuity	Transparency and ethics in purchasing Reducing supplier risk
Production			
Relocation of facilities (low labor cost)	Just in time operations	Sustainable products and process design	Environmnetal conscious design (ECD) and Life-cycle analysis (LCA)
Layout optimization	Outsourcing Waste management Lean production	Postponement Green manufacturing and remanufacturing Lean-Green synergies	Closed-loop SC Source-reduction and Pollution prevention Lean-Green operations
W&T			
Reduction of the number of warehouses	Inventory reduction	Inventory reduction and proper storage of hazardous materials	Donation of excess or obsolete inventory
Full truck deliveries	Cross-docking	Recyclable and reusable packaging and containers	Reverse logistics, minimizing traffic and reducing noise pollution
	Intermodal transportation Efficient delivery (standardization in packaging and delivery units, …)	Clean transportation/fuel efficiency Extending JIT/lean approach to warehouse and transportation	Reward systems linked to sustainability Economic + environmental + safety as a selection criteria of for-hire carriers
Reverse logistics			
Efficient management of returns	Just in time in reverse logistics	Redesigning logistics networks to accommodate returns	Closed-loop supply chains
	Systems of selected collection of materials	ICT as support of RL value measurement	To integrate the financial impact of RL strategies

- Sustainable practices:
  - Reporting on different standards (focused on codes of conduct, product/process-related or management systems and initiatives).
  - Collaborative behaviors with suppliers and customers.

- Collaborating with nontraditional chain members (NGOs, Competitors, Trade groups,…).
- Designing and managing processes to achieve transparency and traceability.
- Addressing the governance structure for SSCM.
- Benchmarking in sustainability.

In Figs. 1 and 2 the frequencies of the two broad categories' occurrence are displayed.

Table 1 shows the evolution of the best practices for each logistics area.

5 Discussion and Conclusions

The confluence of the core-competency and process management movements caused many of the changes in the 1990s. As companies developed their core competencies and included them in their business processes, the tools and concepts of TQM and JIT, and afterward Lean approach, were applied to developing new products and managing the supply chain [7]. But, the literature review of the best practices for effective SCM shows that inter-firm competitive advantage has been mainly focused in profit criteria.

Over the past two decades, increasing pressures from governments, customers and other stakeholders groups have prompted firms to incorporate sustainability issues into their SCM schemes. However, these external pressures on a firm only lead to sustainable supply and production if both the individual firms and the supply chain as a total entity develop the necessary relevant internal resources as prerequisites for implementing SSCM [2]. In this context, most of the companies are still searching for the best way to implement sustainability principles into their supply chain. Our analysis shows that the practices that lead to a more SSCM are equal parts best practices in traditional SCM or slight modifications of existing practices, and innovative practices. However, to gain positive results, the firm must establish management systems and tools that integrate environmental, health and safety metrics with other process metrics within the company and across the SC. In this sense, Rothenberg et al. [10] examined the links between lean and green and found some synergies but also found that harvesting them is not simple. SCM researchers and practitioners are just beginning to face new challenges in integrating sustainability in their areas of interest.

References

1. Andersen M, Skjoett-Larsen T (2009) Corporate social responsibility in global supply chains. Supply Chain Manage: Int J 14(2):75–86
2. Bowen FE, Cousins PD, Lamming RC, Faruk AC (2001) The role of supply chain management capabilities in green supply. Prod Oper Manage 10(2):174–189

3. Carter CR, Rogers DS (2008) A framework of sustainable supply chain management: moving toward new theory. Int J Phy Distrib Logist Manage 38(5):360–387
4. Fink A (2005) Conducting research literature reviews: from the internet to paper. Sage, Thousand Oaks
5. Gold S, Seuring S, Beske P (2010) Sustainable supply chain management and inter-organizational resources: a literature review. Corporate Social Responsibility and Environmental Management, vol 17, pp 230–245
6. Guthrie J, Petty R, Yongvanich K, Ricceri F (2004) Using content analysis as a research method to inquire into intellectual capital reporting. J Intellect Cap 5(2):282–293
7. Kleindorfer PR, Singhal K, Van Wassenhove LN (2005) Sustainable operations management. Prod Oper Manage 14(4):482–492
8. Meredith J (1993) Theory building through conceptual methods. Int J Oper Prod Manage 13(5):3–11
9. Pasukeviciute I, Roe M (2005) Strategic policy and the logistics of crude oil transit in Lithuania. Energy Policy 33(7):857–866
10. Rothenberg S, Pil FK, Maxwell J (2001) Lean, green and the quest for superior performance. Prod Oper Manage 10(3):228–243
11. Seuring S, Müller M (2008) From a literature review to a conceptual framework for sustainable supply chain management. J Clean Prod 16:1699–1710
12. Spens KM, Kovács G (2006) A content analysis of research approaches in logistics research. Int J Phys Distrib Logist Manage 36(5):374–390
13. Stokes P, Harris P (2012) Micro-moments, choice and responsibility in sustainable organizational change and transformation: the Janus dialectic. J Organ Change Manage 25(4):595–611

Supplier Evaluation and Selection: A Review of the literature since 2007

Joan Ignasi Moliné and Ana María Coves

Abstract Supplier selection is currently a subject of great importance to companies. Numerous articles have been published recently, recommending different methods and/or procedures for evaluating and selecting the suppliers with whom the purchasing company will work. The present article reviews a total of 39 articles dealing with this subject, published between 2007 and the present day, in magazines indexed by Journal Citation Reports (in ISI Web of Knowledge). They will be analyzed in order to determine: (i) procedures used in determining criteria, (ii) identification and structure of the criteria under consideration, (iii) methods used to evaluate and select the suppliers and (iv) aims in the selection of suppliers.

Keywords Supplier selection · Supplier criteria

1 Introduction

Supplier selection is one of the strategic elements in managing purchases, as the ability of a company to satisfy its clients, as well as its own continuity, depends to a large extent on its suppliers.

Purchases have a direct and important impact on profits, as the acquired products and services (purchases), have a significant influence on the cost structure of manufacturing companies, ranging from 42 to 79 %. On the other hand, there is a larger framework which defines the general policies of organizations with respect to their relationships with suppliers, some examples being the establishment of partnership deals or global supply chain management.

Supplier selection is basically determined by four decisions, which are: (1) Having the appropriate procedure for determining the criteria and establishing their structure and ranking. (2) Identifying the criteria with which the suppliers will be evaluated. (3) Selecting the most suitable suppliers from those available. (4) Obtaining a list of suppliers, the Suppliers Panel, with whom orders will be placed.

A. M. Coves (✉) · J. I. Moliné
Department of Management, Institute of Industrial and Control Engineering,
Universitat Politècnica de Catalunya, Av. Diagonal 647, 08028 Barcelona, Spain
e-mail: anna.maria.coves@upc.edu

C. Hernández et al. (eds.), *Managing Complexity*, Lecture Notes in Management and Industrial Engineering, DOI 10.1007/978-3-319-04705-8_25,

This process involves different types of criteria (strategic, tactical, and operational) which are cohesive, as they originated in corporate strategy and are in consonance with it.

The described situation has motivated the present article, which reviews a total of 39 articles dealing with this subject, published between 2007 and the present day, in magazines indexed by Journal Citation Reports (in ISI Web of Knowledge). They all deal with the subject of supplier evaluation and selection.

This article is organized in such a way that the following section includes an analysis of the principal aspects and, finally, Sect. 3 shows the conclusions which can be drawn from this review of the latest developments.

2 Analysis of the Latest Developments (Since 2007)

The most recent theories on supplier selection are organized according to the 4 decisions mentioned in the introduction, in a way that Sect. (2.1) analyzes the procedures for determining the criteria, and Sect. (2.2) studies the criteria used for supplier selection. Sect. (2.3) describes the methods used to evaluate and select the suppliers, while Sect. (2.4) analyzes the aims of the different works on the subject of supplier selection.

2.1 Procedures for Determining Criteria

The criteria used for evaluating and selecting suppliers depend on the procedures used for determining them. In this context it can be seen that the most frequently used source is consultation with experts (19 articles, 57.5 %), either as a unique source (15 articles, 45.4 %) or reinforced by a review of the available literature (4 articles, 12.1 %). A detailed analysis shows that some articles described the methodology used, whereas others do not specify how the information was obtained. The experts may either be employees of the purchasing company or others who are familiar with the industry.

The second most used source is the review of literature; 6 (18.2 %) articles are used as a unique source as well as the previously mentioned case of 4 (12.1 %) in conjunction with the experts. Not one article describes how the review was carried out.

There is just one article [12] which describes a specific method, the QFD (Quality Function Deployment) which translates shareholders' needs into criteria for supplier selection. There are seven articles (21.21 %) that do not specify the procedure used.

2.2 Identifying the Criteria for Obtaining the Panel

The panel of suppliers is made up of those suppliers who have best satisfied the criteria of the purchasing company. These criteria can be grouped and ranked according

Table 1 Classification Family vs. Typology using distinct criteria

		Typology				
		Strategic	Operational	Tactical	Total general	
Family	Assets and infrastructure	78	10	18	106	56.4%
	Cost	7	20	4	31	16.5%
	Logistics	1	9	7	17	9.0%
	Quality	12	19	3	34	18.1%
	Total general	98	58	32	188	100%

Table 2 Classification Family vs. Typology counting all repetitions

		Typology				
		Strategic	Operational	Tactical	Total general	
Family	Assets and infrastructure	124	12	50	186	48.9%
	Cost	12	37	23	72	18.9%
	Logistics	1	13	35	49	12.9%
	Quality	21	28	24	73	19.2%
	Total general	158	90	132	380	100%

to families. Analysis of the articles shows how each author opts for different ways of structuring them, only coinciding in one case. The present study uses the families proposed by Erdem and Göçen [7], "Assets and Infrastructure, Costs, Logistics and Quality"; these authors propose an exhaustive classification obtained from reviewing the literature and interviewing experts. The criteria can also be grouped according to their typology: strategic, tactical and operational.

Table 1 classifies the criteria of the articles according to their family and typology, taking each criterion as distinct (not counting the number of times it is repeated in different articles). Table 2 includes the repeats of each criterion, the number of times it is cited in the different articles.

Tables 1 and 2 show that a total of 380 criteria were mentioned, of which 188 are distinct. Detailed analysis of the distinct criteria shows a very disparate level of detail, some criteria being very generic (e.g. technique) and others which are more specific (e.g. the number of Rejected items at entry quality level).

Regarding Typology, the most common are strategic criteria (98 criteria, 52.13 %), although if the total number of articles citing them is taken into account the number drops (158 criteria, 41.58 %). There are 32 criteria (17 %) of a tactical nature, representing 34.74 % of citations. Finally, there are 58 operational criteria (30.85 %) which make up 23.68 % of all citations.

To determine the level of criteria standardization, the ratio between the number of citations and the number of distinct criteria is determined, resulting in: "Assets and Infrastructure" (1.75), "Cost" (2.32), "Logistics" (2.88) and Quality (2.15). The family with the highest number of criteria is "Assets and Infrastructure" with 1.75 citations per criterion. This is due to the fact that 75 of the criteria have only one citation; this family reflects the specific cases of the company concerned, leading

Table 3 Relation of criteria with most citations

Criteria	Total	Family
Delivery performance	21	Logistics
Price	20	Cost
Quality performance	19	Quality
Production capacity	16	Assets and infrastructure
General demand	10	Assets and infrastructure
Financial stability	8	Assets and infrastructure
Communication openness	7	Assets and infrastructure
Location	7	Assets and infrastructure
Transportation	5	Cost
design capability	5	Assets and infrastructure
Quality management practices and systems	5	Quality

to fewer repeated criteria. The other three families have fewer distinct criteria and higher levels of repetition, which indicates that their criteria are used by different authors; there is more consensus and a certain standardization.

Table 3 shows the most mentioned criteria, a total of 11, which have five or more repeats, and which show a high degree of concentration, as the 11 criteria (5.85 %) have a total of 123 citations (32.36 %).

There are 156 criteria (82.97 % of the total) that have two or more citations, representing 49.47 % of the criteria of all the articles.

2.3 Methods of Evaluating and Selecting Suppliers

The methods used to evaluate and select suppliers are very diverse, and Table 4 includes a correlation of all those found in the analyzed articles and the total number of times each one has been used.

Analysis of Table 4 shows that there is a great variety of methods, finding 25 different methods in a total of 35 articles. The principal ones are: Analytic Hierarchy Process (AHP) + mathematical programming (four articles) and those based solely on mathematical programming (five articles). A detailed study of those that include mathematical programming shows how the authors opted for different procedures and so, as each one is practically unique; there are 34 different methods and only one repetition. It can also be seen that 17 of the articles (48.6 %) use the AHP or its variations (ANP, Fuzzy AHP, Fuzzy ANP).

2.4 Objectives in Selecting Suppliers

The selection of suppliers, in general, has the aim of determining a number N of suppliers and forming a panel of suppliers. An analysis of the current situation shows that the articles suggest different methods:

Table 4 Methods/procedures used to evaluate and select suppliers

Methods/procedures	Total	References
AHP (Analytic Hierarchy Process)	1	[12]
AHP + Mathematical Programming	4	[7];[15];[21];[32]
AHP + CFPR(Consistent Fuzzy Preference Relations)	1	[6]
ANP (Analytic Network Process)	1	[10]
ANP + Mathematical Programming	3	[14];[27];[31]
ANP + TOPSIS+ Mathematical Programming	1	[20]
FAHP	2	[13];[16]
FAHP + Mathematical Programming	1	[25]
Fuzzy AHP + Fuzzy TOPSIS + DEA	1	[34]
FANP	1	[29]
FANP + Mathematical Programming	1	[19]
DEA (Data Envelopment Analysis)	1	[26]
DEA + Decision Trees (DT) + Neural Network (NN)	1	[31]
Discret Choice Analysis (DCA)	1	[28]
Fuzzy logic	1	[2]
Grey ralational analysis (GRA)	1	[11]
Influence diagram + Fuzzy logic	1	[9]
MultiAlternative proposal	1	[4]
NN	1	[1]
Mathematical Programming	5	[5];[8];[22];[23];[24]
Stochastic Dynamic Programming (SDP)	1	[18]
Stochastic programming model (SP)	1	[17]
Weighted additive fuzzy programming	1	[33]
Supplier evaluation system (utilitza PROMETHEE)	1	[3]
Vague sets theory	1	[35]

1. Articles which recommend a ranking of suppliers (23 articles), normally in descending order according to their weight. In general the values of their weight show the degree to which each supplier satisfies the client's needs. Their value is composed of the relative weight of the criteria for the client and the degree to which each supplier satisfies each criterion.
2. Articles that classify the suppliers in binary fashion (3 articles), only including those who are suitable in the panel. The result of the selection is as follows: supplier selected/not selected [1], supplier efficient/inefficient [34].
3. Articles that propose different solutions to the anterior:

   3.1 Articles that select suppliers without creating a panel (8 articles); directly assigning orders to the cloud of suppliers [23], or obtaining a ranking which is not based on weight or binary classification; one example is the classification of suppliers according to their partnership consideration [3].

   3.2 Two articles that analyze other aspects arising from the selection process: a comparison of the weight of criteria in different countries [28] and the analysis to determine whether the relationship with suppliers should be long or short term [18].

3 Conclusions

Purchasing management is strategic and supplier selection one of the most decisive processes. In this paper, a review of current thinking in selecting suppliers demonstrates the inexistence of a general model for determining the panel of suppliers, while showing a great diversity in the methods used for creating the panel. Each proposal is almost unique, with those articles that use the AHP or its variations standing out.

There is no standardization in the criteria themselves nor in the ways of classifying or ranking them. In general, the criteria are determined in two ways: consultation with experts and reviewing the literature, and a detailed analysis shows the wide variety of procedures used. Therefore it can be stated that there is a lack of standards for determining and classifying criteria, which currently depends on the decision-maker's experience.

Almost all the articles deal with different industrial sectors (automobiles, electronics, …), and only one was found which dealt with the service sector [8], presenting a method of decision making, in the case of an airline company, to solve the problem of supplier selection in subcontracting services.

Future research works may include: (i) expand the scope of the review focusing on criteria and validate the lack of standards detected: (ii) implement the stages of supplier selection in the area of services.

Acknowledgements The authors gratefully acknowledge the partial support of grant DPI2010-15614 (Ministerio de Economia y Competitividad, Spain).

References

1. Aksoy A, Öztürk N (2011) Supplier selection and performance evaluation in just-in-time production environments. Expert Syst Appl 38(5):6351–6359
2. Amin SH, Razmi J, Zhang G (2011) Supplier selection and order allocation based on fuzzy SWOT analysis and fuzzy linear programming. Expert Syst Appl 38(1):334–342
3. Araz C, Ozkarahan I (2007) Supplier evaluation and management system for strategic sourcing based on a new multicriteria sorting procedure. Int J Prod Econ 106(2):585–606
4. Ben-David A, Gelbard R, Milstein I (2012) Supplier ranking by multi-alternative proposal analysis for agile projects. Int J Proj Manage 30(6):723–730
5. Burke GJ, Carrillo JE, Vakharia AJ (2007) Single versus multiple supplier sourcing strategies. Eur J Oper Res 182(1):95–112
6. Chen YH, Chao RJ (2012) Supplier selection using fuzzy preference relations. Expert Syst Appl 39(3):3233–3240
7. Erdem AS, Göçen E (2012) Development of a decision support system for supplier evaluation and order allocation. Expert Syst Appl 39(5):4927–4937
8. Feng B, Fan ZP, Li Y (2011) A decision method for supplier selection in multi-service outsourcing. Int J Prod Econ 132(2):240–250
9. Ferreira L, Borenstein D (2012) A fuzzy-Bayesian model for supplier selection. Expert Syst Appl 39(9):7834–7844
10. Gencer C, Gürpinar D (2007) Analytic network process in supplier selection: a case study in an electronic firm. Appl Math Modell 31(11):2475–2486

11. Golmohammadi D, Mellat-Parast M (2012) Developing a grey-based decision-making model for supplier selection. Int J Prod Econ 137(2):191–200
12. Ho W, Dey PK, Lockström M (2011) Strategic sourcing: a combined QFD and AHP approach in manufacturing. Supply Chain Manage Int J 16 (6):446–461
13. Kilincci O, Onal SA (2011) Fuzzy AHP approach for supplier selection in a washing machine company. Expert Syst Appl 38(8):9656–9664
14. Kirytopoulos K, Leopoulos V, Mavrotas G, Voulgaridou D (2010) Multiple sourcing strategies and order allocation: an ANP-Augmecon meta-model. Supply Chain Manage Int J 15 (4):263–276
15. Kokangul A, Susuz Z (2009) Integrated analytical hierarch process and mathematical programming to supplier selection problem with quantity discount. Appl Math Modell 33(3):1417–1429
16. Lee AHI (2009) A fuzzy supplier selection model with the consideration of benefits, opportunities, costs and risks. Expert Syst Appl 36(2):2879–2893
17. Li L, Zabinsky ZB (2011) Incorporating uncertainty into a supplier selection problem. Int J Prod Econ 134(2):344–356
18. Li S, Murat A, Huang W (2009) Selection of contract suppliers under price and demand uncertainty in a dynamic market. Eur J Oper Res 198(3):830–847
19. Lin RH (2009) An integrated FANP-MOLP for supplier evaluation and order allocation. Appl Math Modell 33(6):2730–2736
20. Lin CT, Chen CB, Ting YC (2011) An ERP model for supplier selection in electronics industry. Expert Syst Appl 38(3):1760–1765
21. Mafakheri F, Breton M, Ghoniem A (2011) Supplier selection-order allocation: a two-stage multiple criteria dynamic programming approach. Int J Prod Econ 132(1):52–57
22. Mansini R, Savelsbergh MWP, Tocchella B (2012) The supplier selection problem with quantity discounts and truckload shipping. Omega Int J Manage Sci 40(4):445–455
23. Mendoza A, Ventura JA (2012) Analytical models for supplier selection and order quantity allocation. Appl Math Modell 36(8):3826–3835
24. Ng WL (2008) An efficient and simple model for multiple criteria supplier selection problem. Eur J Oper Res 186(3):1059–1067
25. Shaw K, Shankar R, Yadav SS, Thakur LS (2012) Supplier selection using fuzzy AHP and fuzzy multi-objective linear programming for developing low caron supply chain. Expert Syst Appl 39(9):8182–8192
26. Toloo M, Nalchigar S (2011) A new DEA method for supplier selection in presence of both cardinal and ordinal data. Expert Syst Appl 38(12):14726–14731
27. Ustun O, Demirtas EA (2008) An integrated multi-objective decision-making process for multi-period lot-sizing with supplier selection. Omega Int J Manage Sci 36(4):509–521
28. Van der Rhee B, Verma R, Plaschka G (2009) Understanding trade-offs in the supplier selection process: the role of flexibility, delivery, and value added services/support. Int J Prod Econ 120(1):30–41
29. Vinodh S, Ramiya RA, Gautham SG (2011) Application of fuzzy analytic network process for supplier selection in a manufacturing organisation. Expert Syst Appl 38(1):272–280
30. Wu D (2009) Supplier selection: a hibrid model using DEA, decision tree and neural network. Expert Syst Appl 36(5):9105–9112
31. Wu WY, Sukoco BM, Li CY, Chen SH (2009) An integrated multi-objective decision-making process for supplier selection with bundling problem. Expert Syst Appl 36(2):2327–2337
32. Xia W, Wu Z (2007) Supplier selection with multiple criteria in volume discount environments. Omega Int J Manage Sci 35(5):494–504
33. Yücel A, Güneri AF (2011) A weighted additive fuzzy programming approach for multicriteria supplier selection. Expert Syst Appl 38(5):6281–6286
34. Zeydan M, Çolpan C, Çobanoglu C (2011) A combined methodology for supplier selection and performance evaluation. Expert Syst Appl 38(3):2741–2751
35. Zhang D, Zhang J, Lai KK, Lu Y (2009) An novel approach to supplier selection based on vague sets group decision. Expert Syst Appl 36(5):9557–9563

Logistic Management Employing Tabu Search and Neural Network Algorithms: A Case Study

Isidro Peña García-Pardo and Fausto Pedro García Márquez

Abstract This paper presents a real case study of a routing problem in a Spanish firm leader in the fresh food industry. The main objective is to improve profits and competitiveness based on logistic operations, minimizing the transportation cost employing the tabu search and neural network meta-heuristics algorithms. The simplest case considered is the Traveling Salesman Problem (TSP). The real case study presented in this paper there are capacity restrictions and different demands at each node, therefore the problem is solved as a Capacitated Vehicle Routing Problem (CVRP).

Keywords CVRP · Neural network · Tabu Search · Logistic Management

1 Introduction

Logistic is an essential area in business competitiveness. The delivery of goods from a warehouse to local customers is a critical issue and therefore it should be considered as a source of competitive advantage [5]. Logistics management is of particular relevance in highly competitive industries such as the food industry, in wich logistics costs constitute an important percentage of the total costs and the product must be supplied with extreme punctuality [17]. The food industry is interested in reducing transport costs maintaining the quality of the services, especially for fresh and perishable products. For a range of goods labelled as perishables, particularly food, their quality degrades with time is high. It can be mostly reduced with low temperatures, but it takes time, costs and a good coordination for the shipment, and every delay can have negative consequences. An inefficient distribution system leads that the product will not reach to be served on time, and the volume

I. P. García-Pardo (✉)
Dpto. de Administración de Empresas, Facultad de Derecho y Ciencias Sociales,
Universidad de Castilla-La Mancha, 13071 Ciudad Real, Spain
e-mail: isidro.pena@uclm.es

F. P. García Márquez
INGENIUM GROUP, ETSI Industriales, Universidad de Castilla-La Mancha,
13071 Ciudad Real, Spain
e-mail: faustopedro.garcia@uclm.es

C. Hernández et al. (eds.), *Managing Complexity*, Lecture Notes in Management and Industrial Engineering, DOI 10.1007/978-3-319-04705-8_26,

ordered reduces storage space and raises costs considerably, because many products are returned by traders, and this complicates the process of making the optimal logistic decision [16]. It is therefore essential to determine optimal routes with the principal objective of minimizing the distance, costs and time, maintaining the quality of products [7]

The food industry has already been considered in the last few years in many research studies (i.e. [6, 10, 12, 14, 17]). The objective function of these problems usually involves distances, costs, number of vehicles and delivery time [7]. The authors have not found in the literature any research work that can solve the problem presented in this paper taking into account the purchase volume, frequency of orders, type and prices of the products demanded by the customers.

2 Case Study and Problem Description

This case study was conducted in an important firm in the fresh food industry with a large commercial structure, including a logistics department that has expanded and become consolidated in the Spanish market and has a relevant presence in the EU market, Portugal, Italy and France. In Spain, the firm has segmented the market in the following areas in order to maximize its control of operations: Central Area, Northern Andalucía, Western Andalucía, Eastern Andalucía, Extremadura and Castilla y León. This research is focused on a real case study in the region of Extremadura.

Many managerial problems, as routing problems, facility location problems, scheduling problems, network design problems, can be modeled as a Vehicle Routing Problem (VRP). The VRP is one of the combinatorial optimization problems most frequently studied owing to both to its practical relevance and its considerable difficulty [18]. A VRP problem is proposed in order to find a collection of K simple routes (corresponding to vehicle routes) with the minimum cost, defined as the sum of the costs of the arcs belonging to the routes, and such that:

i. each route visits the depot vertex.
ii. each vertex is visited by exactly one route.
iii. each city has a specific demand.

In this study, the routing design process of the firm must optimize the number of vehicles to involve and find out the routes of each vehicle. The problem to solve can be described as a Capacitated Vehicle Routing Problem (CVRP) (e.g. [6]), with the following characteristics: (a) unknown fleet size; (b) homogeneous fleet (refrigerator trucks loading a fix volume of products); (c) single depot; (d) deterministic demand; (e) objective: minimizing distances.

The problem is formulated as a Hamiltonian cycle, i.e. the problem is defined by the graph $\mathbf{G}=(\mathbf{V}, \mathbf{E})$, where $\mathbf{V} \in R^2$ is a set of n cities, and $\mathbf{E}$ is a set of arcs connecting these cities. Under these conditions, the problem can be formulated as:

$$\text{Minimize: } \sum_i \sum_j c_{ij} x_{ij},$$

where x_{ij} is the binary decision variable that in case of $i<j$ has the following values:

$$x_{ij} \begin{cases} 1 \text{ if the arc joining cities i and j is used in solution} \\ 0 \text{ otherwise} \end{cases},$$

being **c** is the associated cost matrix of **E**, composed by the elements c_{ij} that represents the "distance" (expressed as physical distance) between the cities i and j. In this paper, the cost matrix is asymmetric and subjects to the following constraints:

$$\sum_{i=1} x_{ij} = 1, \text{ for all } j \in \mathrm{V} \setminus \{0\} \tag{1}$$

$$\sum_{j=1} x_{ij} = 1, \text{ for all } i \in \mathrm{V} \setminus \{0\} \tag{2}$$

$$\sum_{j=1} x_{0j} = K \tag{3}$$

$$\sum_{i=1} x_{i0} = K \tag{4}$$

$$\sum_{i=1} \sum_{i=1} x_{ij} > \gamma(S) \text{ For all } \mathbf{S} \subseteq \mathbf{V} \setminus \{0\}, \ \mathrm{S} \neq \emptyset \tag{5}$$

The indegree and outdegree constraints (1)–(2) impose that exactly one arc enters and leaves each vertex associated with a customer. Constraints (3)–(4) impose the vehicles requirements from the depot. All available vehicles must be used in one route. This number of vehicles is not smaller than the minimum number of vehicles needed to serve all the customers' demands. Constraint (5) imposes the connectivity of the solution and the vehicle capacity requirements, i.e. each route (V/S, S) defined by a vertex set S, is crossed by a number of arcs bigger than γ (S) (minimum number of vehicles needed to serve customer set S, see eq. (6)) [18].

$$\sum_i \sum_j st_i^k x_j^k + \sum_i \sum_j t_{ij\,ij}^k x_{ij}^k \leq \mathrm{T}^k \tag{6}$$

where s_i is the time to serve a customer, and t_{ij} the time needed to travel from city i to city j. Both must be less than total time (T) for each vehicle k.

Finally, subtour elimination constraints (SECs; 7) are required in order to prevent undesirable subtours that are degenerate tours formed between intermediate nodes and not connected to the origin [2]. The restriction of Miller et al. [13] is used (7).

$$u_i - u_j + Cx_{ij} \leq C - d_j \text{ for all } i, j \in \mathrm{V} \setminus \{0\}; i \neq j;$$
$$\text{s. t. } d_i + d_j \leq Cd_i \leq u_i \leq C \text{ for all } i\ \mathrm{V} \setminus \{0\};$$

where u_i; $i \in V\backslash\{0\}$, is an additional continuous variable representing the load of the vehicle after visiting the customer, and d_i is the demand of each customer. When $x_{ij}=0$ the constraint is not binding because $u_i \leq C$ and $u_j \geq d_j$, whereas when $x_{ij}=1$ they impose that $u_j \geq u_i + d_j$ [18].

Different algorithms can be used to find optimal routes, but none are feasible for large variables because they all grow exponentially in terms of computational cost [8]. The practical importance of the CVRPprovides the motivation for the effort involved in the development of metaheuristics algorithms [1]. Several metaheuristics have been proposed for the CVRP in recent years. The six main types of metaheuristic that have been applied to the CVRP are: Simulated Annealing (SA), Deterministic Annealing (DA), Tabu Search (TS), Genetic Algorithms (GA), Ant Systems (AS), and Neural Networks (NN) [9]. In this paper the problem has been solved by using TS and NN algorithms.

2.1 Neural Network

The NN algorithm originated in the 50s, and was used for optimization. NN is usually employed in combinatorial optimization problems to calculate routes for the structure and characteristics of the case study presented in this paper [11]. NN algorithm is based on concepts related to the human brain mechanism, and they are inspired by the neuron system.

NN consists of processing units, called artificial neurons, which are organized in layers. One of the advantages of NN is that it requires a minimum computational cost in comparison to that required by exact algorithms and other heurist algorithms. NN can learn from a reference solution set. This process is done in an adaptive manner, where the connection between neurons is adjusted in order to implement the desired behaviour (which in this research work will be to minimize the total distance covered). New advantages have been found in self organization, in which NN organizes the information received during training time, and it can work online, and the computational operations are carried out in parallel, thus minimizing the information lost during the process [15].

The NN can be single layered, with an input and output layer, or multiple layered, i.e. with an input, hidden and output layer- net [15]. A multiple NN is employed in this paper because it allows consistent results to be obtained and solves this type of problems efficiently (Fig. 1).

After an NN has been developed and trained in a learning phase, NN will provide new results by itself. The number of neurons in the input and output is determined automatically. However, the number of hidden layers and neurons in each hidden layer has been defined by multiple combinations tested for NN. The NN approach in this paper is based on the following functions:

- *Pdv Function*: This function creates a weight matrix, which gives an importance weight to the entries. It also creates a route pattern, generating a vector path or solution.

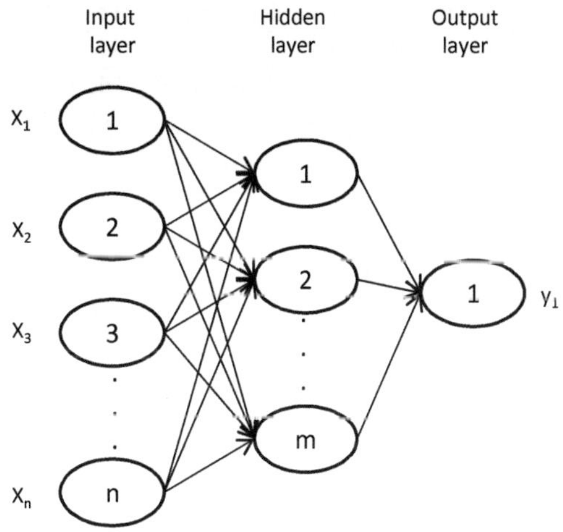

Fig. 1 Neural network with three layers

- *Geper Function*: This generates the permutations necessary for the pdv function.
- *Dpdv function*: This founds the path, starting from the vector map generated by a function called pos.
- *Fc function*: Fc does the learning in order to obtain the reference optimal solutions.

The function command is fc (N, A), where N is the number of nodes and A represents the matrix of distances between the different nodes

2.2 *Tabu Search*

The TS algorithm is based on artificial intelligence using the concept of memory. It has been implemented through simple structures. TS is used in empirical research in CVRP with excellent results [3]. Courdeau and Laporte [4] noted that: *"While the success of any particular method is related to its implementation features, it is fair to say that tabu search (TS) clearly outperforms competing approaches"*.

TS explores the solutions by moving in each iteration from a reference solution to the best solution in its neighborhood. A neighborhood to a reference solution is defined as any solution that is obtained by a pair wise exchange of any two nodes in the solution. This always guarantees that any neighborhood to a feasible solution is a feasible solution. The new solution may deteriorated when is moved from one iteration to another [8]. This can be avoided by using a list that contains recent modifications that have occurred during the transformation from the current solution, called a tabu list. The tabu list only permits new solutions if the method improves the best solution obtained. The TS approach in this paper uses the following primary functions:

Table 1 Distances and runtimes found by NN and TS

	Route 1		Route 2		Route 3	
	Distance (km)/time (h)	Runtime (s)	Distance (km)/time (h)	Runtime (s)	Distance (km)/time (h)	Runtime (s)
Before	1237.3/ 20,15		1353.5/ 21,23		1442.6/ 20, 56	
Neural network	922/ 15,45	1.875	804/ 13,55	14.25	874.5/ 12,45	5.219
Tabu search	925.8/ 14,54	0.187	804/ 13,55	0.125	874.5/ 12,45	0.110

- *Tabu Function*: This function is set for doing all the possible permutations for finding the minimum route.
- *List Function*: This function contains the tabu list which includes the solutions that have been explored in order to not consider these solutions.

TS also employ the *tabu* function (**A**, **path**), where **path** is a vector composed by the initial solutions.

Table 1 shows the total distances corresponding to the routes defined by the firm. The proposed algorithms provided reliable and practical solution with considerable improvements in operational performance, reducing total distance and time in comparison with previous scenario given by the company for the same routes.

It can be seen that the solution generated by TS and NN methods is almost the same, but solution given by NN is more feasible than that obtained by the TS based on the location of populations. In both algorithm, time constrains are satisfied (two work days, 16 h), and it allows delivery with actual vehicle availability and capacity.

However, the computational cost is different due to the TS algorithm uses a heuristic method that is fixed, however NN employs a learning process. NN algorithm is more useful than the TS algorithm when the number of nodes is big, and the NN method generates possible solutions is the same to the number of entries elevated, whereas in the method of the TS generated number of possible solutions raised according to the input nodes.

3 Conclusions

This paper shows that implementing the Tabu Search and Neural Network algorithms for solving various instances of CVRP can significantly reduce the transportation costs in delivery process. The computational performance of these algorithms produces good results in a short computational time. The company's management was satisfied with the performance of the algorithm and decided to use it in their weekly distribution.

In accordance with the findings of this study, new opportunities for research are presented. The first opportunity for research deals with the reconfiguration of customer portfolio based on profitability criterion. Secondly, is possible to optimize the transport costs minimising the distance covered using algorithms set to solve the CVRP.

Acknowledgements The work reported here with has been financially supported by the Spanish Ministerio de Economía y Competitividad, under Research Grant DPI2012-31579.

References

1. Baldacci R, Toth P, Vigo D (2007) Recent advances in vehicle routing exact algorithms, 4OR. Q J Oper Res 5:269–298
2. Bektas T (2006) The multiple traveling salesman problem: an overview of formulations and solution procedures. Omega 34:209–219.
3. Brandão J (2011) A tabu search algorithm for the heterogeneous fixed fleet vehicle routing problem. Comput Oper Res 38:140–151
4. Cordeau JF, Laporte G (2004) Tabu search heuristics for the vehicle routing problem. In: Rego C, Alidaee B (eds) Metaheuristic optimization via memory and evolution: Tabu search and scatter search. Kluwer, Boston, pp 145–163
5. Euchi J, Chabchoub H (2010) A hybrid Tabu Search to solve the heterogeneous fixed fleet vehicle routing problem. Logist Res 2:3–11
6. Faulin J (2003) Applying MIXALG procedure in a routing problem to optimize food product delivery. Omega 31:387–395
7. Faulin J, Juan A, Lera F, Grasman C (2011) Solving the capacitated vehicle routing problem with environmental criteria based on real estimations in road transportation: a case study. Proced-Soc Behav Sci 20:323–334
8. Ganesh K, Nallathambi AS, Narendran TT (2007) Variants, solution approaches and applications for vehicle routing problems in supply chain: agile framework and comprehensive review. Int J Agil Syst Manage 2:50–75
9. Gendreau M, Hertz A, Laporte G (1994) A tabu search heuristic for the vehicle routing problem. Manage Sci 40:1276–1290
10. Hsu CL, Feng S (2003) Vehicle routing problem for distributing refrigerated food. J East Asia Soc Transp Stud 5:2261–2272
11. Leung KS, Jin HD, Xu ZB (2004) An expanding self-organizing neural network for the traveling salesman problem. Neurocomputing 62:267–292
12. Ma H, Cheang B, Lim A, Zhan L, Zhu Y (2012) An investigation into the vehicle routing problem with time windows and link capacity constraints. Omega 40:336–347
13. Miller CE, Tucker AW, Zemlin RA (1960) Integer programming formulation of traveling salesman problems. J Assoc Comput Mach 7:326–329
14. Prindezis N, Kiranoudis CT, Kouris DM (2003) A business-to-business fleet management service provider for central food market enterprises. J Food Eng 60:203–210
15. Sivanandam SN, Dumathi S, Deepa SN (2006) Introduction to neural networks using MATLAB 6.0. McGraw Hill, Delhi, pp 487–512
16. Tarantilis CD, Kiranoudis CT (2001) A meta-heuristic algorithm for the efficient distribution of perishable foods. J Food Eng 50:1–9
17. Tarantilis CD, Kiranoudis CT (2002) Distribution of fresh meat. J Food Eng 51:85–91
18. Toth P, Vigo D (2002) Models, relaxations and exact approaches for the capacitated vehicle routing problem. Discret Appl Math 123:487–512

Multiagent Model for Supply Chain Management

Borja Ponte Blanco, Raúl Pino Díez, Isabel Fernández Quesada, Nazario García Fernández and Manuel Monterrey Meana

Abstract There are several circumstances which, in the last two decades, have granted the Supply Chain Management (SCM) a strategic role in the search for competitive advantage. Thus, this paper applies multiagent methodology to optimize the management. We represent the supply chain as a Global Multiagent System, composed of four Multiagent Subsystems, which replicate the behavior of the different levels of the supply chain. Thereby, each member has its own decision-making capacity and seeks to optimize the performance of the supply chain. We will tackle the problem from two complementary perspectives: reducing the Bullwhip Effect, which can be considered as one of the main sources of inefficiencies in the SCM, and minimizing management costs, both from a non collaborative approach, where each level seeks the best solution for himself, and from a collaborative approach, where each level negotiates with the rest looking for the best solution for the whole supply chain.

Keywords Supply chain · Bullwhip effect · Multiagent systems · Time series forecasting

B. Ponte Blanco (✉) · R. Pino Díez · I. Fernández Quesada · N. García Fernández · M. Monterrey Meana
Grupo de Ingeniería de Organización, Dpto. de Administración de Empresas, Escuela Politéc, de Ingeniería de Gijón, Universidad de Oviedo, Campus de Viesques S/N, 33204, Gijón, Spain
e-mail: borjaponte87@gmail.com

R. Pino Díez
e-mail: pino@uniovi.es

I. Fernández Quesada
e-mail: ifq@uniovi.es

N. García Fernández
e-mail: ngarciaf@uniovi.es

M. Monterrey Meana
e-mail: manuelmonterrey@uniovi.es

C. Hernández et al. (eds.), *Managing Complexity*, Lecture Notes in Management and Industrial Engineering, DOI 10.1007/978-3-319-04705-8_27,

1 Introduction

There have been several changes in the last two decades in the macro environment of the companies that have set up a new competitive perspective, where the production function is considered to have a strategic importance which it did not have. From this perspective, practices related to the SCM currently represent one of the main concerns of business.

This situation leads to talk about the Bullwhip Effect. Analyzing the supply chain, [4] noted that small changes in customer demand are amplified along the same, leading to larger variations in demand supported by the different levels, as they are further away from consumer. This is called the Bullwhip effect and, according to [3], it can be considered a major source of inefficiency in the supply chain, as it causes a significant increase in labor, stockout, storage, obsolescence, and transport costs.

In this context, this paper proposes the application of Artificial Intelligence techniques to the SCM in order to create a tool oriented to optimize the management. More specifically, Distributed Intelligence is applied to the problem through a multiagent system that allows finding an optimal solution to the problem.

This document is divided into five sections, besides this introduction. Section 2 is a review of the most relevant and recent literature on the subject. Section 3 describes the model, with the different agents, the structure which includes them and their relationships. Section 4 presents the results, after testing the model with time series obtained from the literature. Section 5 describes the conclusions.

2 Background: Solutions for Supply Chain Management

[6] demonstrated that the transfer of distorted information from one end of a supply chain to the other is the reason for the generation of the Bullwhip Effect. Thus, the authors identified four main causes: Errors in demand forecasting, Order batching, Price Fluctuation, and Rationing for fear of stockout.

2.1 Traditional Solutions

To reduce the Bullwhip Effect and optimize the supply chain, it is necessary to insist on the accuracy of the information transferred over the same. This fact determines that collaboration is the key to improving the SCM. However, the data transfer is not free and those who facilitate it require compensations. This has created a wide range of ways of collaboration. These are called the traditional solutions to the Bullwhip Effect, and they are practices that take place in some companies for two decades: Use of Information Technologies and Communication, Postponement, Efficient Consumer Response (ECR), Vendor Managed Inventory (VMI) and Collaborative Planning, Forecasting and Replenishment (CPFR).

2.2 *Advanced Solutions*

The SCM is a highly complex problem, conditioned by the interaction of multiple agents, each of which has to serve a large number of variables. In turn, it tends to develop in increasingly uncertain environments, marked by the difficulty of forecasting some parameters which affect it in a crucial way. For this reason, in recent years, they have looked for different ways to optimize the SCM using techniques based on Artificial Intelligence. They are called the advanced solutions to the Bullwhip Effect.

Mainly, multiagent systems have been used. They simulate the behavior of each one of the main members of a supply chain through an intelligent agent, which interact among them. [5] were pioneers in the proposal of the supply chain as a network of intelligent agents. Later, [9] created Metamorph, oriented to companies. With the turn of the decade, there was the development of these tools. Thus, [7] were based on the famous "Beer Game" [10] to devise the "Quebec Wood Supply Game". In [13] introduced fuzzy logic. Other relevant are recent works are the ones of [12] and [8].

The work presented in this paper continues this line, using the multiagent methodology not only to minimize the Bullwhip Effect, but also to minimize management costs, in order to analyze the relationship between the two alternatives.

3 Description of the Model

To prepare the base model, we have considered a traditional supply chain with linear structure, which consists of four main levels (Shop Retailer, Retailer, Wholesaler and Factory) plus Consumer.

We modeled the supply chain as a Global Multiagent System (MAGS), which consists of four main Multiagent Subsystems (MASS 1, 2, 3 and 4), each one of them associated with one of the levels of the supply chain. Each MASS interact with the next ones, and it is able to make intelligent decisions, aimed at optimizing the SCM around predefined objectives. Figure 1 shows the general outline of the proposed model. To implement it, we used NetLogo 5.0.1 [11].

Each MASS replicates the behavior of one of the main levels of the supply chain. In turn, it will consist of several interconnected intelligent agents, as shown in Fig. 1. The four MASS have identical structure.

Information Agent contain the information issued by the other agents, and it stores it in the database. It also responds to the information needs of the rest.

Communication Agent communicates the MASS with its environment. It works, thus, as a spokesman, mainly to collect demand from the lower level, and to transmit the order to the higher level.

Forecasting Agents are the core of the system. Each one of them will perform the demand based on a different method. 1–1 Agent estimates the demand in a period as the previous demand. MM Agent forecasts the demand as the average of the last

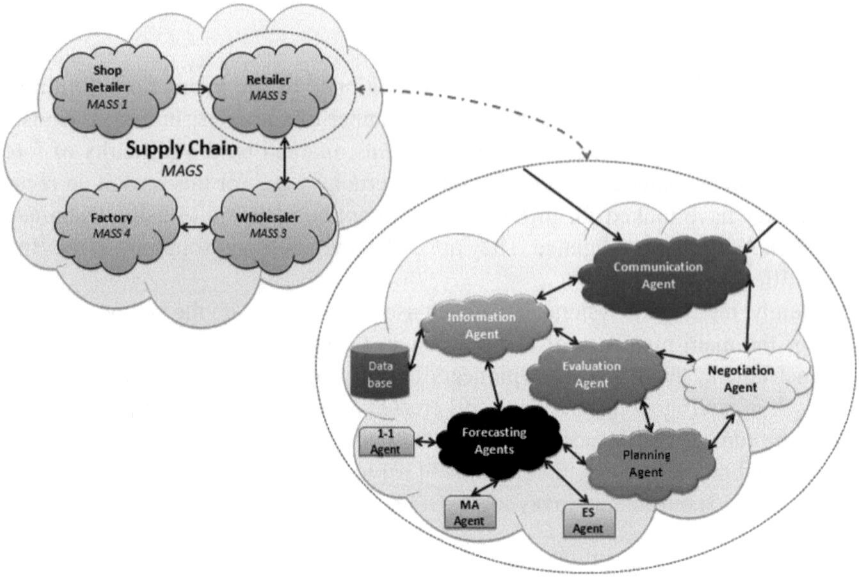

Fig. 1 Multiagent system model used

n demands (n from 2 to 15). Finally, AE agent is based on a weighted average with coefficient α (α from 0.1 to 0.9) between the demand and the forecast in the previous period. In the latter two cases, the system selects as the best forecasting these which generates lower Mean Square Error (MSE).

Planning Agent uses the three forecastings for processing three plannings, through the calculation of four parameters for each period to consider: initial inventory, final inventory, deliveries and orders to be made. Evaluation Agent compares the three plannings, based on specified criteria from the following options:

1. Minimize the Bullwhip Effect in the level, understood it as the ratio between the variance of demand transmitted and received by the level (Eq. 1).

$$BW_n = \frac{\sigma_{dt}^{2\ n}}{\sigma_{dr}^{2\ n}} \tag{1}$$

2. Minimize the total cost of inventory management, stated as the sum of stockout and storage costs (Eq. 2), because, with the assumptions we have made, all other costs would be independent of the forecasting method.

$$C_{im}^{\ n} = C_{so}^{\ n} + C_{st}^{\ n} \tag{2}$$

Negotiation Agent, finally, will be enabled or disabled by the user. When enabled, it activates the information sharing between Shop Retailer and Retailer, on one hand, and Wholesaler and Manufacturer, on the other.

Table 1 Time series data used for testing multi-agent model

Database	Name	Data	Content
[1]	AL03	106	Electricity consumption
[1]	AL04	108	Car sales
[1]	AL11	106	Gas consumption
[2]	BJ02	369	Price of IBM shares
[2]	BJ06	100	Wolfer sunspots
[2]	BJ08	144	Airline company passengers

Table 2 Results with the model oriented to the minimization of the Bullwhip Effect

Test	Series	BW1	BW2	BW3	BW4
A-1	AL03	65,90	26,29	1,54	1,22
A-2	AL04	48,70	12,41	1,32	1,11
A-3	AL11	13,74	7,42	6,00	2,88
A-4	BJ02	4,20	1,21	1,12	1,05
A-5	BJ06	15,41	7,54	4,18	3,35
A-6	BJ08	12,28	3,31	1,25	1,18

4 Numerical Application

To evaluate the performance of the developed tool, we have used six series obtained from databases typically used in forecasting models. Table 1 lists the main data of each series.

4.1 Model Oriented to Minimize the Bullwhip Effect

First, we detail some tests carried out on the model when Evaluation Agent is aimed at minimizing the Bullwhip Effect. Table 2 summarizes the results for each series: Bullwhip Effect generated if all the levels used 1–1 method (BW1) and a three period moving average (BW2) for demand forecasting; and Bullwhip Effect generated with the developed multiagent model without activating Negotiation Agent (BW3) and activating it (BW4).

The results demonstrate the efficiency of the model in reducing the Bullwhip Effect. In all cases, its solution (BW3) significantly improves the result if all the levels used the 1–1 method (BW1) or a moving average (BW2). Thus, the use of simple forecasting methods, coordinated by a multiagent system, allows a great improvement in the SCM, minimizing the impacts of Bullwhip Effect. Finally, by comparing the results when Negotiation Agent is active (BW4) and when it is not (BW3), we can conclude the effectiveness of cooperation in improving the SCM.

By way of example, Fig. 2 shows, with AL03 series, the different demands along the supply chain if all levels use the 1–1 method and Fig. 3 represents the solution offered by the developed model. It is easy to observe the high performance in mitigation the demands variability along the supply chain.

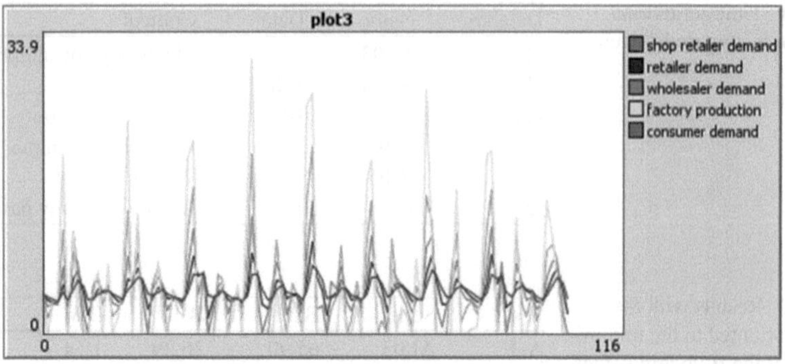

Fig. 2 Variations of orders for the test A-1 if all levels using the method 1–1

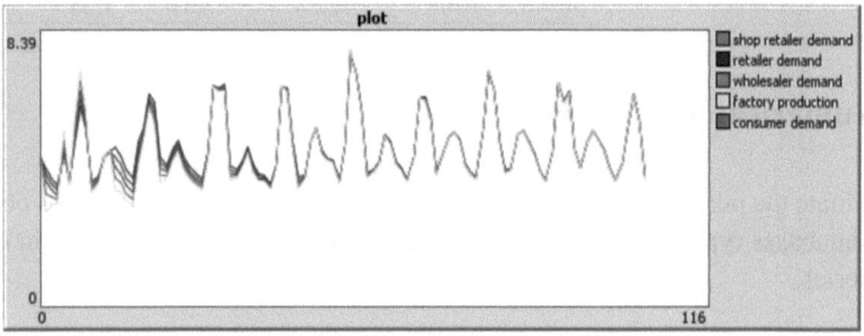

Fig. 3 Solution offered by the model to variations multiagent orders

Table 3 Results with the model oriented to the minimization of the management costs

Test	Series	BW3	BW5	BW6	BW7
B-1	AL03	1,54	4,20	4,20	4,20
B-2	AL04	1,32	2,85	8,32	48,70
B-3	AL11	6,00	6,47	6,47	13,74
B-4	BJ02	1,12	1,16	1,16	1,16
B-5	BJ06	4,18	4,22	7,07	15,41
B-6	BJ08	1,25	1,29	1,96	1,96

4.2 Model Oriented to Minimize Management Costs

Secondly, we have oriented Evaluation Agent to the minimization of management costs and we have tested it with the same series. Table 3 presents the results, collecting the solution provided by the model multiagent when it is oriented to minimize the Bullwhip Effect (BW3), and when it is oriented to minimize management costs if stockout costs and storage are equal (BW5), if the first are four times greater than the second (BW6) and if the first are ten times higher (BW7).

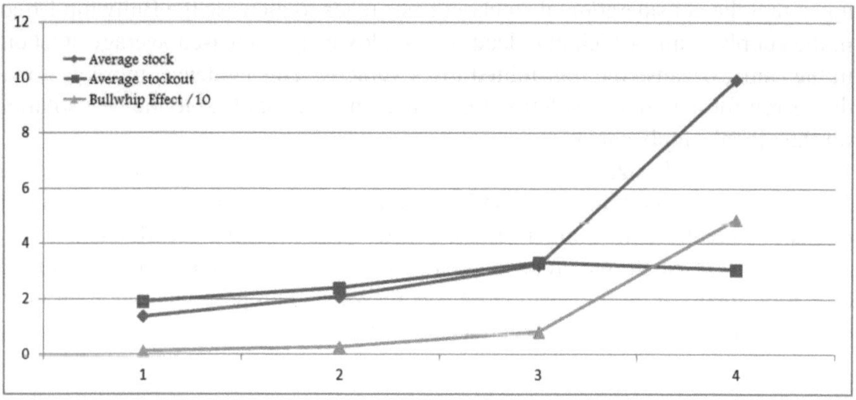

Fig. 4 Detailed results of test B-5

The results show that the stockout costs contribute to generate the Bullwhip Effect. These costs motivate decisions in the different members of the supply chain that, in view of the data, which tend to increase the variability of transmitted demands.

The results show that the stockout costs contribute to generate the Bullwhip Effect. These costs motivate decisions in the different members of the supply chain that, in view of the data, which tend to increase the variability of transmitted demands.

Analyzing the results in more detail, we can draw conclusions of great interest. The consideration of stockout costs at each level, paradoxically, tends to increase the average stockout generated in the supply chain. That is, if each level tends to reduce its own stockout, the stockout generated in the supply chain tends to increase. This occurs because, in this way, each level increases the distortion of the information transmitted to the supply chain, generating a high Bullwhip Effect, which hurts the other levels. This phenomenon is presented in Fig. 4, which summarizes the results of the BJ06 series.

5 Conclusions and Future Research Lines

Tests performed on time series demonstrate the importance of minimizing Bullwhip Effect in managing the supply chain, as a major source of inefficiency thereof. The use of simple forecasting methods without clear coordination leads to a large amplification of the variability of the demands transmitted along the supply chain. In these circumstances, the application of the developed multiagent model significantly reduces the Bullwhip Effect in the supply chain.

In another analysis, the model has been oriented to the minimization of management costs through different coefficients between stockout and storage costs. We noted that the consideration of stockout costs tends to increase the Bullwhip Effect in the supply chain, which may lead to, paradoxically, increased average stockout in the same, because the transmitted loses veracity. This evidence, moreover, that the search for the optimal solution for each agent does not lead to the best solution for the global supply chain.

Thus, we emphasize in cooperation as an effective alternative to optimize the SCM but it is also easy to understand the main barrier to implementation: each participant is interested in taking decisions that do not benefit the overall chain.

As future research, we propose to use multiagent methodology for even closer to reality, focusing on the NegotiationAgent. We also propose incorporating more complex forecasting methods with the aim of further improving the management.

References

1. Abraham B, Ledolter J (1983) Statistical methods for forecasting. Wiley, New York
2. Box GEP, Jenkins GM (1976) Time series analysis: forecasting and control. Holden Day, San Francisco
3. Disney SM, Farasyn I, Lambrecht M, Towill DR, Van de Velde W (2005) Taming the bullwhip effect whilst watching customer service in a single supply chain echelon. Eur J Oper Res 173(1):151–172
4. Forrester JW (1961) Industrial dynamics. MIT, Cambridge
5. Fox MS, Chionglo JF, Barbuceanu M (1993) The integrated supply chain management system. Internal Report, Dept. of Industrial Engineering, Univ. of Toronto. http://citeseerx.ist.psu.edu/viewdoc/download?doi=10.1.1.69.6293&rep=rep1&type=pdf
6. Lee HL, Padmanabhan V, Whang S (1997) The bullwhip effect in supply chains. Sloan Manage Rev 38(3):93–102
7. Moyaux T, Chaib-Draa B, D'Amours S (2004) An agent simulation model for the Quebec forest supply chain. Lect Notes Artif Intell 3191:226–241
8. Saberi S, Nookabadi AS, Hejazi S (2012) Applying agent-based system and negotiation mechanism in improvement of inventory management and customer order fulfilment in multi echelon supply chain. Arab J Sci Eng 37(3):851–861
9. Shen W, Xue D, Norrie DH (1998) An agent-based manufacturing enterprise infrastructure for distributed integrated intelligent manufacturing systems. In: Proceedings of PAAM, vol 98, pp 533–548
10. Sterman JD (1989) Modelling managerial behaviour: misperceptions of feedback in a dynamic decision making experiment. Manage Sci 35(3):321–339
11. Wilensky U (1999) NetLogo. Northwestern University, Evanston, IL: the center for connected learning and computer—based modeling. http://ccl.northwestern.edu/netlogo/
12. Wu SN, Gan WH, Wei FM (2011) Analysis on the bullwhip effect based on ABMS. Procedia Eng 15:4276–4281
13. Zarandi MH, Pourakbar M, Turksen IB (2008) A fuzzy agent-based model for reduction of bullwhip effect in supply chain systems. Expert Syst Appl 34(3):1680–1691

Part IV
Methods and Applications

Revisiting the SEC Classification Framework in Traditional and Electronic Commerce

Ángel F. Agudo-Peregrina, Julián Chaparro-Pélaez and Félix J. Pascual-Miguel

Abstract The recent growth of e-commerce sales has its cause not only on the increase of products sold, but also on the incorporation of new products which were previously not selling well on the online channel. An outstanding example of these products in Spain is clothing, which has rocketed from a marginal position in online sales to the fifth place of the top selling products over the Internet. This is an interesting case because clothing has conventionally been classified as an Experience product within the SEC (Search, Experience, Credence) classification framework, which proposes product segmentation based on the ability of consumers to identify the characteristics and attributes of the products before and after their purchase and use. This situation raises the question whether the SEC classification is still valid today and if there are changes in consumer perceptions about which segment the different products are categorized into. In order to answer these questions, a selection of 26 products was made by e-commerce experts; then, 204 undergraduate and graduate students were asked to classify those products within the SEC framework, and to declare their purchase intentions and actual shopping behavior for each product, both in the online and traditional channel. The findings from this study suggest that the SEC classification is still valid in electronic commerce but not in traditional retail stores. Moreover, the study detected actual changes in the customer perceptions of the nature of some products within the SEC classification.

Keywords E-commerce · Product · Segmentation · Search · Experience · Credence · Intention · Shopping behavior

Á. F. Agudo-Peregrina (✉) · J. Chaparro-Pélaez · F. J. Pascual-Miguel
Departamento de Ingeniería de Organización, Administración de Empresas y Estadística, Universidad Politécnica de Madrid Av, Complutense 30, 28040 Madrid, Spain
e-mail: af.agudo@upm.es

J. Chaparro-Pélaez
e-mail: julian.chaparro@upm.es

F. J. Pascual-Miguel
e-mail: felixjose.pascual@upm.es

C. Hernández et al. (eds.), *Managing Complexity*, Lecture Notes in Management and Industrial Engineering, DOI 10.1007/978-3-319-04705-8_28,

1 Introduction

Electronic commerce has experienced a strong growth in recent years in Spain. Online sales have raised by 84.7% between 2007 and 2011 [24]. This growth has been caused by an increase of total product units' sales, but also by a higher variety of products sold. A widely accepted classification of products is the SEC (Search, Experience, Credence) classification framework introduced by Nelson [22]. Within this classification, *Search* products are those whose attributes and characteristics can be identified by buyers before purchase; *Experience* products are those whose attributes and characteristics can not be identified until the buyer has bought the product; and *Credence*, those for which the buyer needs to experience longer use before forming a complete opinion of the product [7].

According to these definitions, the Internet channel has been considered particularly suitable for selling Search products, due to the lack of physical examination when buying from web stores [1]. On the other hand, other products such as clothing, have traditionally been viewed as Experience products and not considered especially adequate for e-commerce, since consumers place great value on the ability to touch and inspect the product [19].

However, according to data from the Spanish Telecommunications Market Commission, since 2010 clothing are among the top ten best selling products online, up to the fifth position in late 2011 [6]. In fact, brands like Zara or Mango, Spanish textile distribution giants, have focused their efforts on developing their online channels—with current sales between 2.5 and 5% of their annual total sales, and annual growth rates of nearly 100% [7].

From these sales data, it is clear that online shoppers show a different behavior than what could be expected according to the SEC framework, and this raises the question whether this framework is still appropriate for classification of online products. Moreover, given the changes in attitudes of shoppers since the original proposition of the SEC framework, its suitability to traditional commerce should also be assessed. On the other hand, should the framework still be proven valid, it would be of high interest to evaluate whether there have been changes in shoppers' perceptions of products attributes, so that some products may have shifted from one classification group to another.

To achieve this research objective, the paper is organized as follows: Sect. 2 makes a review of the literature on the SEC framework; research methodology, procedures and sample description are shown in Sect. 3, data analysis and results are presented in Sect. 4, and in Sect. 5 a discussion of the results and findings from the experiment are presented.

2 Theoretical Background

One of the most accepted classifications for product segmentation is the SEC classification in which, initially, products were classified as search and experience [22]; later on Darby [7] added the category of Credence products. Norton and Norton

[22] extended the framework by dividing experience products into durable and non-durable, and Klein [17] proposed a division of Experience products into Experience 1—similar to the original definition of Experience, and Experience 2—those in which the search cost is bigger than the experience cost. Although the SEC-product classification framework was developed for the traditional channel, it has also been used in electronic commerce e.g. [10, 11].

In order to study the influence of product type differences in shoppers' purchase behavior, purchase intention has been usually considered a proxy for actual purchase behavior, according to the Theory of Reasoned Action [9]. Some studies have confirmed the existence of such behavioral differences [4], and others have found that, in fact, search products are better suited for the online market than experience products [21] due to reduction of search costs in Internet, and that Search products show higher purchase intention than Experience products [5]. Moreover, Credence and Experience-2 products have the lowest online purchase intentions [11, 18]. For the traditional channel, there are differences in purchase intentions between different types of products too, although these differences are smaller [10]. It should be noted that consumer characteristics, such as age [3], education [12] or experience [17, 26] may also influence purchase intentions across product types.

However, some authors suggest that the SEC framework is not adequate for e-commerce [25]. Due to the reduction of the search costs [13] and the increase of new features that improve how information is displayed on Internet—3D, high quality videos—[20], there are less differences between Search and Experience products, and they are now based on the type of information needed by buyers for final purchase decision [14].

3 Research Methodology

The empirical study has been divided in two stages. First, a group of ten experts in e-commerce proposed a list of 105 representative goods and services from SEC literature, and selected 26 based on differences between them according to other segmentation classifications, in order to achieve higher product diversity.

Then, a survey was distributed among 370 Spanish graduate and postgraduate students, with a total of 204 valid responses. Sample selection criteria aimed to ensure sample homogeneity: all respondents were between 18 and 35 years, 99 % of them used Internet more than 5 h a week and only 1 % of them had never bought online. The survey was divided in two parts: in the first one, students had to classify the 26 products within the SEC framework; in the second part, respondents were asked about their purchase intention through virtual and traditional retail stores, as well as their real shopping behaviors in both channels. The 5-point Likert scale used in the survey has wide acceptance in purchase intention literature [10, 15].

4 Data Analysis and Results

First, we show the classifications of the selected products made by respondents in the Search, Experience and Credence categories, as well as their purchase intentions and shopping behavior in traditional and electronic commerce (Table 1). In 12 of the products—35 %, marked with an asterisk in Table 1—the opinion of the majority of respondents ("Sur." column) does not match the experts' proposed product classification ("Exp." column). In 8 products (30.8 %), respondents did not even reach a consensus, since none of the categories reached at least 50 % of the responses.

As it may be observed in Table 1, there are some changes in the categories in which products have been traditionally classified. An ANOVA was performed to test if SEC-products were different depending on consumer purchase intentions and actual shopping behaviors, using consumer purchase intention and actual shopping behavior as dependent variables and product class as the categorical independent variable (see Table 2). In contrast to other studies where the opinion of the majority was chosen as independent variable [10, 15], and since we were measuring differences in individual perceptions, the product classification stated by each respondent was considered the independent variable for this study.

With regard to differences between groups, from the ANOVA analysis there is a significant difference between groups in all cases—at $p<0.05$ for actual purchase in the traditional channel, and $p<0.01$ in all other cases. From highest to lowest, the means of purchase intentions and actual shopping behavior in e-commerce corresponded to Search, Experience and Credence products. However, in traditional commerce, the resulting order for purchase intention was the opposite—Credence, Experience and Search. Regarding actual shopping behavior in the traditional channel, the order from highest to lowest was Experience, Search and Credence.

After observation of the Levene statistic, homoscedasticity of data distribution could not be assured, and therefore a Games-Howell multiple comparisons procedure was chosen for the post-hoc analysis. The results indicate that in the Internet channel, for both purchase intention and actual shopping behavior, the three groups are significantly different from each other ($p<0.01$). However, in the case of purchase intentions in the traditional channel, there are significant differences between Search and Credence ($p<0.01$) and Search and Experience ($p<0.05$). Regarding actual shopping behavior, significant differences were found only between Experience and Credence products ($p<0.05$) in traditional stores.

From the results, the SEC classification would still be suitable for product segmentation in the online channel—all the groups were significantly different from each other, both for purchase intention and actual shopping behavior. However, the differences between products seem to have blurred in the case of the traditional channel, especially for actual shopping behavior, where only significant differences between the Experience and Credence products were found.

Table 1 Search/experience/credence classification, purchase intention and shopping behavior

Product	Exp.	Sur.	S (%)	E (%)	C (%)	Online int.	Trad. int.	Online shop	Trad. shop
MP3 player	S	S	*81.9*	13.2	4.9	3.89	3.76	2.2	3.65
E-book	S	S	*70.1*	18.1	11.8	4.59	2.31	3.8	1.32
Clothing*	E	S	*62.7*	29.9	7.4	3.09	4.32	1.87	4.07
Online music*	E	S	*61.3*	32.4	6.3	4.53	1.84	3.91	1.35
Insurance	S	S	*59.8*	19.6	20.6	3.82	3.28	3.13	2.02
Hosting	S	S	*57.8*	29.4	12.8	4.46	2.21	2.64	1.3
Flowers*	E	S	*56.9*	39.7	3.4	2.67	4.34	1.51	4.24
Car rental	S	S	*55.4*	33.8	10.8	4.04	3.32	2.79	2.05
Antivirus*	C	S	*54.4*	20.6	25	4.28	2.74	3.48	1.91
Game	S	S	*50*	40.7	9.3	4.52	2.2	3.51	1.72
Handicraft	S	S	*49.5*	41.2	9.3	2.2	4.31	1.23	3.75
Photoshop*	S	–	*47.1*	*47.1*	5.8	4.07	2.81	2.69	1.43
Photo retouch	E	E	23.5	*65.7*	10.8	3.59	3.03	2.09	1.59
Hotel lodging	E	E	21.6	*63.7*	14.7	4.44	3.15	3.93	2.07
Housekeeping	E	E	22.5	*54.9*	22.6	2.24	4.22	1.25	2.77
Fruit	E	E	40.2	*52.9*	6.9	1.77	4.51	1.28	4.56
Telemedicine*	C	E	15.2	*48.5*	36.3	3.37	2.74	1.27	1.3
Social network	E	E	30.9	*47.1*	22	4.59	1.87	4.13	1.24
Astrology*	C	E	13.2	*46.6*	40.2	2.25	2.8	1.23	1.26
Web programming*	S	E	34.8	*46.1*	19.1	4.16	2.47	2.28	1.32
Crowd funding*	S	E	28.4	*41.2*	30.39	4.08	2.3	1.59	1.23
Lotion	C	C	6.9	10.3	*82.8*	2.46	3.91	1.21	2.4
Custom medicine	C	C	29.4	17.6	*53*	2.24	4.04	1.3	2
Tutoring*	E	C	7.4	42.2	*50.4*	2.54	4.05	1.37	3.59
Physiotherapy*	E	C	6.9	43.1	*50*	2.48	4	1.42	2.84
Gym*	E	C	34.3	30.9	*34.8*	3.02	3.95	1.3	3.96

5 Discussion of Results

The results of the study show that the SEC classification framework is still valid for analysis purposes in e-commerce, but changes in shoppers' perceptions of SEC products classification in this channel should be taken into account in the future. Furthermore, Search products are still more likely to succeed in the online market. On the other hand, changes in perceptions about products types may have impacted perceived differences in the traditional channel, where product type seems to be no longer significant in terms of purchasing behavior.

We have found that some products which were previously seen as Experience products, e.g., clothing, online music and flowers, are now perceived by consumers as Search products. In all three cases online sales have increased, as well as the quantity and quality of online information available about these products. This increase of information availability and the introduction of new methods to display the information allow customers to access all the information they need before they make their online purchase decisions, giving them a high degree of product experience without even having any physical contact with the product [14].

Table 2 ANOVA and Games-Howell post-hoc analysis results

	Class	Mean	SD	N	F-*Sig*	Product (A)	Product (B)	Mean-diff	Std.Err
Online purchase intention	S	3.74	1.29	2085	123.126[b]	S	E	0.34[b]	0.04
	E	3.40	1.36	1992		S	C	0.75[b]	0.05
	C	2.99	1.42	1227		E	C	0.41[b]	0.05
Traditional purchase intention	S	3.15	1.38	2085	10.844[b]	S	E	−0.12[a]	0.04
	E	3.27	1.35	1992		S	C	−0.23[b]	0.05
	C	3.38	1.31	1227		E	C	−0.11	0.05
Online shopping behavior	S	2.55	1.69	2085	94.884[b]	S	E	0.32[b]	0.05
	E	2.23	1.59	1992		S	C	0.78[b]	0.05
	C	1.77	1.38	1227		E	C	0.46[b]	0.05
Traditional shopping behavior	S	2.34	1.61	2085	3.449[a]	S	E	−0.04	0.05
	E	2.38	1.66	1992		S	C	0.11	0.06
	C	2.23	1.69	1227		E	C	0.15[a]	0.06

[a] $p < 0.05$.
[b] $p < 0.01$.

Experience and Credence products have lower values of actual purchase through Internet [10]. This means that some barriers still exist for certain products which prevent buyers to perform a thorough analysis of a product's characteristics based on the information received through the online channel. Interestingly enough, this is not the case of two of the products on the list—social networks and hotel lodging—which, although being identified as Experience products, have good online sales performance; but nonetheless, social networking is a digital service that can only be accessed via the Internet, and in the case of hotel lodging, this service has generally achieved a high degree of standardization and has been constantly improving the quality and quantity of online information available, which has in turn contributed to increase their online sales [2]. These results suggest that there may be a shift in customers' perceptions of these services, which might become Search products in the near future. However, even though the adequacy of SEC classification framework has been previously confirmed [12], it may require further analysis in order not to group together products with very different characteristics. Therefore, it is also recommended to take into account other classification variables when using the SEC framework, such as standardization [25] or digitizability [15].

When comparing the traditional and the online channels, the findings from this study contend previous results, where the SEC classification was considered adequate for analysis of the traditional channel [10]; the results from this study suggest that the growth of e-commerce may be contributing to blur the differences of individuals' perceptions of the products in traditional shopping.

Thus, while there is still room for improvement and sales increases in the online channel through a process of "searchabilization" of Experience products, there seems to be no option to further differentiate products in the traditional channel. This "searchabilization" process might require increases in quality and quantity of information offering, implementation of multichannel strategies, enhancement of transaction processing and product delivery, as well as focusing on the intensification of other elements that contribute to products being perceived as Search products [16].

References

1. Alba J, Lynch J, Weitz B, Janiszewski C (1997) Interactive home shopping: consumer, retailer, and manufacturer incentives to participate in electronic marketplaces. J Market 61:38–53
2. Beldona S, Morrison AM, O'Leary J (2005) Online shopping motivations and pleasure travel products: a correspondence analysis. Tourism Manage 26(4):561–570
3. Bhatnagar A, Ghose S (2004) Segmenting consumers based on the benefits and risks of Internet shopping. J Bus Res 57(12):1352–1360
4. Brown M, Pope N, Voges K (2003) Buying or browsing? An exploration of shopping orientations and online purchase intention. Eur J Mark 37(11/12):1666–1684
5. Chiang K, Dholakia RR (2003) Factors driving consumer intention to shop online: an empirical investigation. J Consumer Psycho 13(1-2):177–183
6. Comisión del Mercado de las Telecomunicaciones (2011) Informe sobre el Comercio Electrónico en España a través de entidades de medios de pago. Cuarto trimestre 2011. Madrid, Spain: CMT
7. Darby MR, Karni E (1973) Free competition and the optimal amount of fraud. J Law Econ 16(1):67–88
8. Elizalde I, Ruiz R (2012) Inditex duplicará las ventas de su negocio online el año que viene. *Expansión*. p 3. http://expansionpro.orbyt.es/2012/10/14/tmt/1350234607.html. Accessed 1 March 2013
9. Fishbein M, Ajzen I (1975) Belief, attitude, intention, and behavior: an introduction to theory and research. Wesley, MA
10. Girard T, Dion P (2010) Validating the search, experience, and credence product classification framework. J Bus Res 63(9-10):1079–1087
11. Girard T, Korgaonkar P, Silverblatt R (2003) Relationship of type of product, shopping orientations, and demographics with preference for shopping on the internet. J Bus Psychol 18(1):101–120
12. Girard T, Silverblatt R (2002) Influence of product class on preference for shopping on the internet. J Comput-Mediat Commun 8(Oct):1–18
13. Hoffman DL, Novak TP (1996) Marketing in hypermedia computer-mediated environments: conceptual foundations marketing in hypermedia computer-mediated environments. J Market 60(1):50–68
14. Huang P, Lurie NH, Mitra S (2009) Searching for experience on the web: an empirical examination of consumer behavior for search and experience goods. J Market 73(March):55–69
15. Kiang MY, Ye Q, Hao Y, Chen M, Li Y (2011) A service-oriented analysis of online product classification methods. Decision Support Syst 52(1):28–39
16. Kim EY, Kim Y-K (2004) Predicting online purchase intentions for clothing products. Eur J Mark 38(7):883–897
17. Klein LR (1998) Evaluating the potential of interactive media through a new lens: search versus experience goods. J Bus Res 41:195–203
18. Korgaonkar P, Silverblatt R, Girard T (2006) Online retailing, product classifications, and consumer preferences. Internet Res 16(3):267–288
19. Levin AM, Levin IP, Heath CE (2003) Product category dependent consumer preferences for online and offline shopping features and their influence on multi-channel retail alliances. J Electron Commerce Res 4(3):85–93
20. Li H, Daugherty T, Biocca F (2002) Impact of 3-D advertising on product knowledge, brand attitude, and purchase intention: the mediating role of presence. J Advert 31(3):43–57
21. Liang T, Huang J (1998) An empirical study on consumer acceptance of products in electronic markets: a transaction cost model. Decision Support Syst 24:29–43
22. Nelson P (1970) Information and consumer behavior. J Polit Econ 78(2):311–329
23. Norton SW, Norton S Jr (1988) An economic perspective on the information content of magazine advertisements. Int J Advert 7:138–148

24. Urueña A, Valdecasa E, Ballestero MP, Antón P, Castro R, Cadenas S (2012) Estudio sobre Comercio Electrónico B2C 2011 (edición 2012). Observatorio Nacional de las Telecomunicaciones y de la Sociedad de la Información (ONTSI). Madrid, Spain
25. Peterson RA, Balasubramanian S, Bronnenberg BJ (1997) Exploring the implications of the internet for consumer marketing. J Acad Mark Sci 25(4):329–346
26. Shim S, Eastlick MA, Lotz SL, Warrington P (2001) An online prepurchase intentions model: the role of intention to search. J Retail 77:397–416

A Roadmap to Establish Collaboration Among SMEs Belonging to Non-Hierarchical Networks

Beatriz Andres and Raul Poler

Abstract The importance of collaboration has increased in supply networks; and thus, the number of so called non-hierarchical manufacturing networks (NHN). When establishing collaborative processes appear a number of barriers that companies must face. This paper proposes a roadmap, *NHNmap*, to support researchers on the migration process towards the long-term vision to establish collaborative relationships in NHN. This NHNmap phases are briefly described and depicted in a chart in order to identify the roadmap time plan.

Keywords Roadmap · Collaboration · Non-hierarchical networks · SMEs

1 Introduction

SMEs modus operandi is evolving towards complex value chains [8]. A variety of collaborative networks have emerged during the last years, collaborative non-hierarchical networks (NHN) represent a network topology evolved from the centralised view, in which a node makes the decision optimising the objectives of the entire network, (hierarchical networks—HN) towards the decentralised decision making perspective, in which individual entities make its own decisions to optimise their own objectives, requiring greater exchanges of information and commitment of all companies [1]. Participation in NHN implies sharing responsibilities, organisations actively participation in the decision-making, jointly treatment of problems and equally consideration of partners. Many factors may cause failures when establishing collaborative relations within networks [25]. Andrés and Poler [1] highlight the problems that SMEs have to face when they decide to participate in collaborative NHN, admitting that the participation in collaborative NHN brings new challenges

B. Andres (✉)
Research Centre on Production management and Engineering (CIGIP),
Universitat Politècnica de València (UPV), Plaza Ferràndiz y Carbonell, 2, 03801 Alcoy, Spain
e-mail: beaanna@cigip.upv.es

R. Poler
Centro de Investigación de Gestión e Ingeniería de Producción (CIGIP),
Universitat Politècnica de València, Camino de Vera s/n, 46022 Valencia, Spain
e-mail: rpoler@cigip.upv.es

C. Hernández et al. (eds.), *Managing Complexity*, Lecture Notes in Management and Industrial Engineering, DOI 10.1007/978-3-319-04705-8_29,

Table 1 Classification criteria in the literature review

	Criteria	Description
Context	Network level	Roadmap developed at network level
	SME level	Roadmap developed at enterprise level
Roadmap type	Inter-enterprise	Identification, evaluation and promotion of collabotive projects. Network goals establishment
	Technological	Gaps identification in technological arena
	Product	Product line description. Introduction of new products
	Product-technology	Technology needs and to develop a product
	Project	Strategies and project planning. Projects integration
Roadmap evaluation		Not included
Features to support the establishment of collaborative relationship in networked SMEs	*	Includes in a weak way
	**	Includes in a moderate way
	***	Includes features in a strong way

that they have to overcome. In the light of this, a roadmap to deal with these SMEs challenges is developed. A literature review is carried out to identify roadmaps addressing collaborative network issues. The literature provides roadmaps facing collaborative problems focused on the centralised decision making perspective, and specific issues, such as technology, interoperability, virtual organisations, collaboration, alignment, etc. The work proposed in the literature is classified to obtain a better understanding in terms of the (i) roadmap context, (ii) roadmap type and (iii) evaluation criteria (Table 1). Considering the classification criteria, Table 2 chronologically shows the roadmaps provided in the literature; each work is briefly described and analysed. From the literature review, it is concluded that the proposed roadmaps are not developed from the NHN context; this highlights the need to develop a roadmap to specifically deal with the establishment of collaborative relationships within NHN.

2 Baseline and NHNmap Vision

With the emergence of NHN arises the need to establish the foundations for a new investigation regarding the needs that appear in the establishment of new collaborative relationships of SMEs within the same NHN. In order to guide researchers in the transition path towards enterprises' collaboration, the proposed roadmap allows to identify SMEs needs and challenges when establish collaborative processes. Two important elements are considered in the roadmap development (i) the SMEs current state characterised by establishing non-collaborative relationships and (ii) the SME desired vision of establishing collaborative relationships and processes

Table 2 Literature review

Author	Work description	Context	Roadmap type	Evaluation
[23]	Provides the Semantic Grid roadmap focusing on e-Science and GRID infrastructure needs	Network level	Technological	*
[9]	Roadmap to address interoperability of enterprise applications and software	SME level	Technological	**
[15]	Assembly-net roadmap: discusses research challenges in advanced collaborative manufacturing systems	Network level	Project	**
[24]	Roadmap focused on virtual communities and cooperative environments (COCONet)	Network level	Project	**
[5]	Roadmap for advanced virtual organizations (VOmap). Identification of research challenges needed to fulfil the vision of the European initiative on collaborative dynamic virtual organizations. A modelling approach considering different perspectives of environmental characteristics, life cycle modelling and collaborative networks	Network level	Project inter-enterprise	**
[20]	Brings together different partners of the collaborative network, giving the opportunity to exchange information and point of views and provides a vehicle for holistic consideration of problems, opportunities and new ideas	Network level	Project technological	*
[19]	Reflects the inherent complexities in the innovation and development of new products and offers a view of the future management tools. Includes technology roadmaps and the impact on decision making	Network level	Product technology	*
[22]	Roadmap for planning support to integrate technology markets, products and technologies	SME level	Product technological	-
[11]	Roadmap to address the vision of extended enterprises from the centralized decision making context to integrate all processes to connect two supply chains	Network level	Inter—enterprise	**
[21]	Integration of systems and interoperability to operate in virtual enterprises	Network level	Technological	*
[27]	Identification of critical elements associated with the transition from a not collaborative state to collaborative context	Network level	Inter—enterprise	***

Table 2 (continued)

Author	Work description	Context	Roadmap type	Evaluation
[13]	Analysis for design concepts to address problems appearing in collaborative business processes	Network level	Technological	**
[2]	Identification of SMEs capacities availability, skills and abilities of the system to establish collaborative relationships within the network	SME level	Project inter-enterprise	***
[6]	Strategic Research Plan covering social, organizational and technological perspectives in collaborative networks	Network level	Project	**
[12]	Roadmap for integrating the supply chain developed at SMEs level and network level in order to progress together towards integrated business solutions, and propose a cultural changes	Network Level SME level	Project	***
[26]	Comprehensive vision and roadmaps for the future construction of information technologies in the industry	SME level	Technological	-
[3]	Evaluation and implementation to facilitate innovation in collaborative environments	Network level	Technological	*
[11]	Relate the SMEs needs and emerging models of collaborative networks discipline (GloNet project)	SME level	Inter—enterprise	***
[16] [4]	Multidisciplinary approach involving technological perspective to face collaboration	Network level	Technological	*
[17]	Collaborative network approach to develop a business concept of integrated services in VO	Network level	Inter—enterprise	**
[18]	Roadmap for defining exchange of resources and capabilities and to develop behaviours	Network level	Inter—enterprise	*
[25]	Roadmap to identify conflicts, select collaborative members that best fit and assign roles and rights within collaborative supply network	Network level SME level	Inter—enterprise	***
[14]	Roadmap based on the establishment of web services system to provide interoperability. Change management systems that provide information to support decision making	Network level	Technological	*
[10]	Roadmap to acquire teamwork knowledge and necessary knowledge regarding the tools for collaboration modelling	SME level	Inter—enterprise	**

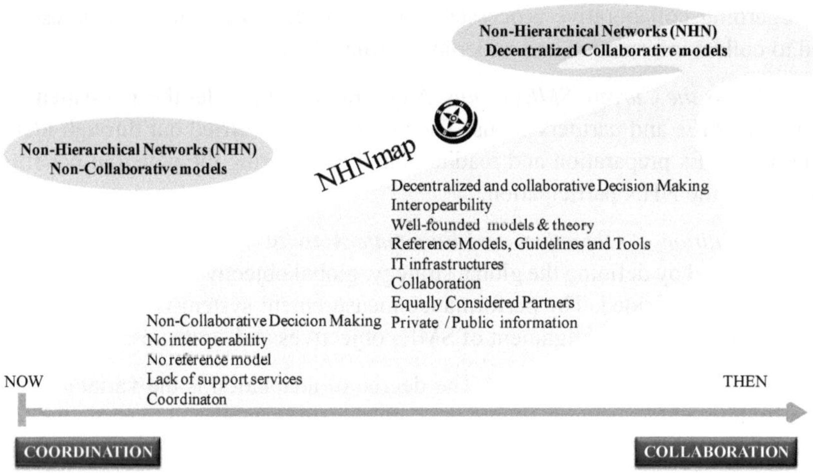

Fig. 1 NHNmap

in non-hierarchical manufacturing networks. Figure 1 represents the vision of the provided roadmap, *NHNmap*, which develops a guide for SMEs on the transition path towards the future state of collaborative relationships establishment in NHN. Participation in NHN enables SMEs to create new opportunities to be more competitive and innovative. The NHNmap main contribution is devoted to guide and support on the SMEs adaptation path towards collaborative NHN.

3 NHNmap Phases

The NHNmap implementation helps SMEs, who decided to participate in a NHN, determine what to do to start decentralised collaborative relationships. NHNmap allows the alignment of partners to propose common objectives and guiding, by sharing information, the adaptation process towards collaborative NHN, continuously evaluating and reviewing the results. The NHNmap is structured through four focus areas (i) collaboration establishment, (ii) performance evaluation, (iii) models, guidelines and tools proposals to address potential problems in collaborative processes and (iv) Information Technology and Systems to support decentralised decision making. NHNmap is multi-layered (i) at network level, where collaborative NHN requirements are defined, and (ii) at SMEs level, where SMEs resources and capabilities are identified. The Roadmap consists on ten stages, briefly explained below:

Phase 1 Defining the Scope of Collaboration SME assignment of, at least, one process in the NHN so that each of the NHN partners identifies responsibilities and roles regarding collaborative processes. An agenda for future meetings is established to collect information and undertake further stages.

Phase 2 Assess the Current SMEs Status A questionnaire enables the assessment of the current SMEs and partners status. The assessment is carried out through identifying the SMEs preparation and readiness and considering the potential benefits derived from the NHN participation.

Phase 3 Definition of Objectives and Economic Activity The collaborative agreement is achieved by defining the global strategy, global objective and the sub-objectives in which is divided. The performance measurement system is also designed. The aim is to achieve the alignment of SMEs objectives and strategies.

Phase 4 SMEs Needs Identification The degree of adaptation is the variable used to identify how SMEs features fit into the requirements established to participate in collaborative NHN. This phase allows networked partners to identify the problems they have to overcome to fulfil the collaborative NHN requirements.

Phase 5 Preparing for Collaboration Potential partners are trained. Technological change is promoted to meet the exchange of information (technology, platforms, information management, visibility, standards, and synchronisation) and the processes connection (reengineering, collaborative transactions, platforms).

Phase 6 Solutions Identification Considering the analysis results of phase 2 and the identified needs in phase 4, this phase provides solutions to overcome a set of defined collaborative problems. Solutions are proposed in form of models, methodologies and tools.

Phase 7 Solution Implementation Enables collaborative partners to identify the extent to which the solution proposed in the literature can be particularly applied to each SME, taking into account the resources and capabilities they have.

Phase 8 Collaboration Implementation The start up process to participate in a collaborative NHN is initiated. Initial results and feedback from NHN partners are set out so that partners can discuss appropriate adjustments to be further performed in the NHN. Future information and technology systems and software modifications are identified to improve the collaborative processes.

Phase 9 Results Assessment Relations within partners, business processes and technologies used to achieve collaboration within the NHN are evaluated. The performance evaluation loop is periodically repeated. Reporting the assessment results of collaboration is an opportunity to identify achieved progress.

Phase 10 Improvements Identification and Implementation Improvements are identified in order to establish future collaborative relationships. Following steps regarding the collaborative NHN are agreed and the involved organisations plan to deploy additional collaborative initiatives to be developed. A report with results is distributed to NHN partners to contribute to recommendations for expansion.

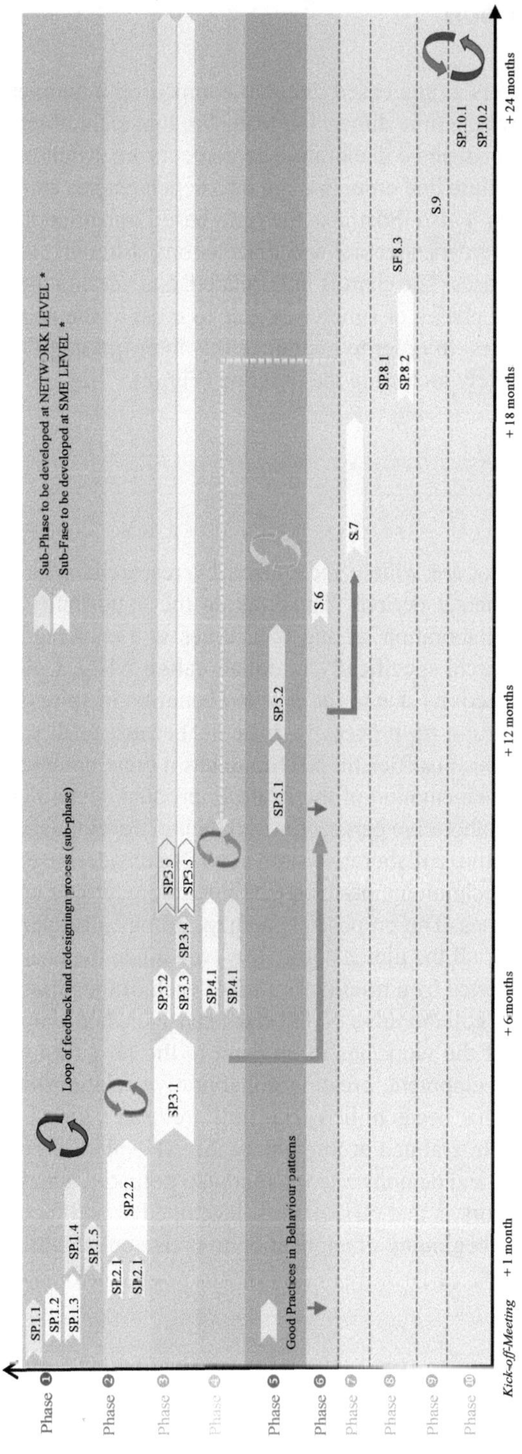

Fig. 2 Roadmap background and vision

3.1 NHNmap Chart

The generic roadmap is a time based diagram comprising a number of layers. The NHNmap chart shows the time dimension to ensure that technology, products, processes and organisational and collaborative agreements are synchronised. NHNmap chart exemplifies the time line order and the interdependencies among the different phases and subphases, it also illustrates the time-based transition steps within each phase showing the interdependencies and links among different subphases, as well as among different phases. The chart is inter related, considering the transition steps needed from different phases or subphases, due to most of the subphases need the results of previous ones. In order to guarantee this inter-linking, the subphases defined are to be accurately located in the timeline (Fig. 2).

4 Conclusions

Collaborative networks are widely recognised by researchers for SMEs survival in turbulent environmental periods. Thus, taking the opportunity to structure the evolution towards collaboration among companies is a challenge for the associated community research, specifically in collaborative NHN. Collaborative NHN involve a significant activity due to the reached benefits. In spite of the benefits of collaboration, the starting up process is difficult, because SMEs capabilities and resources are sometimes insufficient. NHNmap has a great potential to support the development and implementation of the strategy, product, technology and partners relations from the collaborative perspective. NHNmap develops the migration path to reach the desired future collaborative state in NHN considering different perspectives: collaboration, technology, performance measurement and solutions to overcome SMEs weaknesses. The proposed roadmap, graphically represented ensures the synchronisation of all the dimensions. However, since the chart is synthesised, the NHNmap is supported by a brief definition of each of the phases.

Research work, in collaborative NHN, has been launched only few years ago. Therefore the most of the work has to be done in the next future, and we are in the early stage of development, creating collaborative relationships. NHNmap is a dynamic construct that needs to be periodically revised taking into account new trends. Future research is aimed at implement the NHNmap in order to complete an external validation and demonstrate the roadmap potential benefits. The original contribution of this work is that NHNmap is developed at both network and SMEs levels and serves as a beginning of application to create collaborative NHN.

References

1. Andrés B, Poler R (2012) Relevant problems in collaborative processes of non-hierarchical manufacturing networks. In: Prado JC, García J, Comesaña JA, Fernández AJ (eds) 6th international conference on industrial engineering and industrial management. Vigo, pp 90–97
2. Barton R, Thomas A (2009) Implementation of intelligent systems, enabling integration of SMEs to high-value supply chain networks. Eng Appl Artif Intell 22:929–938
3. Budweg S, Schaffers H, Rulanda R, Kristensenc K, Prinz W (2011) Enhancing collaboration in communities of professionals using a living lab approach. Prod Plan Control 22:594–609
4. Caetano M, Amaral DC (2011) Roadmapping for technology push and partnership: a contribution for open innovation environments. Technovation 31:320–335
5. Camarinha-Matos L, Afsarmanesh H (2003) A RoadMap for strategic research on virtual organizations. In: Camarinha-Matos LM, Afsarmanesh H (eds) Processes and foundations for virtual organizations. Kluwer, Boston, pp 3–16
6. Camarinha-Matos LM, Afsarmanesh H (2010) Active ageing roadmap—a collaborative networks contribution to demographic sustainability. In: Collaborative networks for a sustainable world, IFIP Adv in Inf and Com Tech. Springer, Berlin, pp 46–59
7. Camarinha-Matos LM, Afsarmanesh H (2011) Active aging with collaborative networks. IEEE Technology and Society Magazine, pp 12–25
8. Camarinha-Matos LM, Afsarmanesh H, Galeano N et al (2008) Collaborative networked organizations—concepts and practice in manufacturing enterprises. Comput Ind Eng 57:46–60
9. Chen D, Doumeingts G (2003) European initiatives to develop interoperability of enterprise applications-basic concepts, framework and roadmap. Annu Rev Control 27:153–162
10. Gallardo J, Bravo C, Redondo MA (2012) A model-driven development method for collaborative modeling tools. J Netw Comput Appl 35:1086–1105
11. Hunt I, Wall B, Jadgev H (2005) Applying the concepts of extended products and extended enterprises to support the activities of dynamic supply networks in the agri-food industry. J Food Eng 70:393–402
12. Hvolby HH, Trienekens JH (2010) Challenges in business systems integration. Comput Ind 61:808–812
13. Ku KC, Kaob HP, Gurumurthy ChK (2007) Virtual inter-firm collaborative framework-An IC foundry merger/acquisition project. Technovation 27:388–401
14. Lin HW, Nagalingam SV, Kuik SS et al (2012) Design of a global decision support system for a manufacturing SME: towards participating in collaborative manufacturing. Int J Prod Econ 136:1–12
15. Onori M, Oliveira J, Lastra J et al (2003) European precision assembly -roadmap. Assembly-Net. ISBN 91-7283-637-7
16. Osório L, Camarinha-Matos LM, Afsarmanesh H (2011a) Cooperation enabled systems for collaborative networks. In: Camarinha-Matos LM, Afsarmanesh H, Koelmel B (eds) 12th IFIP WG 5.5 working conferences on VE, PRO-VE 2011. Springer, Sao Paulo, 263–270
17. Osório L, Afsarmanesh H, Camarinha-Matos LM (2011b). A service integration platform for collaborative networks. Stud Inform Control 20:19–30
18. Perks H, Moxey S (2011) Market-facing innovation networks: how lead firms partition tasks, share resources and develop capabilities. Ind Mark Manage 40:1224–1237
19. Petrick IJ, Echols AE (2004) Technology roadmapping in review: a tool for making sustainable new product development decisions. Technol Forecast Soc Change 71:81–100
20. Phaal R, Farrukh CJP, Probert DR (2004) Technology roadmapping—a planning framework for evolution and revolution. Technol Forecast Soc Change 71:5–26
21. Rezgui Y, Zarli A (2006) Paving the way to the vision of digital construction: a strategic roadmap. J Constr Eng Manage 132 (7)767–776
22. Rinne M (2004) Technology roadmaps: Infrastructure for innovation. Technol Forecast Soc Change 71:67–80

23. Roure D, Jenning N, Shadbolt N (2001) Research agenda for the semantic grid: a future e-science infrastructure. EPSRC/DTI core e-science programme
24. Schaffers H, Ribak A, Tschammer V (2003) COCONET: a roadmap for context-aware cooperation environments. Processes and foundations for virtual organizations. Kluwer, Boston
25. Shadi M, Afsarmanesh H (2011) Addressing behavior in collaborative networks. In: Camarinha-Matos LM, Afsarmanesh H, Koelmel B (eds) 12th IFIP WG 5.5 working conferences on VE, PRO-VE 2011. Springer, Brazil, pp 263–270
26. Shen W, Hao Q, Mak H et al (2010) Systems integration and collaboration in architecture, engineering, construction, and facilities management: a review. Adv Eng Inform 24:196–207
27. Spekman RE, Carraway R (2006) Making the transition to collaborative buyer-seller relationships: an emerging Framework. Ind Mark Manage 35:10–19

Incorporating the Work Pace Concept into the *MMSP-W*

Joaquín Bautista, Rocío Alfaro, Cristina Batalla and Alberto Cano

Abstract This work proposes an extension for the *MMSP-W* (Mixed-Model Sequencing Problem with Work overload Minimization) with variable processing times by the incorporation of the work pace or work speed concept. A computational experience, linked to a case study of Nissan Powertrain plant in Barcelona, is carried out to compare the performance of the reference model with the new proposed model.

Keywords Mixed model assembly line · Sequencing · Work factor · Work overload · Linear programming

1 Introduction

The product variety, that is demanded today, forces manufacturers to have mixed-product assembly lines. These lines are composed by a set of workstations (K) arranged serially. Each workstation is characterized by its workload, or set of task assigned to it, and the available standard (normal activity or normal pace) time or cycle time (c) to process these tasks.

Because of that type of assembly lines treat several product types and each one may require different processing times and resources it is necessary to determine the manufacturing order of the product units to avoid (1) high stock levels in the production system, and (2) the inefficiency of the line.

J. Bautista (✉)
Dpto. Organización de Empresas, ETSEIB, UPC Catalunya, 08028 Barcelona, Spain
e-mail: joaquin.bautista@upc.edu

R. Alfaro · C. Batalla · A. Cano
Cátedra PROTHIUS, Dpto. de Organización de Empresas, Universitat Politècnica de Catalunya, Avda. Diagonal, 647, 7th floor, 08028, Barcelona, Spain
e-mail: rocio.alfaro@upc.edu

C. Batalla
e-mail: cristina.batalla@upc.edu

A. Cano
e-mail: alberto.cano-perez@upc.edu

C. Hernández et al. (eds.), *Managing Complexity*, Lecture Notes in Management and Industrial Engineering, DOI 10.1007/978-3-319-04705-8_30,

Thus, the sequencing problems [3] can be focused on (A) minimizing the work overload or lost work, and (B) minimizing the stock levels.

In this work, we study a problem focused in objective (A), the *MMSP-W* (Mixed-Model Sequencing Problem with Work overload Minimization). That problem consist of sequencing T products, grouped into a set of I product types, of which d_i are of type$I$ ($I=1,\ldots,|I|$). Each product unit requires a processing time at normal work pace, $p_{i,k,}$ for each homogeneous processor at each workstation ($k=1,\ldots,|K|$). If the cycle time is not sufficient to complete the required work by a product unit at the workstation k, the product unit can be held at the station for a time, called the time window, equal to l_k, which is longer than the cycle time ($l_k>c$), reducing the available time of the workstation for processing the next product unit.

When it is not possible to complete all of the work required, it is said that an overload is generated. The objective of the *MMSP-W* is to maximize the total work completed [7], or minimizing the total work overload generated [6], being equivalent both objectives (see Theorem 1 in [1]).

Usually, deterministic and fixed operation processing times, $p^o_{i,k}$, are considered in sequencing problems. These times are initially determined, through the *MTM* system (Methods and Time Measurement) in *JIT* (Just In Time) and *DS* (Douki Seisan) manufacturing environments, and correspond to the time required by an average skilled operator, working at normal pace or normal speed, to perform a specified task using a prescribed method, allowing time for personal needs, fatigue, and delay. Therefore, these operation processing times correspond to the predetermined standard times [4].

The present study aims the extension of reference models for the *MMSP-W* considering variable processing times of the operations regarding the operator activation or work pace. In particular, we focus on minimizing the work overload, increasing the work pace at certain intervals of the workday, taking into account the usual conditions of the automotive companies and the relationship between the performance of the operator and his level of activation or stress level.

2 *MMSP-W* with Variable Work Pace

Large automotive companies negotiate with labor unios a set of work conditions once established the processing times of operations according to the *MTM* system with a work speed 100, *MTM_100*. Among these conditions we found the selection of work pace considered as normal for the company.

Typically, the processing times accepted are the corresponding to *MTM_110*, which are obtained as follows:

$$p\,(MTM\_110) = p\,(MTM\_100)\,.\frac{100}{110} = p^o.\frac{100}{110} \qquad \forall i \in I \quad \forall k \in K \quad (1)$$

Obviously, from the reference times determined with the normal work pace accorded by the company, *MTM_110*, we can establish a correspondence between the processing times regarding any pair of work paces. To do this, we define the work pace factor, α, of an operation as the division between the processing times measured at normal work pace (p) and those at the work pace that is carried out ($\hat{p}$); that is: $\alpha = p/\hat{p}$.

Therefore, if we consider that the processing times obtained through *MTM_100* correspond with the standard work pace and the times *MTM_110* with the normal pace, we can determine $\alpha^0 = 0.\widehat{90}$ (standard) and $\alpha^N=1$ (normal), respectively.

Similarly, the companies set an activity or optimum work speed, which typically is 20% higher than the normal work pace ($\alpha^*=1.2$, *MTM_132*). This activity is considered the maximum work pace that a worker can develop without loss of working life, working 8 *h* a day.

Thus, considering that the work pace can vary throughout the workday and therefore operations can expand or contract over time, we can establish a new model for the *MMSP-W* from the reference model [2]. The parameters and variables of the model are:

Parameters	
K	Set of workstations ($k=1,\ldots,\lvert K\rvert$).
b_k	Number of homogeneous processors at workstation k.
I	Set of product types ($i=1,\ldots,\lvert I\rvert$)
d_i	Programmed demand of product type i.
$p_{i,k}$	Processing time required by a unit of type i at workstation k for each homogeneous processor (at normal pace or activity).
T	Total demand; obviously, $\sum_{i=1}^{\lvert I\rvert} d_i = T$.
t	Position index in the sequence ($t=1,\ldots,T$).
c	Cycle time, the standard time assigned to workstations to process any product unit.
l_k	Time window; maximum time that each processor at workstation k is allowed to work on any product unit, where $l_k-c>0$ is the maximum time that the work in progress (*WIP*) is held at workstation k.
$\alpha_{k,t}$	Work pace factor associated with the tth operation of the product sequence ($t=1,\ldots,T$) at workstation k ($k=1,\ldots,\lvert K\rvert$).
α_t	Work pace factor associated with the period t ($t=1,\ldots,T+\lvert K\rvert-1$) of the extended workday that includes T manufacturing cycles (total demand) more $\lvert K\rvert-1$ additional cycles needed to complete the required work by all the units at all workstations.
	Note that if we associate the same factor to each moment of workday in all workstations, we have: $\alpha_{k,t}=\alpha_{t+k-1}$ $(\forall k, \forall t)$.
	If $\alpha_{k,t}=1$ (normal work speed) we will use the MTM_110 scale; for $\alpha_{k,t}=1.1$ and $\alpha_{k,t}=1.2$ we will use MTM_121 and MTM_132, respectively.
Variables	
$x_{i,t}$	Binary variable equal to 1 if a product unit i ($i=1,\ldots,\lvert I\rvert$) is assigned to the position t ($t=1,\ldots,T$) of the sequence, and to 0 otherwise.
$\hat{s}_{k,t}$	Positive difference between the start instant and the minimum start instant of the tth operation at workstation k.
$w_{k,t}$	Overload generated for the tth unit of the product sequence at station k for each homogeneous processor (at normal activity); measured in time.
$v_{k,t}$	Processing time applied to the tth unit of the product sequence at station k for each homogeneous processor at normal work pace or activity.
$\hat{v}_{k,t}$	Processing time reduced by a work pace factor $\alpha_{k,t}$. We impose here: $v_{k,t} = \alpha_{t+k-1} \cdot \hat{v}_{k,t}$

Model $M\_4 \cup 3\_\alpha 1$:

$$Min\,W = \sum_{k=1}^{|K|} \left(b_k \sum_{t=1}^{T} w_{k,t} \right) \Leftrightarrow Max\,V = \sum_{k=1}^{|K|} \left(b_k \sum_{t=1}^{T} v_{k,t} \right) \quad (2)$$

Subject to:

$$\sum_{t=1}^{|T|} x_{i,t} = d_i \quad (\forall i = 1, \ldots, |I|) \quad (3)$$

$$\sum_{i=1}^{|I|} x_{i,t} = 1 \quad (\forall t = 1, \ldots, |T|) \quad (4)$$

$$v_{k,t} + w_{k,t} = \sum_{i=1}^{I} p_{i,k} x_{i,t} \quad (\forall k = 1, \ldots, |K|); \quad (\forall t = 1, \ldots, T) \quad (5)$$

$$\alpha_{t+k-1} \cdot \hat{v}_{k,t} - v_{k,t} = 0 \quad (\forall k = 1, \ldots, |K|); \quad (\forall t = 1, \ldots, T) \quad (6)$$

$$\hat{s}_{k,t} \geq \hat{s}_{k,t-1} + \hat{v}_{k,t-1} - c \quad (\forall k = 1, \ldots, |K|); \quad (\forall t = 2, \ldots, T) \quad (7)$$

$$\hat{s}_{k,t} \geq \hat{s}_{k-1,t} + \hat{v}_{k-1,t} - c \quad (\forall k = 2, \ldots, |K|); \quad (\forall t = 1, \ldots, T) \quad (8)$$

$$\hat{s}_{k,t} + \hat{v}_{k,t} \leq l_k \quad (\forall k = 1, \ldots, |K|); \quad (\forall t = 1, \cdots, T) \quad (9)$$

$$\hat{s}_{k,t}, v_{k,t}, \hat{v}_{k,t}, w_{k,t} \geq 0 \quad (\forall k = 1, \ldots, |K|); \quad (\forall t = 1, \ldots, T) \quad (10)$$

$$x_{i,t} \in \{0.1\} \quad (\forall i = 1, \ldots, |I|); \quad (\forall t = 1, \ldots, T) \quad (11)$$

$$\hat{s}_{1,1} = 0 \quad (12)$$

In the model, the equivalent objective functions (2) are represented by the total work performed (V) and the total work overload (W). Constraint (3) requires that the programmed demand be satisfied. Constraint (4) indicates that only one product unit can be assigned to each position of the sequence. Constraint (5) establishes the relation between the processing times applied to each unit at each workstation and the overload generated in each unit at each workstation. Constraint (6) reduces the processing times applied regarding the work pace factor. Constraints (7)—(9) constitute the set of relative start instants of the operations at each station

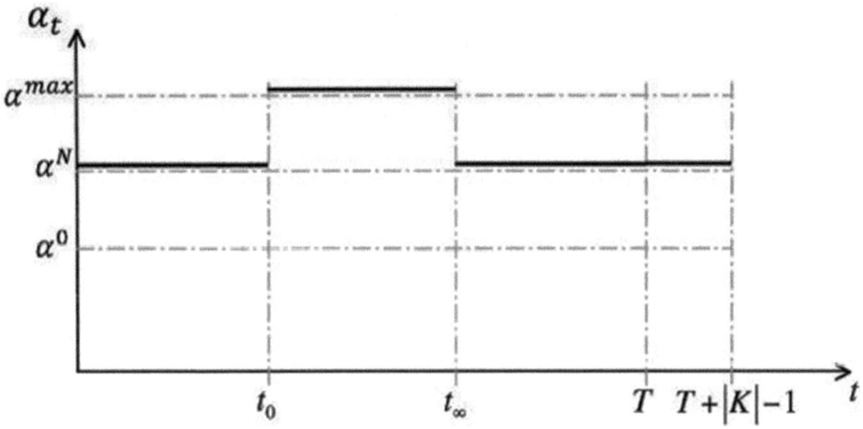

Fig. 1 Function of work pace factor

and the processing times reduced to the products for each processor. Constraint (10) indicates the non-negativity of the variables. Finally, constraint (11) requires the assigned variables to be binary, and the equality (12) fixes the start instant of the operations.

2.1 *Function of the Work Pace Factor throughout the Workday*

The relationship between an operator's performance and his or her level of "activation" or "arousal", reflected in his level of stress, can be considered curvilinear [5]. The "Yerkes–Dodson law" argues that it is an inverted-U.

From this idea, in this work, we associate the operator's efficiency with the work pace by a direct correlation between the work pace factor and the stress. Thus, considering the Yerkes-Dodson's optimum stress curve, we determine a function of the work pace factor throughout time (see Fig. 1). In this way, on one hand, the first and last product-units sequenced will be processed with less activation of stress and, therefore, with a work speed similar to normal work pace. While, on the other hand, the time periods in which the operator reaches the routine, stress is increased by increasing the work pace factor until reach the fatigue that is a characteristic of the end of the workday.

In our case, we define a step function of work pace factor, where α^0 is the activity factor corresponding with the *MTM*_100, α^N with *MTM*_110 (work pace established as normal by companies) and α^{max} with *MTM*_121. Thus we set the values of the work pace factor over time. Specifically:

$$\alpha_t = \begin{cases} \alpha^N = 1.0 & if\left(1 \le t \le t_0\right) V \left(t_\infty + 1 \le t \le T + |K| - 1\right) \\ \alpha^{max} = 1.1, & if\left(t_0 + 1 \le t \le t_\infty\right) \end{cases} \tag{13}$$

Table 1 NISSAN-9ENG instances and demand plans

Demand plan	4×4			Vans		Trucks			
	p_1	p_2	p_3	p_4	p_5	p_6	p_7	p_8	p_9
1	30	30	30	30	30	30	30	30	30
2	30	30	30	45	45	23	23	23	23
3	10	10	10	60	60	30	30	30	30
6	50	50	50	30	30	15	15	15	15
9	70	70	70	15	15	8	8	7	7
12	24	23	23	45	45	28	28	27	27
18	60	60	60	30	30	8	8	7	7

3 Computational Experience

To evaluate the influence of the incorporation of the work pace factor into the *MMSP-W* on the total work overload generated, we compare the obtained results by the new model $M\_4 \cup 3\_\alpha 1$ with the obtained results by the reference model $M\_4 \cup 3$.

To do this, we use a case study that corresponds to an assembly line from Nissan's plant in Barcelona. That line has 21 workstations with an effective cycle time of $c = 175\ s$, a time window of $l_k = 195 s \forall k$, and an identical number of homogeneous processors $b_k = 1 \forall k$. The line assembles nine types of engines $(p_1, \ldots, p_9)$ grouped into three classes: 4×4 $(p_1, \ldots, p_3)$; vans (p_4, p_5); trucks $(p_6, \ldots, p_9)$. Each engine class has different processing times, therefore we use several instances corresponding to different demand plans associated with a single workday of 13.127 h with 2 shifts. Each one of these instances has a total demand of 270 engines with different production mixes. Specifically, for this manuscript we have selected 7 instances that correspond, each one, to a representative situation of the demand (see Table 1).

In addition, considering the function of work pace defined in section 2 and the conditions of Nissan, in the computational experience we fixed the values of work pace factor as follows (see figure 2):

To implement the two models, the Gurobi v4.6.1 solver was used on a Apple Macintosh iMac computer with an Intel Core i72.93 GHz processor and 8 GB of RAM using MAC OS X 10.6.7. The solutions from this solver were obtained by allowing a maximum CPU time of 7,200 s for each model and for each of the 7 demand plans selected from the NISSAN-9ENG set. Table 2 shows the obtained results by each model.

From Table 3, we can see how the incorporation of work pace factor decreases the obtained value of over all work overload, regarding the obtained value by the model $M\_4 \cup 3$, at all instances tested. In particular, we see that the fact of passing from a factor of 1 to a factor of 1.1, in the second third of the work shift, reduces the work overload a 96.88 %, 82.39 %, 69.44 % and 70.80 % in instances #2, # 6,

Table 2 Work overload (W) of the 7 NISSAN-9ENG instances selected given by Gurobi for models $M\_4 \cup 3$ and $M\_4 \cup 3\_\alpha 1$ with a execution time of 7,200 s. Optimal solutions are marked with *

# instance	#1	#2	#3	#6	#9	#12	#18
V_0 [a]	807,420	807,370	807,260	807,505	807,615	807,360	807,535
$W_{M\_4 \cup 3}$	228	384	425	477	782	326	678
$W_{M\_4 \cup 3\_\alpha 1}$	0*	12	0*	84	239	0*	198
$\Delta W(\%)$	100	96.88	100	82.39	69.44	100	70.80

[a] The total work performed can be calculated as $V = V_0 - W$

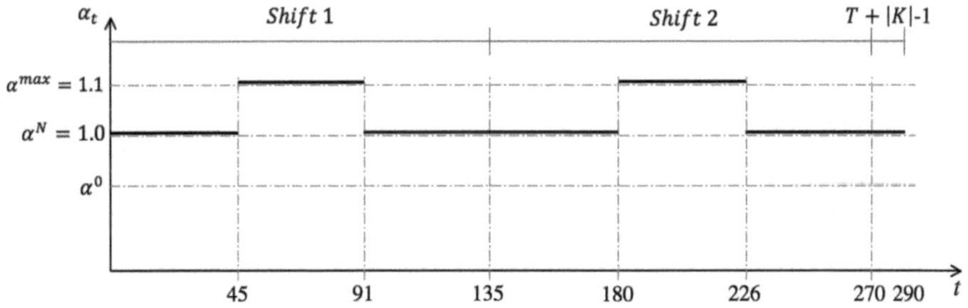

Fig. 2 Function of Nissan's work pace factor

#9and#18, respectively, and a 100 % in instances #1, #3 and #12, reaching, in these three cases, the optimal solution.

4 Conclusions

In this paper we have presented a new model, the $M\_4 \cup 3\_\alpha 1$, from the reference model, $M\_4 \cup 3$, proposed by [2]. This new model for the *MMSP-W* incorporates variable processing times of operations according to the work pace factor or activity. Specifically, it has set a staggered function of this factor throughout the workday, based on Yerkes-Dodson's optimal stress function. This function sets the normal work pace, fixed by the company $\alpha^N = 1$ at the beginning and end of the work shift, and increases this value to $\alpha^{max} = 1.1$ in intermediate moments of the shift.

Defined the model and work pace function, we have performed a computational experience linked to the assembly line of Nissan Powertrain plant in Barcelona. We have selected 7 instances that are representative of different demand plans that can be found, and the results obtained by the new model have been compared with those obtained with the reference model.

After computational experience, we have observed that an increase in the work pace of 10 %, for a third of the work shift reduces the total work overload generated, on average, 88.5 %, a maximum of 100 % and a minimum of 69.44 %. In addition, by the incorporation of work speed concept into the *MMSP-W*, we have reached the optimal solution in three instances, the #1, #3 and #12, in which the overload is 0.

Moreover, if we consider that the loss of an engine supposes a cost of the 4000 €/*year*, cycle time is 175 *s* and the work schedule contains 225 workdays, we can obtain savings by a maximum of 2.79, a minimum of 1.17 and an average of 2.03*M* €/*year*.

Acknowledgements This work is supported by the Spanish Ministerio de Educación y Ciencia under Project DPI2010-16759 (PROTHIUS-III) including EDRF fundings.

References

1. Bautista J, Cano A (2011) Solving mixed model sequencing problem in assembly lines with serial workstations with work overload minimisation and interruption rules. Eur J Oper Res 210(3):495–513
2. Bautista J, Cano A, Alfaro R (2012) Models for MMSP-W considering workstation dependencies: a case study of Nissan's Barcelona plant. Eur J Oper Res 223(3):669–679
3. Boysen N, Fliedner M, Scholl A (2009) Sequencing mixed-model assembly lines: survey, classification and model critique. Eur J Oper Res 192(2):349–373
4. Lawrence S Aft (2001) Measurement of work. In: Maynard's industrial engineering handbook, fifth edition. McGraw-Hill, New York, pp 5.3–5.22
5. Muse LA, Harris SG, Feild HS (2003) Has the inverted-U theory of stress and job performance had a fair test? Hum Perform 16(4):349–364
6. Scholl A, Klein R, Domschke W (1998) Pattern based vocabulary building for effectively sequencing mixed-model assembly lines. J Heuristics 4(4):359–381
7. Yano CA, Rachamadugu R (1991) Sequencing to minimize work overload in assembly lines with product options. Manage Sci 37(5):572–586

Applying the LEGOstics concept in formal education at Technical University of Cartagena

Peter Bajor, María Victoria de la Fuente Aragón and Lorenzo Ros-McDonnell

Abstract The LEGOstics training programs are under continuous development in Industrial Management Laboratory at UPCT (Spain), working together with Szabó-Szoba R&D Laboratory at Széchenyi University (Györ, Hungary). The main purpose of these innovative simulation projects is to construct special real-life environments for modeling logistics systems and phenomena (the nature of product and information flow in a supply chain, the meaning of delivery in time shipments, material handling and order picking processes of a warehouse or a factory, flexible manufacturing, work-in-process inventory management, lean thinking, etc). During the learning-by-doing LEGOstics trainings participants can get practical knowledge and develop many innovative skills to be able to construct, design and re-engineer sustainable and efficient logistics processes, and feel the responsibility of making decisions. In our paper we present how we implemented LEGOstics concept into the formal education focusing on the experiences we got during 2011–2013 at Technical University of Cartagena (Spain).

Keywords Logistic processes simulation · Learning-by-doing · LEGOstics

1 LEGOstics Modelling of Logistics Processes

The collective noun LEGOstics is coming from the LEGO and LOGISTICS words: LEGO products are very suitable for modeling such semi-virtual (in most cases non-semi, but real) logistics environments.

The bricks and parts are very popular, and considered as high quality innovative products to help in development of constructive skills.

It is important to highlight that we are using these products in totally new applications. What is really innovative in LEGOstics Laboratory: "How to use?", not "Which type of products to use?", so in education environments we can use also metal construction boards or any other wood-bricks—the reason why LEGO

M. V. de la Fuente Aragón (✉) · P. Bajor · L. Ros-McDonnell
Grupo GIO. ETSII, Universidad Politécnica de Cartagena,
C/Dr. Fleming, 30202 Cartagena, Spain
e-mail: marivi.fuente@upct.es

C. Hernández et al. (eds.), *Managing Complexity*, Lecture Notes in Management and Industrial Engineering, DOI 10.1007/978-3-319-04705-8_31,

Fig. 1 The GrEta standardized production management training

products were chosen was: the students are very familiar with, and these modelled environments are also "mobile" applications [1].

The goal is to develop different environments on the same platform: analyzing and developing the processes according to the technology and real nature of logistics systems (warehousing, material handling, production and transportation) [2, 3].

The present work shows the two projects developed at present course in UPCT: the GRETA and the TRUDI project.

2 The GrEta Project

GrEta is a self-developed, non-official LEGO product, based on our intention to construct a relatively simple model, what is possible to build with many and totally flexible ways. The result is a nice car: GrEta (Fig. 1).

At the GrEta BoardGame there are four participants around the plotting board, each of them are assembly workers, responsible to fulfil a given assembly process and manage the material flow. During the game, the students are working together on the same model, assembling a few parts of the model, and usually there is a "master" place for final assembling.

GrEta car has 8 separate functional parts: chassis, wheels, engine, engine hood, seat, computer unit, cabin, lamps. Inside the parts there are several ways again to construct, so constructing GrEta for four people has a lot of possible strategies and production system structures, according to the decisions of the team based on the different personal attributes [2, 3].

They can get experience in production teamwork (allocate procedures according to different features of a given workstation—speed, accuracy, quality checking, etc.), and also in process analysis (after a round they discuss their observations, make some changes, do it again, evaluate the consequences, etc.)

The main questions are:

1. What is the real meaning of efficiency at a given case?
2. How to improve capacity utilization?
3. What to measure and how to measure?
4. How to fit different processes following each other on the best way?

Using GrEta with many types of groups (students, logistics experts, assembling workers, etc.), the participants don't compete with each other, because their job is to improve productivity and provide high-level quality checking at the same time. Our experience is during the round by round optimization results in fast time reduction (initial average is 15 min, then gradually lowers to 8 and finally 3–5 min). The discussions generally considering:

- initial strategy—better sharing of the assembling tasks
- flow control—better managing of material flow (prior actions, match-mismatch of concurrent tasks)
- improving the individual assembling speed, parallel assembling

During spontaneous discussions a lot of theoretical knowledge is coming into practice:

- Controlling viewpoint—not just financial, but time-based indicators
- Sensing viewpoint—automatic detection of bottlenecks and inactive times
- Performance measurement with comparing the current strategy to the previous solutions

3 The TRUDI Project

The Trudi LEGOstics plotting board initially was constructed to help in the selection of new employees at a hungarian manufacturer company. With the Trudi Project the research group is developing a special mass-production system, focused on several functions: bottleneck problem analysis approach, work-in-process inventory management viewpoint, lean philosophy implementation problems, etc. [5].

The plotting board is separated for 4 parts as work-stations, each sector has an operator. The 4 types of roles as system operators are: Picking, Parts assembling, Final assembling, Disassembling and Selection. The system architecture is modular and convertible. The materials easily flow across the workstations. The shelving system is infinitely variable and can be placed anywhere on the table. Between the work places we use boxes for material handling (Fig. 2).

3.1 Measurements and Scientific Aspects of TRUDI LEGOstics Simulation

Measurement of logistics performance is one of the critical factors for being able to develop a well-balanced supply chain [3, 4]. It is possible to measure the takt-time

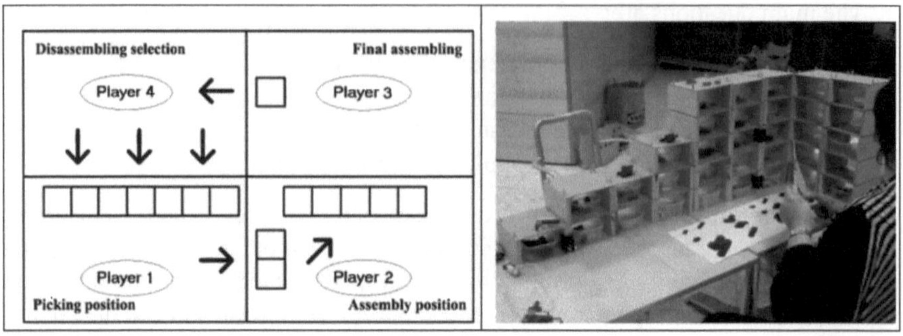

Fig. 2 The TRUDI plotting board in operation

Table 1 Future research areas in laboratory of LEGOstics

Inventory management	Production system re-engineering
Bullwhip-effect	MRP
WIP inventory problems	JIT
Order policy	KANBAN
Outsourcing: Make or Buy decisions	Lean thinking (different colours of product—parts can stay in the same box!)

of each operator, and build-up a systematic and coherent performance measurement system. We realized that it's not enough to synchronize the different processes based on their time requests. It's a possible direction of development and new results can come if we apply bar-codes on the boxes—so we can automatically monitor the products and performance of the operators during the process.

As concluded during the trainings, LEGOstics simulations have many promising insights for scientific research—as we can highlight in Table 1:

4 Preparation for Implementation—Create a LEGOstics Classroom

Our goal was to implement the LEGOstics method into the formal education in the frame of an international collaboration between the Széchenyi István University (Györ, Hungary) and the Technical University of Cartagena—UPCT (Spain).

The UPCT provided the classroom and the necessary equipments. For the first step the interactive environment was created, where we prepared the plotting boards and elaborated the course of the trainings together with the teachers of UPCT and a couple of UPCT students. When everything was prepared (trainings, trainers, and infrastructure) we integrate the LEGOstics trainings into the schedule supplementing the theoretical teaching with some practical aspects.

5 Scenario of a Lesson

The lessons start with a short introduction about the learning-by-doing method and the LEGOstics, presenting the plotting boards, working processes, what the trainings are representing. Later, the participants start playing the GrEta or Trudi simulations.

During the trainings, students discuss their experiences round-by-round, they describe the problems appeared. They have to make decisions and search for possible solutions: how to organize the working processes, arrange the layouts, assign the time and allocate the tasks.

On each lesson, the trainers or some voluntary observers filled the report; all data of the processes (time, ideas, and problems) was registered.

5.1 Scenario of the Trudi Training

The Trudi plotting board itself is rather flexible than the GrEta Board. In this simulation it is difficult to recognize the task of the operators for external observers and also for the participants.

According to individual skills the bottleneck can appear at different work stations, they can get the experience about the mechanism of push and pull systems this way (Fig. 3).

There is no need for central control, the process automates itself. The operators have to communicate and collaborate, they can use visual signs to help each other and handle the system.

The steps of the Trudi training are the following:

- Presenting the work stations tasks (selection, order picking, assembling, disassembling)
- Playing one or two sample round, being familiar with the process
- Playing, measuring, evaluating, optimization

5.2 Further Direction of Development in Trudi Model

Next stage of development in the Trudi plotting board is automatic measurement, or measurement with informatics support. On the trainings we realised that registering the details of the processes may get too complicated. Using some electronic devices can be an effective solution: the electronic devices with the proper programs can collect and analyze the data from this LEGOstics environment.

The bests are barcodes or QR codes. The QR code can store more data and easier to read it. So we chose to use tablets, because it has the features we need: a camera, being mobile, with rechargeable batteries, appropriate screen, input possibilities, and a proper operation system, software for reading QR codes and collect the data.

Fig. 3 The visual guides, playing and measurement of TRUDI

We will put the codes to the boxes of Trudi, and use the tablet to read them, during the game. The software will save the time of reading, and the code of the item. It will be saved to the memory, in a.csv file, we can export for further use. The operator's task consists only two easy steps: first read the codes, than save. Take much less time than the old type paper based administration.

6 Experiences

The LEGOstics method was implemented in the formal education from first grade of Bachelor and final grade of Master level. In that period in Cartagena (Spain) there were more than 10 trainings, with the participation of about 100 students. We experienced that the students were really motivated to learn the logistics phenomena in practice; they were opened for collaboration and communication.

During the GrEta part of the lesson, it was interesting to realize cultural, habital aspects of work organization. In our previous Hungarian experiences we concluded that the so called "general manager" usually owns the initial parts of assembling. Spanish participants usually preferred the finishing procedure and quality checking in this position. We intend to organize trainings in international groups.

When the students made the Trudi part of the class, they discussed more. It was a great experience, than the operators concluded the importance of real-time

Fig. 4 Measurement and paper-based administration

communication during the process. The biggest problems were, that in the plotting board the boxes has fixed place, but they wanted to send the boxes also with the items. The visual guides gave them an inevitable help during the game (Fig. 4).

The observers measured the time, but the administration was not sufficient. First we just gave them a paper to register the times, but it was not in the manageable form for analysis. We made a documentation form and their task got easier, just fulfilling it—than discussion of results got faster.

6.1 Students Opinions

The students enjoyed the trainings, they were interested about every aspects of it, and they made a huge effort to reach better result in the games. They communicated and worked together, so in this way they were able to optimize the processes. Generally the students said they had a good time, during and after lessons. They told us it was very interesting to utilize in practice what they learned in the courses about different logistics processes, and most of the times it's better to see and experience theoretical things in practice. They also told that in this way, they can understand better the importance of communication and process optimization.

7 Conclusions

The implementation of the LEGOstics method into the formal education was successful. Many students took part in the lessons, despite it was not compulsory. The students enjoyed the trainings; they had fun and study a lot at the same time. They could use most of their theoretical knowledge and use them during the games. They understood why it is so important to work as a team, why we put the emphasis on the communication. They were able to realize the problems and trough the discussions they tried to find the solutions. The students measured the times, and they could decide what worth to measure, what measurement points and data can be good for the further analysis, focusing on the performance. The LEGOstics in the formal education is a necessary practical part of the classes.

References

1. Bajor P, Bódis T (2011) Qualiy thinking in LEGOstics laboratory innovations. In: Conference proceedings, pp 351–356. 7th research/expert conference with international participation, Neum, Bosnia & Herzegovina, 1–4 June 2011
2. Haapasalo H, Hyvönen J (2001) Simulating business and operations management—a learning environment for the electronics industry. Int J Prod Econ 73:261–272
3. Holweg M, Bicheno J (2002) Supply chain simulation—a tool for education, enhancement and endeavor. Int J Prod Econ 78:163–175
4. Lewis MA, Maylor HR (2007) Game playing and operations management education. Int J Prod Econ 105:134–149
5. Szander N, Makrai Z, Bajor P (2012) Development of LEGOstics environments in Szabó-Szoba laboratory. In: Conference proceedings: 2nd international students symposium in logistics and international business, Bremen, Germany 28th November–2nd December 2011

Neural Network Application for Risk Factors Estimation in Manufacturing Accidents

Jesús Antonio Carrillo-Castrillo, José Guadix Martín, Rafael Grosso de la Vega and Luis Onieva

Abstract In occupational safety, when a neural network is trained, it is possible to predict the outcome given a combination of risk factors. Risk assessment is probably the most important issue in occupational safety. Risk assessment facilitates the design and prioritization of effective prevention measures. Neural network were applied for predicting the severity of accidents, which is important to assess risks. Data sets were obtained from the official accident notifications in the manufacturing sector of Andalusia in 2011. The results confirm that neural networks are useful in risk factor estimation. Association analysis was used to identify the most important risk factors within the predicting variables. Diagnostic array analyses show that for preventive purposes it is better to use a reduced data set with a case-control approach in order to improve the specificity and the sensitivity.

Keywords Neural networks · Occupational accidents · Risk factors · Sensitivity analysis · Manufacturing sector

1 Introduction

Occupational safety is a complex issue. The analysis of accidents deals with multiple causes and circumstances. In most cases, the accident occurrence implies multiple failures in the preventive and protective barriers [9]. Models of accident causation seek to explain the contribution of each risk factor.

Most of research has been focused on determining causal relationships expressed as injury rates, relative risk, odds ratios or quantitative risk estimation. However,

J. A. Carrillo-Castrillo (✉) · J. Guadix Martín · R. Grosso de la Vega · L. Onieva
Dpto. de Organización Industrial y Gestión de Empresas II. ETSI, Universidad de Sevilla, Camino de los Descubrimientos, s/n. Isla de la Cartuja, 41092 Seville, Spain
e-mail: jcarrillo@io.us.es

J. Guadix Martín
e-mail: guadix@us.es

R. Grosso de la Vega
e-mail: rgrosso@us.es

L. Onieva
e-mail: onieva@us.es

C. Hernández et al. (eds.), *Managing Complexity*, Lecture Notes in Management and Industrial Engineering, DOI 10.1007/978-3-319-04705-8_32,

because of the socio-technical nature of accident causation, most of those models have little predictive ability.

From a public policy perspective, the objective is to identify risk factors at the macro level. For Public Administrations the main concern is identifying intervention areas.

At the same time, safety practitioners need an initial estimation for the risk assessment. Expected severity and likelihood of accidents based on the registered accidents in small and medium companies is not feasible due to the low rate of occurrence, and quantitative risk analysis is an expensive and time consuming alternative [2010].

Data mining is an interdisciplinary subfield of computer science focusing on the discovery of patterns in large data sets. Data sets of accidents are an excellent candidate to test if data mining techniques can provide meaningful information.

Many data mining techniques are available, such as logistic regression, classification trees, and Chi-squared automatic interaction detection, among others [3]. One of the most powerful tools for mining of complex data is using neural networks. In fact, when comparing the predictive ability of the different data mining techniques for risk analysis, neural networks show better results [11].

Previous studies have used neural networks for different purposes such as for classification of industrial jobs in terms of the risk of low back disorders [12], for the prediction of occupational injury risk [3] or for prediction of the severity in traffic accidents [6].

In occupational safety, when a neural network is trained, the outcome of a combination of risk factors can be predicted. Risk assessment may be considered the most important issue in occupational safety, used to facilitate the design and prioritization of effective prevention measures.

Previous studies in the manufacturing sector of Andalusia (Spain) have shown that a number of individual worker characteristics can affect likelihood of occupational safety injuries [5] and in case of injury these characteristics also affect the severity of injuries [4].

Prediction of the expected severity of accidents associated with certain exposure variables is a quantitative tool that identifies risk factors at the macro level. At the company level, depending on the prevention measures adopted, that initial estimation at the macro level may need to be modified according to the real working conditions. However, from a public policy point of view, the initial estimation of risk factors for the severity of accidents at the macro level is useful for designing effective intervention programs.

2 Data and Methods

2.1 Accident Data

The Data used is from the accident notification database. Accident notification is mandatory for every accident with at least one day of absence from work. Data is

Table 1 Exposure (predicting) variables for the full dataset: Categories and their distribution

Variable	Categories	Abbreviation	Number of examples
Sex	Male	S1	19,442
	Female	S2	2,181
Nationality	Spanish	N1	20,559
	Foreign	N2	1,064
Age	Young (less 25 years)	A1	7,804
	Normal (from 25 to 54 years)	A2	9,116
	Senior (more than 54 years)	A3	4,703
Job	Non-manual	J1	621
	Qualified manual	J2	16,513
	Non-qualified manual	J3	4,489
Experience in company	Low (less than 4 months)	E1	4,832
	Medium (from 4 to 12 months)	E2	4,695
	High (more than 12 months)	E3	12,096
Contract	Permanent	P1	11,656
	Temporary	P2	9,967
Company size	Micro (less than 10 workers)	C1	4,508
	Small (from 10 to 49 workers)	C2	8,385
	Medium (from 50 to 249 workers)	C3	6,367
	Big (more than 249 workers)	C4	2,363
Workstation	Usual	WT1	20,987
	Unusual	WT2	636
Working Environment	Industrial	P1	18,868
	Non-industrial	P2	2,755
Work in process	Production	W1	15,636
	Construction	W2	575
	Agricultural	W3	71
	Services	W4	300
	Maintenance, auxiliary	W5	3,852
	Transport	W6	451
	Other	W9	738
Physical activity	Operating machines	A1	1,902
	Working with hand-held tools	A2	4,061
	Driving	A3	369
	Handling objects	A4	7,276
	Carrying by hand	A5	2,786
	Movements	A6	4,550
	Other	A9	679

coded according to the European Statistics on Accidents at Work Methodology—ESAW [7]. In every accident notification there are two separate types of variables from the prevention point of view: the variables of exposure (before the accident); and the variables of result (after the accident).

Given the objective of this paper, to use the data to predict the possible accidents and their severity, we will only use exposure variables to train the neural network. Those exposure variables are presented in Table 1 (total number of cases 21,623). Exposure variables are those not related to the accident occurrence or results.

Because in the data set there is a very high proportion of a slight accident, an alternative data set was prepared using a case-control approach. In that approach, a sample of slight accidents is randomly selected. Therefore in the second data set there are a similar number of examples with slight and non-slight severity (total number of cases 164).

2.2 Use of Neural Network for Severity Prediction

The modern usage of the term often refers to artificial neural networks, which are composed of artificial neurons or nodes. Artificial neural network algorithms attempt to achieve good predictive ability with a low generalization error.

During supervised error-back propagation training, input patterns are presented sequentially to the system along with the correct response. The network learns by comparing the targeted correct response with the actual response. This process is continued until all examples from the training set are learned within an acceptable error. Then the network is ready to operate in a feed-forward manner, attempting to classify accurately situations not encountered in training. In order to validate the usefulness of the network, examples with real output results are used. In modern software that process is integrated.

Finally, a separate set of real examples are used for querying. These are for testing the accuracy of the predictions of the trained network.

2.3 Neural Network Design and Learning Process Development

The tool used in this paper allows the selection of networks with one, two or three hidden layers. Once the network architecture is set, the learning process can start.

There are several controls for learning what needs to be set. As we want to avoid over fitting, the learning stops when the average validating error increases [2]. Other criteria for stopping are when average error is less than 0.01 or when 100 % of the validating examples are within the 50 % of the desired outputs.

2.4 Evaluation of the Predictive Ability of a Neural Network

The main evaluation parameter for the predicting purpose of the neural network is the number of examples correctly classified. For that purpose, the best way of expressing the results is to use a diagnostic array (see Table 2).

In relation to the analysis of the risk factors, there are two important parameters for each variable: importance and sensitivity. Importance shows the sum of the absolute weights of the connections from the input node to all the nodes in the first hidden layer. Sensitivity indicates how much an output changes when the inputs

Table 2 Diagnostic array[a]

		Real examples	
		True	False
Observed examples	True	True positive (TP)	False positive (FP)
	False	False negative (FN)	True negative (TN)

[a] Results for a diagnostic array:
Sensitivity: TP/(TP+FN)
Specificity: TN/(TN+FP)
Positive prediction: TP/(TP+FP)
Negative prediction: TN/(TN+FN)
Efficiency: (TP+TN)/(TP+FP+FN+TN)

Table 3 Association rules

Variable1	Variable2	Strength Full data set	Strength Case-control data set
Severity	Workstation	7.9	2.1
Severity	Nationality	–	1.6

are changed. The inputs are all set to the median values and then each in turn is increased from the lowest value to the highest value. The change in the output is measured as each input is increased from lowest to highest to establish the sensitivity to change.

3 Results

3.1 *Associations*

The first analysis looks at associations between variables. This analysis offers a data mining initial pattern of the strength of the relationship between each pair of variables (see Table 3). In terms of risk assessment, variables that have strong association with the *Severity* outcome variable should be considered risk factors.

For the full data set, the *Severity* output variable only has association with one predicting variable: *Workstation*. For the case-control data set, there is also association with *Nationality*.

3.2 *Neural Network Analysis Based on the Full Data Set*

Two models were developed for the full data set, one with one hidden layer and the other with two hidden layers. Both of them include the eleven variables of exposure that are available. Errors and results are presented in Table 4 and diagnostic array in Table 5.

Table 4 Neural networks based on the full data set

Heading	Validating error (%)	Validating OK (%)	Cycles	Hidden layers	Hidden Nodes
Model 1	0.28	99.6	1,000	1	6
Model 2	0.25	99.7	1,000	2	6+8

Table 5 Diagnostic array for query examples of the neural networks based on the full data set

		Real outcome	
	Predicted outcome	Slight	Severe
Model 1 hidden layer	Slight	4,983	14
	Severe	0	3
Model 2 hidden layers	Slight	4,986	14
	Severe	0	0

Table 6 Neural networks based on the case-control data set

Heading	Validating error(%)	Validating OK(%)	Cycles	Hidden layers	Hidden Nodes
Model 1	0.00	86.2	2,400	1	9
Model 2	0.93	86.2	2,560	2	9+6

Table 7 Diagnostic array for query examples of the neural networks based on the case-control data set

		Real outcome	
	Predicted outcome	Slight	Severe
Model 1 hidden layer	Slight	13	2
	Severe	4	11
Model 2 hidden layers	Slight	14	3
	Severe	3	10

3.3 *Neural Network Analysis Based on a Case-Control Data Set*

Two models were developed for the case-control data set, one with one hidden layer and the other with two hidden layers. Both of them include the eleven variables of exposure available. Errors and results are presented in Table 6 and diagnostic array in Table 7.

4 Discussion

The neural networks were built with the variables that are available. Other risk factors such as the level of training, protection measures implemented or risk level of the task are not available. As previous researchers have proposed [8] the notification system in Europe should include other relevant data.

The use of these neural networks to assess the expected severity of accidents for a set of exposure variables is a very useful practice and can provide safety practitioners with an initial estimation of the severity of the accidents for a group of tasks in the manufacturing sector.

The results of the neural networks based on the two data sets are very different. In terms of efficiency, the full data set is higher, whereas in terms of specificity the case-control is more precise. From de preventive point of view, what really matters to safety practitioners understands the possible risk factors of severity, and for that purpose the case-control approach is more useful [10].

These neural networks can be easily applied by safety practitioners. Given the levels of the variables in a specific job based on the real data from the enterprise, it is possible to use trained neural networks based on the accidents notified to predict the severity of accidents. Ultimately, that initial estimation should be complemented with the analysis of the prevention measures and the working conditions in the risk assessment process in order to assess the risk of severe accidents.

References

1. Aneziris O, Papazoglou I, Doudakmani O (2010) Assessment of occupational risks in an aluminum processing industry. Int J Ind Ergon 40(3):321–329
2. Asensio-Cuesta S, Diego-Mas JA, Alcaide-Marzal J (2010) Applying generalised feedforward neural networks to classifying industrial jobs in terms of risk of low back disorders. Int J Ind Ergon 40(6):629–635
3. Bevilacqua M, Ciaparica FE, Giacchetta G (2010) Data mining for occupational injury risk: a case study. Int J Reliab Qual Saf Eng 17(4):351–380
4. Carrillo JA, Onieva L (2012) Severity Factors of Accidents: Analysis of the Manufacturing Sector in Andalusia. In: Occupational Safety and Hygiene—SHO 2012. Arezes et al. Portuguese Society of Occupational Safety and Hygiene, Guimaraes, pp 111–115
5. Carrillo JA, Gómez MA, Onieva L (2012) Safety at work and worker profile: analysis of the manufacturing sector in Andalusia in 2008. In: Occupational Safety and Hygiene—SHO 2012. Arezes et al. Portuguese Society of Occupational Safety and Hygiene, Guimaraes, pp 116–120
6. Delen D, Sharda R, Bessonov M (2006) Identifying significant predictors of injury severity in traffic accidents using a series of artificial neural networks. Accid Anal Prev 38(3):434–444
7. European C (2002) European statistics on accidents at work (ESAW)—Methodology (ed. 2001). Office for Official Publications of the European Communities, Luxembourg
8. Jacinto C, Canoa M, Guedes Soares C (2009) Workplace and organizational factors in accident analysis within the food industry. Saf Sci 47(5):626–635
9. Reason J (2000) Human errors: models and management. Br Med J 320:768–770
10. Sorock GS, Courtney TK (1997) Advancing analytic epidemiologic studies of occupational injuries. Saf Sci 25(1-3).29–43
11. Wang Y-M, Elhag THS (2007) A comparison of neural network, evidential reasoning and multiple regression analysis in modelling bridge risks. Expert Syst Appl 32:336348
12. Zuruda J, Karwowski W, Marras WS (1997) A neural network-based system for classification of industrial jobs with respect to risk of low back disorders due to workplace design. App Ergon 28(1):49–58

A New Constructive Heuristic for the Fm|block|ΣT

Ramon Companys Pascual and Imma Ribas Vila

Abstract This paper deals with the blocking flow shop problem and proposes new constructive procedures for the total tardiness minimization of jobs. The heuristic has three-phases to build the sequence; the first phase selects the first job to be scheduled, the second phase arranges the remaining jobs and the third phase uses the insertion procedure of NEH to improve the sequence. The proposed procedures evaluate the tardiness associated to the sequence obtained before and after the third phase in order to keep the best of both because the insertion phase can worsen the result. The computational evaluation of these procedures against the benchmark constructive procedures from the literature reveals their good performance.

Keywords Flow shop · Blocking · Tardiness

1 Introduction

In a flow shop, there are n jobs that have to be processed in m machines. All jobs follow the same route in the machines. The processing time of job $i \in \{1,2,\ldots,n\}$ on machine j, $j \in \{1,2,\ldots,m\}$, is $p_{j,i} > 0$. In the traditional version of the problem, it is assumed that there are buffers of infinite capacity between consecutive machines, where jobs, after being processed by the previous machine, can wait until the subsequent machine is available. However, in many industrial systems this supposition cannot be made, since the capacity of buffers is zero, due to the characteristics of the process [6]. In this type of productive configuration, a machine can be blocked by the job already processed if the buffer is full or absent. Therefore, an accurate scheduling is necessary to avoid or to minimise the blocking time of machines. Some examples can be found in the production of concrete blocks where storage is not allowed in some stages of the manufacturing process [4]; in the iron and steel industry [3]; in the treatment of industrial waste and the manufacture of metallic

R. Companys Pascual (✉) · I. Ribas Vila
Dpto. de Organización de Empresas, Escola Tècnica Superior d'Enginyers Industrials de Barcelona, Universitat Politècnica de Catalunya, BarcelonaTech, Avda. Diagonal, 647, ed. H, P.7, 08028 Barcelona, Spain
e-mail: ramon.companys@upc.edu

I. Ribas Vila
e-mail: imma.ribas@upc.edu

C. Hernández et al. (eds.), *Managing Complexity*, Lecture Notes in Management and Industrial Engineering, DOI 10.1007/978-3-319-04705-8_33,

parts [7]; or in a robotic cell, where a job may block a machine while waiting for the robot to pick it up and move it to the next stage [11].

The tardiness criterion has been less studied than the make span or total flow time criteria, despite the fact that scheduling according to this performance measure helps companies offer a high service level to their customers, which is essential for survival in the market. In particular, to the best of our knowledge, only Armentano and Ronconi [2] and Ronconi and Henriques [10] dealt with the blocking flow shop problem for total tardiness minimization. Armentano and Ronconi [2] propose a Tabu Search procedure that uses the LBNEH method proposed in Armentano and Ronconi [1] to obtain the initial solution. Alternatively, in Ronconi and Henriques [10], a constructive method (FPDNEH) and a Greedy Adaptive Search Procedure (GRASP) is proposed for this problem.

In this paper we propose a new constructive heuristic which explores specific characteristics of the problem. The comparison between the most popular constructive heuristic used in the literature reveals its efficacy to find good solutions for the problem dealt with.

The paper is organized as follows: after this brief introduction, the problem is formally defined in Sect. 2. Section 3 describes the proposed constructive heuristics, Sect. 4 shows the computational evaluation and Sect. 5 summarizes the conclusions.

2 Problem Definition

The tardiness blocking flow shop problem denoted as Fm | block | ∑T according to the notation proposed by Graham et al. [5] can be formulated with the following equations, where d_i denotes the due date of job i, $e_{j,k}$ the time in which the job in position [k] starts to be processed on machine j and $c_{j,k}$ is the departure time of this job:

$$e_{j,k} + p_{j,[k]} \le c_{j,k} \quad j = 1,2,\ldots,m \quad k = 1,2,\ldots,n \tag{1}$$

$$e_{j,k} \ge c_{j,k-1} \quad j = 1,2,\ldots,m \quad k = 1,2,\ldots,n \tag{2}$$

$$e_{j,k} \ge c_{j-1,k} \quad j = 1,2,\ldots,m \quad k = 1,2,\ldots,n \tag{3}$$

$$c_{j,k} \ge c_{j+1,k-1} \quad j = 1,2,\ldots,m \quad k = 1,2,\ldots,n \tag{4}$$

$$TT = \sum_{i=1}^{n} \max(c_{m,i} - d_i, 0) \tag{5}$$

$c_{j,0=0} \forall j, c_{0,k} = 0, c_{m+1,k} = 0 \forall k$ are the initial conditions.

If Eqs. (2) and (3) are summarized as (6) and Eq. (1) and (4) as (7), the schedule obtained is semi-active, which is interesting because an optimal solution can be found in the subset of the semi-active set of solutions.

$$e_{j,k} = \max\left\{c_{j,k-}1\ ;\ c_{j-1,k}\right\} \quad (6)$$

$$c_{j,k} = \max\left\{e_{j,k} + p_{j,[k]}, c_{j+1,k-1}\right\} \quad (7)$$

3 Constructive Heuristics for the Fm | block |∑T Problem

The most popular constructive heuristics used to deal with the tardiness criterion consist of using the *early due date* (EDD) rule or *slack* (sl) rule to create an initial sequence, which is then processed by the insertion phase of NEH [8 adapted to the tardiness criterion. We denote to these procedures N_{EDD} and N_{sl}. In both procedures, the creation of the initial sequence is done by assigning an index to each job which is used for their prioritization. However, Ronconi and Henriques [10] proposed a method, named FPD, which can be seen as a two-phase procedure; the first phase selects the job to be scheduled first and the second phase sequences the rest of the jobs. They also proposed to improve the obtained sequence by FPD with the insertion procedure. According to our previous notation, we named to this procedure N_{FPD}.

3.1 New Constructive Heuristics

The proposed heuristics consist of three phases according to the philosophy of the N_{FPD} method. We have implemented four alternatives to choose the first job in the sequence (first phase) and three alternatives to sequence the remaining jobs (second phase). The combination of them has led us to implement twelve methods for creating a sequence which is then processed by the insertion procedure.

We denote as P_i, the sum of processing time of job *i*, and as sl_i the slack of job *i* calculated as the difference between its due date and the sum of its processing time. Next, we describe the rules used in the first phase:

- A1: Select the job with minimum $\lambda \cdot p1_{,i} + (1-\lambda) \cdot d_i$, if $\lambda = 0.5$ this index is exactly the same than the used in FPD.
- A2: Select the job with minimum $\lambda \cdot P_i / \mathrm{m} + (1-\lambda) \cdot sl_i$, to break ties select the job with minimum d_i.
- A3: Select the job with minimum $\lambda \cdot \dfrac{P_i - P_{min}}{P_{max} - P_{min}} + (1-\lambda) \cdot \dfrac{sl_i - sl_{min}}{sl_{max} - sl_{min}}$ where P_{min} and P_{max}, are the maximum and minimum processing time and sl_{min} and sl_{max} the maximum and minimum slack of jobs, respectively.
- A4: Select the job with minimum $\lambda \cdot \dfrac{\mathrm{dif}_i - \mathrm{if}_{min}}{\mathrm{dif}_{max} - \mathrm{dif}_{min}} + (1-\lambda) \cdot \dfrac{\mathrm{sl}_i - \mathrm{sl}_{min}}{\mathrm{sl}_{max} - \mathrm{sl}_{min}}$ where $dif_i = \sum_{j=1}^{m} | p_{i,j} - P_i / m |$, to break ties select the job with minimum d_i.

These procedures are variants of the used in FPD. A1 allows weighting the two factors, A2 considers the important of the total processing time instead of the processing time in the first stage and changes the due date for the slack of jobs, A3 is an evolution of A2 and A4 change the first factor in order to prioritise jobs with a more regular processing time.

The implemented rules for the second phase are:

- B1: the same rule than in FPD.
- B2: select the job with minimum $\mu \cdot \frac{tm_i - tm_{\min}}{tm_{max} - tm_{min}} + (1-\mu) \cdot \frac{dsl_i - dsl_{min}}{dsl_{\max} - dsl_{min}}$, in case of ties choose the job with early due date. tm_i is the idle time generated when job i is scheduled at the end on the partial sequence σ, in position $k+1$, $tm_i = \sum_{j=1}^{m} fj,k+1(i) - fj,k(\sigma) - pj,i)$ and dsl_i is tvhe dynamic slack of job i when it is scheduled at the end of σ, $dsl_i = d_i - c_{m,i}$
- B3: Select, among the jobs with $\frac{\mathrm{dsl}_i - \mathrm{dsl}_{\min}}{\mathrm{dsl}_{\max} - \mathrm{dsl}_{\min}} < \mu$, the job with minimum tm_i.

B2 is an evolution of B1 because, in B1, both terms allow estimating the idle time and the slack time, respectively, whereas in B2 these terms allow calculating the real idle time generated by the schedule job as well as its real slack: i.e. the first term of B2, is the real idle time resulting from scheduling job i in that position. The second term, dynamic slack of job i, is evaluated considering that this job is scheduled at the end of the partial sequence.

B3, instead, uses hierarchical multicriteria decision to select a job.

Finally, the obtained sequence after these two phases is tried to improve by the insertion procedure, adapted to the tardiness criterion. But, as during this research we detected that, in some instances, the tardiness associated to the sequence given by the constructive procedures was smallest than the associated to the sequence obtained after the insertion, the implemented procedures evaluate the sequence before and after the insertion procedure in order to keep the best of both.

3.2 Experimental Parameter Adjustment of the Rules

The proposed procedures have two parameters, λ and μ, that should be adjusted. The calibration of these parameters has been done with 480 instances generated *ad hoc*, 10 instances for each combination of $n = \{20, 50, 100, 200\}$ and $m = \{5, 10, 20\}$ and 4 ranges of due dates, which are named scenarios from now on. The due dates of jobs were uniformly distributed between $LB \cdot (1-T-R/2)$ and $LB \cdot (1-T+R/2)$ as in Potts and Van Wassenhove [9], where T and R are the tardiness factor of jobs and the dispersion range of due dates, respectively. LB is a lower bound of the C_{max} with unlimited buffer in the flow shop [12] problem. Therefore, each of the scenarios correspond to a combination of $R = \{0.6, 1.2\}$ and $T = \{0.2, 0.4\}$. The experiments were carried out on an Intel Core 2 Duo E8400 CPU, with 3 GHz and 2 GB RAM memory. To analyze the experimental results obtained, we measured the relative deviation index (*RDI*), calculated as (8) for each procedure:

$$RDI = \frac{Heur_{h,s} - Best_s}{Worst_s - Best_s} \tag{8}$$

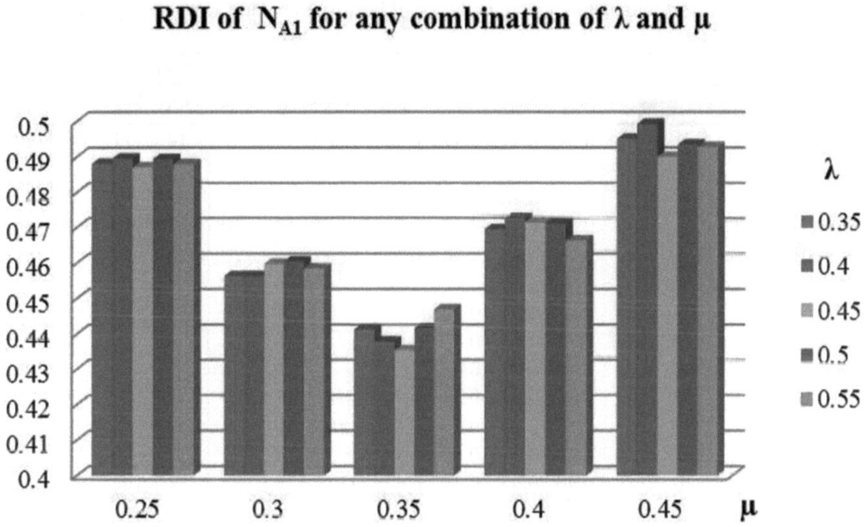

Fig. 1 Overall average of RDI obtained with N_{A1} procedure for any combination of λ and μ

Table 1 Best values of λ and μ per each procedure

–	N_{A1}	N_{A2}	N_{A3}	N_{A4}
λ	0.45	0.1	0.1	0.15
μ	0.35	0.35	0.30	0.35

where $Heur_{h,s}$ is the average of tardiness values obtained by heuristic *h*, in instance *s*, and $Best_s$ and $Worst_s$ are the minimum and maximum value of tardiness obtained for this instance, among all the combinations of parameters.

A preliminary test showed us that any A procedure (first phase) combined with B2 outperformed the procedures that used B1 or B3 in the second phase, which led us to discard B1 and B3. Hence, the adjustment of λ and μ was done for those procedures that use B2 in the second step, which have been denoted as N_{A1}, N_{A2}, N_{A3} and N_{A4}, to indicate the method used in the first step.

As an example, the calibration of N_{A1} is shown in Fig. 1, where the overall average of index RDI obtained, for several values of λ and μ can be seen. Notice, that in this case the lower values of RDI are obtained when $\lambda = 0.45$ and $\mu = 0.35$.

The same experiment was done for each procedure and λ and μ were fixed according to the values showed in Table 1.

4 Computational Evaluation

In this section we compare the proposed procedures with N_{FPD}, N_{EDD} and N_{sl} in order to analyse their performance. This test was done against the 480 instances used in Ronconi and Henriques [10]. The comparison between procedures was done using the index RDI as in [8], where $Best_s$ and $Worst_s$ are the minimum and maximum value of the total tardiness obtained by any of the procedures evaluated in this test.

Table 2 Average RDI value obtained and number of best solutions found by each procedure per *n* and *m* combination

Average RDI								Number of best solutions						
$n \times m$	N_{A1}	N_{A2}	N_{A3}	N_{A4}	N_{EDD}	N_{sl}	N_{FPD}	N_{A1}	N_{A2}	N_{A3}	N_{A4}	N_{EDD}	N_{sl}	N_{FPD}
20×5	0.47	*0.31*	0.49	0.31	0.44	0.41	0.48	11	14	7	14	11	*16*	12
20×10	0.47	*0.43*	*0.43*	*0.43*	*0.43*	0.57	*0.43*	7	9	9	9	*15*	2	14
20×20	0.56	*0.41*	0.45	*0.41*	0.42	0.43	0.60	5	*12*	8	*12*	*12*	8	8
50×5	0.37	0.38	*0.33*	0.38	0.42	0.49	0.65	*11*	10	10	10	10	8	3
50×10	0.51	0.54	*0.41*	0.54	0.49	0.48	0.59	6	6	*12*	6	7	5	8
50×20	0.51	0.54	0.47	0.53	0.46	*0.42*	0.56	2	4	10	4	5	*14*	6
100×5	0.29	0.27	*0.22*	0.27	0.34	0.30	0.74	10	9	11	9	11	*12*	6
100×10	0.33	*0.32*	0.37	*0.32*	0.34	0.33	0.68	*13*	8	4	8	6	4	11
100×20	0.44	0.46	*0.42*	0.46	0.45	0.46	0.56	3	6	6	6	7	6	*13*
200×10	0.33	0.33	0.33	*0.33*	0.47	0.53	0.79	*14*	11	9	12	9	7	0
200×20	*0.32*	0.37	0.44	0.38	0.42	0.54	0.68	10	*12*	8	11	8	1	4
500×20	*0.27*	0.42	0.34	0.42	0.51	0.50	0.73	*18*	14	7	13	3	5	2
All	0.41	0.40	*0.39*	0.40	0.43	0.45	0.62	110	*115*	101	114	104	88	87

Table 2 shows the average RDI values obtained and the number of best solutions found by each procedure in each *nxm* set. The first observation is that the new procedures have better performance than N_{FPD}, N_{EDD} or N_{sl}. However, the difference between them is very small. We can say that, N_{A3} performs slightly better, but its behaviour is worse for $n=20$ and $n=200$, which dilutes the overall average value for this procedure. This can also be seen through the number of best solutions found by each procedure in each set. As the only difference between these procedures is the selection of the first job, we can say that this decision has a great influence in the obtained sequence. Therefore, one line of future research is to explore the convenience of using one of these procedures to select the first job according to the number of jobs to schedule.

Another interesting result is the performance of N_{EDD} and N_{sl} compared to N_{FPD}. The obtained results in this test are much better than those reported in Ronconi and Henriques [10]. This fact can be due to the tie-break criterion used to select the jobs. In N_{EDD}, we break ties by selecting the job with higher P_i whereas, in N_{sl} the job selected is the one with less P_i. This observation indicates that can be interesting to analyse several criteria to break ties because it has an appreciable effect in the obtained results.

5 Conclusions

In this paper we have proposed effective constructive heuristics procedures to deal with the total tardiness flow shop problem with blocking. The presented procedures have three steps; step one selects the first job in the sequence, step two builds the remaining sequence and step three uses the insertion procedure of NEH, adapted to the tardiness criterion, to try to improve the sequence. It has been implemented four procedures for step one and three for step two, which have been combined to build twelve heuristics. The best method for step two was B2 which is focused on the precise evaluation of the idle time and slack time of jobs; instead, the best procedure for the first step is not clear because it seems to be dependent of the number of jobs to schedule. The computational evaluation has revealed that the criterion used to break ties in the constructive procedures has a great influence in the quality of the obtained sequence. Therefore, we recommend to the researchers a detailed analysis of the criterion used to break ties.

References

1. Armentano VA, Ronconi DP (1999) Tabu search for total tardiness minimization in flowshop scheduling problems. Comput Oper Res 26(3):219–235
2. Armentano VA, Ronconi DP (2000) Minimização do tempo total de atraso no problema de flowshop com buffer zero através de busca tabu. Gestao Produçao 7(3):352
3. Gong H, Tang L, Duin CW (2010) A two-stage flow shop scheduling problem on a batching machine and a discrete machine with blocking and shared setup times. Comput Oper Res 37(5):960–969
4. Grabowski J, Pempera J (2000) Sequencing of jobs in some production system. Eur J Oper Res 125(3):535–550
5. Graham RL, Lawler EL, Lenstra JK, Rinnooy Kan AHG (1979) Optimization and approximation in deterministic sequencing and scheduling: a survey. Ann Discret Math 5:287–326
6. Hall NG, Sriskandarajah C (1996) A survey of machine scheduling problems with blocking and no wait in process. Oper Res 44(3):510–525
7. Martinez S, Dauzère-Pérès S, Guéret C, Mati Y, Sauer N (2006) Complexity of flowshop scheduling problems with a new blocking constraint. Eur J Oper Res 169(3):855–864
8. Nawaz M, Enscore EE Jr, Ham I (1983) A heuristic algorithm for the m-machine, n-job flowshop sequencing problem. Omega 11(1):91–95
9. Potts CN, Van Wassenhove LN (1982) A decomposition algorithm for the single machine total tardiness problem. Oper Res Lett 1(5):177–181
10. Ronconi DP, Henriques LRS (2009) Some heuristic algorithms for total tardiness minimization in a flowshop with blocking. Omega 37(2):272–281
11. Sethi SP, Sriskandarajah C, Sorger G, Blazewicz J, Kubiak W (1992) Sequencing of parts and robot moves in a robotic cell. Int J Flexible Manuf Syst 4:331–358
12. Taillard E (1993) Benchmarks for basic scheduling problems. Eur J Oper Res 64(2):278–285

Applying Cluster Analysis to Renewable Energy Emergent Sector at Local Level

Jaso Larruscain, Rosa Río-Belver, Ernesto Cilleruelo, Gaizka Garechana and Javier Gavilanes-Trapote

Abstract This paper aims to provide a brief overview of state-of-the-art methods of cluster analysis and to acknowledge their limitations when applied to local level in renewable energies. This emergent sector is becoming increasingly important within the field of Industrial Organization, with technological and industrial innovation being essential for the competitiveness of future "smart cities". An understanding and analysis of the clusters formed by the different participating actors (public administration, centers of research and knowledge, and businesses) will be the key to safeguarding economic development, especially in their initial stage. As a conclusion, Social Network Analysis (SNA) tools together with Competitive Advantage Analysis (CAA) seem to be the most recommended methods.

Keywords Cluster analysis · Renewable energy · Network analysis · Collaborative networks · Management models

---

J. Larruscain (✉) · R. Río-Belver · J. Gavilanes-Trapote
Foresight Technology and Management (FTM) Group, Department of Industrial Engineering, University of the Basque Country UPV/EHU, calle Nieves Cano 12, 01006 Vitoria, Spain
e-mail: jaso.larruscain@ehu.es

R. Río-Belver
e-mail: rosamaria.rio@ehu.es

E. Cilleruelo
Foresight Technology and Management (FTM) Group, Department of Industrial Engineering, University of the Basque Country UPV/EHU, alameda Urquijo s/n, 48030 Bilbao, Spain
e-mail: ernesto.cilleruelo@ehu.es

G. Garechana
Foresight Technology and Management (FTM) Group, Department of Industrial Engineering, University of the Basque Country UPV/EHU, calle Elcano 21, 48030 Bilbao, Spain
e-mail: gaizka.garechana@ehu.es

J. Gavilanes-Trapote
e-mail: javier.gavilanes@ehu.es

C. Hernández et al. (eds.), *Managing Complexity*, Lecture Notes in Management and Industrial Engineering, DOI 10.1007/978-3-319-04705-8_34,

1 Introduction

In the era of globalization, the tendency to create local clusters is gathering strength as the key factor for the economic development of "smart cities". This is particularly true in the renewable energy (RE) sector, not only in order to comply with sustainability requirements imposed by government organizations, but also to generate economic growth. The competitiveness of a sector can be measured by its clusters. These clusters in turn can be analyzed by the numerous methods used by the scientific community. Whilst the majority of these methods are applied to geographical areas of nations or regions, there is a growing need to be able to apply them to more restricted areas, such as cities.

The purpose of this article is to provide a brief overview of state-of-the-art methods of cluster analysis applied to local level in renewable energy sector. For that purpose, a research methodology is carried out based on three main steps: gathering information about this specific sector, analyzing general identification tools with specific network indicators, as well as highlighting limitations when applying to local level.

1.1 Renewable Energies (Sub)Sector

The objective to increase the percentage of RE in the European energy mix to 20 % by 2020 [2] has caused the scientific production in RE to double in size at European and worldwide levels between 2002 and 2007 [18] and also the boost of a new multi-discipline industrial sector.

Nevertheless, the need to incorporate a greater percentage of different technologies (including not so well-developed ones such as wave, tidal and small wind energies) within the new structures of energy generation and distribution in cities makes local participation increasingly important in terms of industrial development. Investments in this sector will generate a multiplying effect in the economy, and the creation of new organizational structures such as local clusters will become necessary [12].

1.2 Local Clusterisation

According to Porter [14], an industrial cluster is defined as: "a geographic concentration of interconnected businesses and institutions in a particular field, creating a matrix crucial for increasing productivity". Accordingly, clusters have a competitive advantage due to their co-localization [4].

Modern clusterization theories affirm that the experience and know-how shared by actors of a cluster are the greatest source of benefit, as a result of being close by and maintaining local innovation networks [15].

Table 1 Some examples of European Renewable Energy Local Clusters (2011). (Source: http://en.erneuerbare-energien-hamburg.de/, http://www.greencity-cluster.de/, http://www.cphcleantech.com/, and http://www.fomentosansebastian.org/)

Cluster	Creation year	Phase	Members (companies)	Comments (employment)
Hamburg (Germany)	2010	Embryonic	163	Local: 14.563; expected growth (2008–2015): 40 %
Freiburg (Germany)	2009	Established	107	Region: 12.000
Copenhagen (Denmark)	2010	Embryonic	36	Local expected growth (2010–2013): + 1000
San Sebastian (Spain)	2009	Embryonic	85	Local: 1.800

The type of industry is usually a factor that influences the typology of a cluster, which changes from a temporary phase to another, going through the embryonic, established, mature and declining stages. In the case of the RE sector at a local level, they are currently in the first phase and have a mixed typology [8] between the Marshallian, Hub-and-spoke, Satellite platform or State-anchored form [11].

Whilst national and regional clusters have been studied in detail by the scientific community, no specific studies have been carried out at a local level.

At local level, the success of a cluster is largely determined by the growth potential of its small and medium sized industries. For instance, one of the main priorities for policymakers is to promote local enterprise and to allow SMEs to benefit from the availability of the cluster's resources [8].

The extensive and speedy evolution of RE enterprises shows that local clusters as well as industry based on knowledge will grow exponentially in the next few years. Clusterization in this emergent RE sector is an industrial hallmark.

Porter [14] argued that although the role of localization has been ignored in the era of global markets, lasting competitive advantages are to be found in the local characteristics that cannot be matched by far-off competition and that these characteristics, among others the relationships, will have to be studied and analyzed in detail, especially in local new clusters (see Table 1).

2 Cluster Analysis Tools

Cluster analysis is an essential tool in the identification of areas of local-regional economy where there are comparable advantages in terms of the productivity and economic growth of a cluster. Comprehensive analysis of a cluster requires paying attention to concepts such as industrial structure, business strategy, competitiveness between industries and the relationship between knowledge and technology [9].

In general terms, most authors suggest an analysis that takes into account two aspects. On the one hand, the impact indicator (to measure the impact of the network on its members) [13] and, on the other, the size of the network and its average path length (the average number of links between its members).

It has been observed that businesses embedded in alliance networks, which show a high clustering impact and reach (very short lengths of links between businesses), tend to show high innovation performance [19]. With regard to its own growth, this will be directly related to the benefits obtained from being a member of the network, such as economies of scale [4].

2.1 General Methods

Although literature relating to methods is varied and extensive, the methods can be grouped according to the origins of baseline data. **Quantitative** methods are used in the generic analysis of the properties of the network, both at the general and restricted levels [1].The most important methods in this group are: input-output (analyzing an approximation of interdependencies between different areas of the network), cluster dependency (analyzing the dependency ratios which can easily be visualized with the Fuzzy tool), and network analysis (together with the Graph theory used for visualization only) [19]. Alternatively, **qualitative** methods, which provide greater sensitivity regarding the relationship between the actors in the cluster [4], would be: expert opinion, surveys and industry research. A full analysis requires the use of both types of method.

Social Network Analysis (SNA) will specifically provide information that enables possible actors to be aware of the existence, needs and ability requirements of others, including help to develop new alliances [7]. This becomes a key element in the study of relationships at an organizational level, inside and outside the clusters [8, 10].

Its objective is to detect and interpret patterns in the links between the different actors in the network, which are represented as vertices [3]. In addition, attributes (characteristics of the actors, which are not based on their structural position within the network, and calculated statistically) provide added value to the interpretation of the structure.

The most important specific network indicators [3] calculated and visualized by means of certain types of software such as Pajek, Ucinet, Visone, Gephi or Vennmaker [17], are:

- Density: ratio between existing and possible relations
- Cohesion: number of bidirectional choices in relation to the number of dyads
- Multiplexity: share of multiplex relations on all possible relations
- Degree of homophily: describes whether actors with similar attributes are more connected with each other than to actors with different attributes.

The network typology of a cluster is critical from the first moment of its existence, as it will determine its success or failure during the later phase of expansion, growth and development. Analysis of this first phase is essential, particularly to identify the cluster potential [8].

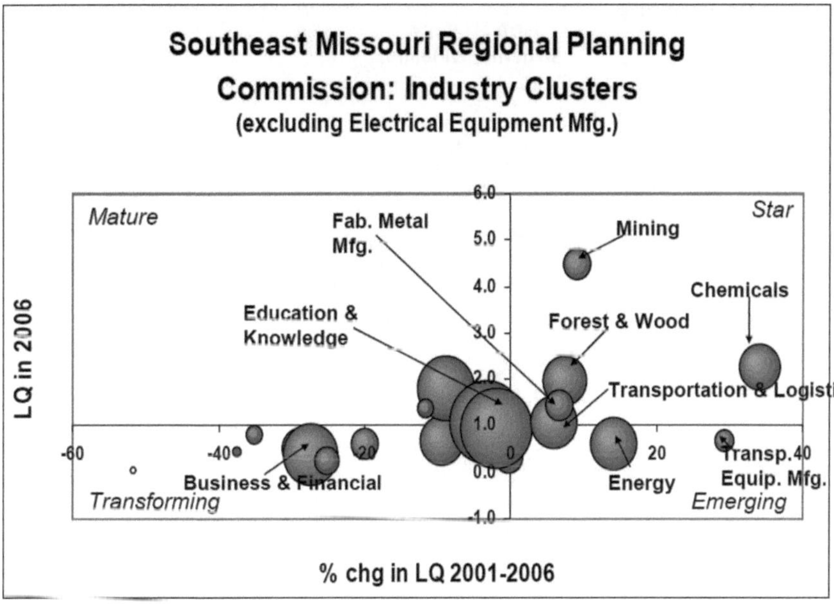

Fig. 1 Tri-dimensional representation of the LQ method using data from 2001 to 2006 [16] to identify emergent clusters in Southeast Missouri

2.2 *Identification Methods and Limitations in RE Sector*

The emergence of a cluster can be attributed to historical circumstances, even before the appearance of the contributing industries themselves, or could even be due to chance or coincidence [14]. It may also be the result of a business opportunity, of the presence of a unique added value, an increase in the influence exercised by a business, an increase in the undertaking or even a change in the policies of a given sector [22]. The latter applies to most RE clusters.

There are numerous methods used in identification, such as: expert opinion, input-output analysis (trade-based and innovation-based), network analysis, surveys or Giniho coefficient of localization [1, 20]. However, the most widely used are explained below.

2.2.1 Location Quotients

This method compares the fraction of the region's variable in a particular industry cluster to the fraction of the nation's variable in the same industry (employment, wages or other economic variable). An example is given in Fig. 1.

It is suitable for the RE sector only as a first approximation due to the fact that although it provides local information, it uses only one analysis factor and it does not allow to predict the behavior of emergent cluster.

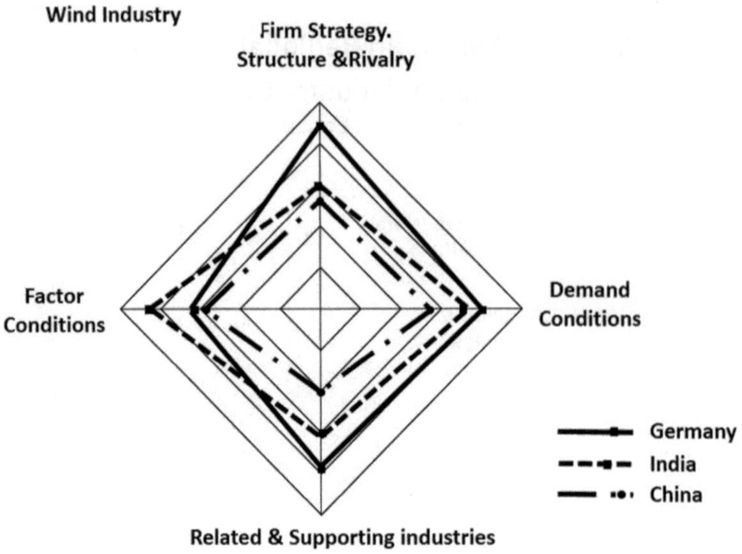

Fig. 2 Competitiveness of RE industries in Germany compared with India and China, using Porter's diamond model adapted to the RE sector [5]

2.2.2 Shift-Share Analysis

It shows easily the influence of the industrial structure and changes in the regional economy on a given variable, categorizing the actual changes into three percentage-effect groups: national, industry cluster and regional [21].

Although it is useful to compare a hypothesis of potential viability of an emerging RE cluster, it can not be corrected with factors dependent on economy (RE is not currently stable) and it is necessary a 5 year time-window.

2.2.3 Ellison and Glaeser Index of Agglomeration

This establishes the degree of agglomeration of a specific industry within a region. In industrial areas, agglomeration occurs in the form of a cluster. The distribution of that industry and the size of the businesses within that sector are both taken into consideration [6]. The lack of detailed information of companies' inner field of activity (RE sector still remains very diversified in industrial activity since RE technology is not mature) complicates accurate results.

2.2.4 Competitive Advantage Analysis (CAA)

This is a mixed method (quantitative and qualitative) based on Porter's Diamond, which calculates the competitiveness of the cluster environment and its components (Fig. 2) and requires the opinion of different managers of the most prominent actors

[20]. It gives useful results in terms of demand, sources, industrial branches and strategies, being sufficient to predict clusters behavior, but no specific studies have been carried out at RE local level.

3 Conclusions

In RE emergent sector, local clusters are currently in their first phases (embryonic, established) being their success largely determined by the growth potential of their small and medium sized industries. As concluded in this research, an analysis of these clusters formed by the different participating actors will be the key to safe guarding economic development. Among the varied and extensive cluster analysis methods studied and which can be grouped according to the origins of baseline data [1, 20]), identification methods appear to be essential and crucial [8] in RE sector where local clusters' births are becoming increasingly important. As a result, the order of applicability efficiency of these could be said to be CAA, Shift-Share analysis and Ellison and Glaeser index of agglomeration [20].

However, limitations appear in two senses. On the one hand, regarding the baseline data, there is still no consensus within the public administration on how to define the sector itself (lack of official and standardized data at local level owing to multidisciplinary diversified industrial activities in no-mature technology), no availability of long time windows (essential to accurate and comparable results) and a difficulty for companies' managers to provide sensitivity information such as their relationships with others within the cluster. On the other hand, applying methods which are used in regions analysis, inaccuracies in several result-factors might appear.

To summarize, and taking into account these available tools and their limitations in RE sector, Social Network Analysis (SNA) together with Competitive Advantage Analysis (CAA) seem to be the most recommended methods.

References

1. Bergman EM, Feser EJ (1999) Industrial and regional clusters: concepts and comparative applications. Regional Research Institute, West Virginia University, West Virginia
2. Communication from the commission to the European parliament, the council, the European economic and social committee and the committee of the regions (2011) A roadmap for moving to a competitive low carbon economy in 2050, COM 2011, 112, Brussels, 8 Mar 2011
3. De Nooy W, Mrvar A, Batagelj V (2011) Exploratory social network analysis with pajek. Cambridge University Press, New York
4. Doeringer PB, Terkla DG (1995) Business strategy and cross-industry clusters. Econ Dev Q 9:225–237
5. Dögl C, Holtbrügge D, Schuster T (2012) Competitive advantage of German renewable energy firms in India and China: an empirical study based on Porter's diamond. Int J Emerg Mark 7(2):191–214

6. Ellison G, Glaeser EL (1997) Geographic concentration in U.S. manufacturing industries: a dartboard approach. J Polit Econ 105:889–927
7. Gulati R, Nohria N, Zaheer A (2000) Strategic networks. Strategic Manage J 21:203–215
8. He J, Fallah MH (2011) The typology of technology clusters and its evolution—evidence from the hi-tech industries. Technol Forecast Soc Change 78(6):945–952
9. Iammarino S, McCann P (2006) The structure and evolution of industrial clusters: transactions, technology and knowledge spillovers. Res Policy 35(7):1018–1036
10. Johannisson B (1995) Paradigms and entrepreneurial networks—some methodological challenges. Entrep Reg Dev 7(3):215–232
11. Markusen A (1996) Sticky places in slippery space: a typology of industrial districts. Econ Geogr 72(3):293–313
12. Marques AC, Fuinhas JA (2012) Is renewable energy effective in promoting growth? Energy Policy 46:434–442
13. Newman MEJ (2001) Scientific collaboration networks: I. Network construction and fundamental results. Phys Rev E 64:016131
14. Porter ME (1998) Clusters and the new economics of competition. Harv Bus rev 76(6):77–90
15. Porter ME (2000) Location, competition and economic development: local clusters in a global economy. Econ Dev Q 14(1):15–34
16. Primont DF, Domazlicky B (2008) Industry cluster analysis for the southeast Missouri region. Center for Economic & Business Research. Missouri State University, Springfield
17. Richards W (2007) Computer programs for social network analysis. http://www.insna.org/. Agosto 2013
18. Romo-Fernández LM, López-Pujalte C, Guerrero Bote VP et al (2011) Analysis of Europe's scientific production on renewable energies. Renew Energy 36(9):2529–2537
19. Schilling MA, Phelps CC (2007) Interfirm collaboration networks: the impact of large-scale network structure on firm innovation. Manage Sci 53(7):1113–1126
20. Stejskal J, Hajek P (2012) Competitive advantage analysis: a novel method for industrial clusters identification. J Bus Econ Manage 13(2):344–365
21. Stimson R, Stough RR, Roberts BH (2006) Regional economic development: analysis and planning strategy, 2nd edn. Springer, New York, p 452. ISBN 978-3-540-34826-34829
22. Su Y, Hung L (2009) Spontaneous vs. policy-driven: the origin and evolution of the biotechnology cluster. Technol Forecast Soc Change 76(5):608–619

How Organizational Cybernetics can Help to Organize Debates on Complex Issues

José Pérez Ríos, Iván Velasco Jiménez and Pablo Sánchez Mayoral

Abstract In this paper we expose how concepts taken from the Systems Thinking field, and in particular Organizational Cybernetics (OC), can help decision makers to study complex issues with the help of the information and communication technologies (ICT). We present some software tools developed within the Systems Thinking and Organizational Cybernetics Research Group (STOCRG) of the University of Valladolid that based on OC concepts use ICTs to help groups of people to study complex issues in a collaborative way through Internet. We discuss some uses of OC to help collective decisions making and we show two international pioneering experiences in which the ICTs have been used in combination with OC concepts. In the first case the purpose was to create a collective scientific book by a group of scientists working at distance and, in the second case, to organize a major academic international event. We end the paper showing an example of specific software (Debates Organizer) developed in the STOCRG through the last 15 years to facilitate any size of groups of people through Internet the organization of debates on complex issues.

Keywords Organizational Cybernetics · Team Syntegrity® · Debates Organizer

1 Introduction

In the last few decades the world has changed dramatically. Many are the interrelated factors involved in that change (socio-economic crisis, globalization, demographic, etc.). The consequence is that we live in a very complex social system. To qualify the level of complexity of a system (or a situation) Ashby [3] proposed the

J. Pérez Ríos (✉) · I. Velasco Jiménez · P. Sánchez Mayoral
Grupo INSISOC, Dpto. de Organización de Empresas y CIM, Escuela Técnica Superior de Ingeniería Informática, Universidad de Valladolid, Campus Miguel Delibes S/N, 47011, Valladolid, Spain
e-mail: rios@uva.es

I. Velasco Jiménez
e-mail: ivan.velasco@uva.es

P. Sánchez Mayoral
e-mail: mayoral@uva.es

C. Hernández et al. (eds.), *Managing Complexity*, Lecture Notes in Management and Industrial Engineering, DOI 10.1007/978-3-319-04705-8_35,

concept of variety (number of possible states of a system) and set in his Law of Requisite Variety that "only variety can destroy variety". At the start of the seventies, Conant and Ashby [8] had argued, in the famous theorem that bears their name, that "a good regulator of a system must be a model of the system" and that the variety of the regulator must be at least equal to the variety of the system that it pretends to regulate. If we think in the role of managers of organizations as "governors" of them then they need models adequate to the task, that means models with requisite variety. The Systems Thinking field and in particular Organizational Cybernetics (OC) provide some models like the Viable System Model (VSM) and tools like Team Syntegrity® (TS)[1] that can help decision takers to tackle the complex problems facing them.

If we focus our attention in how to facilitate the decision-making and communication processes we should consider that, as Pérez Ríos [13, 14] expressed it: "The new frontier of humanity is, at the start of the XXIst century, not so much scientific or technological development as an understanding of the complex social systems in which we are immersed. Such understanding is fundamental for our being able to deal effectively with the problems of social tension facing mankind. We must explore new ways to organize and engage in relations that will enhance the processes of communication and decision-making…certain fundamental challenges which still have not been resolved in a satisfactory way: for example, the development of group-decision processes which are at the same time democratic, creative and efficient, or the replacement of hierarchical organizational structures by other more democratic ones in which all points of view can be effectively taken into consideration." [14, pp. 201–202].

But, at the same time that the world changed and experienced an increase in its dynamic complexity, a whole new set of technological tools related to information and communication technologies (ICT) became available. Also the interest about the role that these can play as a support to higher levels of people participation in discussions and decision-making kept increasing. Examples of those technologies can be found in what is generally known as "groupware", which includes software for planning and programming in groups, computer-assisted cooperative undertakings, and the whole arsenal of tools that have appeared within the so-called Web 2.0 [2, pp. 253–265]. "The conjunction of these two cornerstones—namely, on the one hand, the new conceptual framework for the design of organisational structures and decision processes, and, on the other, the availability of a technological support allowing remote collective inter-communication—opens up new horizons for relations between individuals and institutions." [14, pp. 202].

In this paper we focus our attention on the application of concepts taken from the Systems Thinking field and in particular Organizational Cybernetics [4, 5, 6, 14] to help decision makers to study complex issues with the help of the information and communication technologies. In it we present some software tools developed within the Systems Thinking and Organizational Cybernetics Research Group

[1] Team Syntegrity® and Syntegration® are registered trademarks by Team Syntegrity Internacional Inc. and Malik Management Zentrum St. Gallen.

(STOCRG- INSISOC) of the University of Valladolid (UVA) that based on OC concepts use ICTs to help groups of people to study complex issues in a collaborative way through Internet. The paper is structure as follows. First we make reference to how OC has been applied to help collective decisions making. Next we show two international pioneering experiences in which ICTs have been used in combination with OC concepts. The first case purpose was to create a collective scientific book by a group of scientists working at distance and, in the second case, to organize a major academic international event. In the last part of this paper we show an example of specific software developed in the STOCRG through the last 15 years with the aim of helping any size of groups of people to organize debates on complex issues through Internet.

2 Organizational Cybernetics and Group Decision Making

Among the diversity of conceptual tools that OC can provide to help collective decision-making we will focus our attention on Team Syntegrity®[2] due to its value to facilitate those decision processes. "Team Syntegrity® consists basically of a methodology developed by Stafford Beer [7] with the aim of offering a creative, synergetic and participative platform for studying complex problems…which we might regard as a structured means of creating and communicating a group awareness" [14, pp. 203].

The goals of the TS application can be summarized as follows: "(1) To generate a high level of participation among the individuals concerned; (2) To provide a structure and a system of communication that guarantee the non-hierarchical nature of the process; (3) To benefit from the variety and wealth of knowledge supplied by each individual within the group, putting into practice the synergies derived from the interaction among all its members; (4) To create a collective awareness, if possible shared among all the members of the group, regarding the central issue being considered and analysed" [14, pp. 205].

The main phases in which it is structured the process of application of TS are the following: (a) *Opening question*. The TS application process starts when a question is asked concerning the issue to be studied or discussed. This question is normally (though not necessarily) of a general, open nature; (b) *Explosion of variety (Statements)*. In this stage, each participant prepares statements he/she considers relevant to the central question. The only requirements regarding these statements are that they can be refuted and are not very extensive; (c) *Reduction of Variety and grouping*. After issuing and grouping the statements, we go on to generate Aggregated Statements; (d) *Assigning topics to people*. When the topics for discussion have

[2] A detailed description of the Team Syntegrity protocols can be found in Pérez Ríos [11, 13] and in chapter five of the book: Pérez Ríos [14] Design and Diagnosis for Sustainable Organizations. The Viable System Method, Springer.

been identified, it is necessary to determine which persons among the group are going to take part in the debate on each of them. We need therefore to find out their preferences. Once these preferences have been ascertained, this information is processed with the aid of a computerised assignment algorithm[3], which tries to maximise the degree of satisfaction in the group; (e) *Generation of content.* In this phase the different teams which debate each of the main topics generates the information that clarifies the topic.

In the next section we will see some examples of how TS was used to help two different groups of people to work collaboratively through Internet.

3 Communication and Information Technologies and Group Decision-Making: Two Cases

The fast development and diffusion of ICTs opened new ways to apply elements of the OC to help people to debate complex issues without having to be necessarily in the same place. Two pioneering examples of this are the Stafford Beer Festschrift Project (SBFP) and the Horizonte 2000 Project.

The Stafford Beer Festschrift Project (SBFP) is the first application in the world of TS, using ICTs. The purpose of the project was to set up a collective study in which over 30 cyberneticians (among them J. Pérez Ríos, co-author of this chapter) from four continents and sixteen countries could create a scientific work, revealing the usefulness of S. Beer's different theories for all kinds of organisations and for society in general. This work would be presented to Beer to celebrate his 70th birthday. The presentation in fact occurred on September 25th 1996, at John Moore's University in Liverpool. The undertaking was carried out between October 1995 and July 1996. Almost all of the work, consisting of both identifying the chapters it would include (12) and drawing up the content (more than 600 pages), was done remotely via Internet. This scientific work has been published under the title: *To be and not to be that is the system: A tribute to Stafford Beer,* CD ROM [9]. In the book *Intelligent Organizations*, M. Schwaninger, co-director of the project writes about the elaboration process [15, pp. 123–128].

The second example is the Horizonte 2000 Project. The aim of this project was to promote the cooperation among the universities from Iberoamérica, Filipinas and those of Spanish influence in the U.S.A. The project was presented in the event named "I Encuentro de Rectores de Universidades Hispano-Americano-Filipinas" which, organized by the University of Valladolid, took place in Valladolid (Spain) on 23–27 march 1998. Its purpose was expressed as: "To identify and to start new ways of relationships among the various Spanish speaking universities. It intends, from 1998 on, to open a new historical period of relationships based upon the equality, democracy and mutual trust. To make it possible and to foster this process the

[3] The Systems Thinking and Organizational Cybernetics Research Group at the University of Valladolid developed a set of optimizing algorithms for various TS configurations.

new information and communication technologies will be used." [1, pp. 14–15]. The debates system created to make this possible was based on OC principles and it used a new software tool created for this event (Iberforo-98 Project). We will comment some details of this tool in the next section.

4 Group Decision-Making Software Tools. Debates Organizer

Based on OC principles and in some elements of the first phases of the TS approach we initiated in 1997 within the ST and OC Research Group of the UVA the development of two groups of software tools to support several phases of the process of knowledge capturing and organization of debates. The first group included software tools that helped the application of TS protocols. The second group included software tools oriented to facilitate the collective knowledge capturing and study of complex issues, as well as the realization of debates through Internet. These set of tools were presented in the Fourth Metaphorum Conference in Liverpool in 2006 [12].

The first group of tools (based in TS) includes: (1) Software to configure and organize groups sessions in different size options (number of persons: 30, 24, 28, 12, 6); (2) Software to optimize (maximize participants satisfaction) the assignments (persons to issues) in the physically organized sessions; (3) Software to facilitate through Internet the visualization in 3D of the various TS configurations (view of Topics and Participants, as vertex and struts respectively, in figures corresponding to various groups sizes/configurations as represented by the icosahedron-30, octahedron-12, tetrahedron-6, etc.).

The second group of tools was oriented to help decision makers to study complex issue through Internet. Here we include two different software applications. The first of them (*Col-KCap*) was created to help the members of a group to generate through Internet a causal map of the complex problem object of study. It allowed the incorporation of all kind of information both to the variables and to the relations between them, as well as the realization of a cross-impact analysis and classification of variables in four main groups (active, passive, critical and inert). The second set of software application (*Debates Organizer*) was oriented to facilitate any number of persons the organization of debates on complex issues through Internet (www.debatesorganizer.org).

The advantage of using this Internet modality of debates versus the physical meetings (i.e.: the meetings organized with TS, as mentioned in the previous paragraph) is that the persons who compose the group can be located anywhere in space and can intervene at the time that best fits their needs or availability. Another advantage is that a person is not limited to belong only to two specific teams (Topics), as happens in the physical applications of the TS (the two vertex connected by a strut). A person can participate "virtually" in as many Teams/Topics as she/he likes. Of course there are practical limitations about its number (time availability etc.).

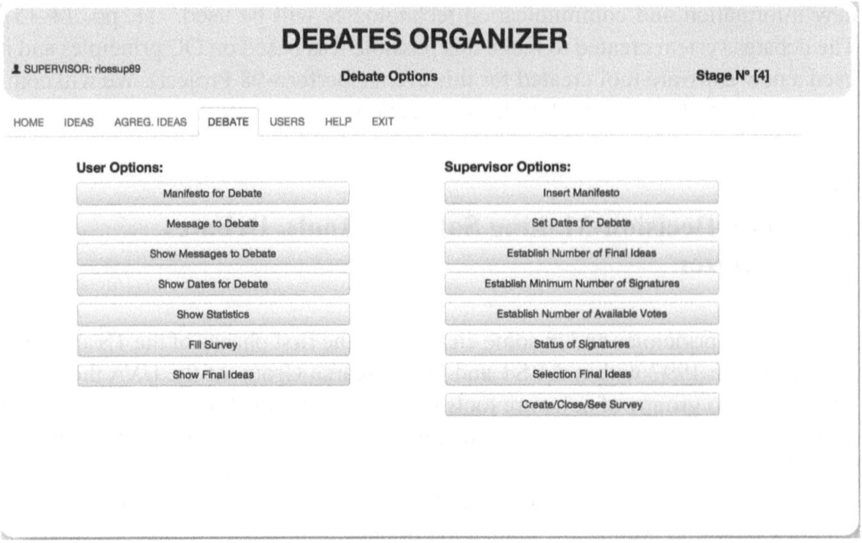

Fig. 1 Debates organizer (Administrator and Users options). www.debatesorganizer.org

The first version of this software was used in the above-mentioned Project Horizonte 2000 [10, 11] to organize the "I Encuentro de Rectores de Universidades Hispano-Americano-Filipinas" mentioned in the Sect. 3 [1]. This project financed in part by the BSCH, was the precursor of the Universia project created in 2000 by the BSCH. The information and communication based software tools used in this project constituted the Iberforo-98 project, later called Iberfora-2000. One of its components was the software to facilitate the organization of debates whose functioning we describe below. The initial version of this software which was used by the rectors who participated in the "I Encuentro de Rectores" included these three main phases of the Internet based debate: (a) Formulation of the Issue to be studied (b) Expansion of Variety and (c) Reduction of Variety and Aggregation of Ideas. An advanced version of this software was presented in the Metaphorum Conference in Liverpool [12]. Since then we have been using it continuously in our teaching activity in the UVA with engineering students (Information Science, Telecommunications etc.). Let us see briefly how it works.

The application of the *Debates Organizer* to study of a complex issue follows the steps that we describe next.

The organization of a debate starts with the identification of the people who is going to intervene in the process (can be located anywhere because their activity will be done through Internet) and the configuration of the debate. In Fig. 1 we can see some of the menu options for the administrator to configure the application (screenshot taken from the software *Debates Organizer*).

Once created the debate the first step is the launching of the question that expresses the issue to be clarified/answered. This question is presented to the group as a debate kick-off in the form of a manifesto to the group. For example in the case of the "I Encuentro de Rectores de Universidades Hispano-Americano-Filipinas" the question was: "How to organize (now and in future) the relationships between the various Universities *Hispano-Americano-Filipinas*, based on a common language and culture, in order to get the maximum benefit for our societies?" Another example is the issue/question proposed to students of Information Science Engineering: What do you think are the effects of the current financial crisis for society and what role ICTs can play in this new scenario? Once known the question the members of the group have a period of time to generate the statements/ideas that each of them considers it relates to answering the issue/question proposed.

In the next step the members of the group propose the aggregation of statements. The purpose of this phase is to consolidate related ideas into a more elaborated "Aggregated Statement". Each of the Aggregated Statement finally agreed would constitute the Final topics which condense the answer of the group to the question/issue under study. This process (generation of ideas, aggregation and final consolidation) is enriched with the comments (visible to the group) to all ideas made by the group members through time. The final step is the Generation of Content for each of the Final Topics. This is done typically in physical meetings but some research is in process about its realization at least partially with the help of Internet.

5 Conclusions

In this paper we presented some examples of applications of OC concepts and tools to help managers to study complex issues in a collaborative way. We commented two international projects (SBFP and Horizonte 2000) in which these have been applied in combination with ICT tools. We exposed as well several software developments done at the STOC research group with the aim of facilitating the organization of debate sessions through Internet. We provide also some information about how one of those -the *debates organizer* software- works. This application can be used not only to help managers of private companies to make group decisions but to all kind of organizations, no matter if they are public, private, big or small, for profit or non-profit etc. In fact one of our current lines of research is to explore how evolved version of this tools can provide support to a more collaborative and efficient citizens participation in public affairs.

References

1. Almuiña C, Martín R, y Pérez Ríos J (eds) (2000) Las Universidades iberoamericanas en la sociedad del conocimiento. Universidad de Valladolid, Valladolid, ISBN:84-8448-03-3
2. Almuiña C, Pérez Ríos J et al (2008) La relevancia de los medios de comunicación en Castilla y León. Consejo Económico y Social de Castilla y León. ISBN:978-84-95308-37-5
3. Ashby WR (1956) An introduction to cybernetics, Vol 2. Chapman Hall, London

4. Beer S (1979) The heart of enterprise. Wiley, Chichester
5. Beer S (1981) Brain of the firm, 2nd edn. Wiley, Chichester
6. Beer S (1985) Diagnosing the system for organizations. Wiley, Chichester
7. Beer S (1994) Beyond dispute: the invention of team Syntegrity. Wiley, Chichester
8. Conant RC, Ashby WR (1970) Every good regulator of a system must be model of that system. Int J Syst Sci 1(2):89–97
9. Espejo R, Schwaninger M et al (1997) To be and not to be that is the system: a tribute to Stafford beer, CD ROM. Carl Auer-Systeme, Wiesbaden
10. Pérez Ríos J (1998) La "sintegración en equipos" y el aprendizaje en las organizaciones. El caso de IBERFORO-98. Key note speech. I Reunión de Rectores de Universidades Hispano-Americanas-Filipinas. Marzo 1998, Valladolid, Spain
11. Pérez Ríos J (2000) Nuevas formas organizativas en sociedades complejas. In: Almuiña C, Martín R, Pérez Ríos J (eds) Las Universidades Iberoamericanas en la sociedad del conocimiento. Universidad de Valladolid, Valladolid, pp 291–317
12. Pérez Ríos J (2006) Information and communication technologies and organizational cybernetics. The fourth metaphorum conference. Liverpool John Moores University, Liverpool, UK, 4–5 May 2006
13. Pérez Ríos J (2008) Diseño y diagnóstico de organizaciones viables. Un enfoque sistémico. IBERFORA 2000. ISBN: 978-84-612-5845–1
14. Pérez Ríos J (2012) Design and diagnosis for sustainable organizations: the viable system method. Springer, Heidelberg
15. Schwaninger M (2006) Intelligent organizations: powerful models for systemic management. Springer, Berlin

Agent-Based Modelling and Archaeological Complexity

David J. Poza and Ricardo del Olmo

Abstract This paper presents a brief overview of social simulation in the field of Archaeology. It has been conducted from an agent-based modelling focus, a very interesting and valid methodology for modelling prehistoric societies. The second part of this work presents an application of this methodology in a specific model that analyses the emergence of ethnicity in a prehistoric society.

Keywords Archaeology · Agent-based modelling · Social simulation · Ethnicity

1 Introduction

Social simulation is a methodology where the modelled system is a social process. Its aim is to generate models of present or past realities that allow creating or validating theories about their regularities and, of course, their exceptions.

Amongst all the modelling paradigms of Artificial Intelligence that allow social simulation, we have opted for Agent-based Modelling (hereafter ABM), which is defined by Gilbert [18] as: "*a computational method that enables a researcher to create, analyse and experiment with models composed of agents that interact within an environment*". The idea is to represent a human society by means of software-objects (the so-called agents) that interact between one another in a virtual environment. There exists a correspondence between the individuals in the real society and the agents in the virtual environment in such a way that observing the evolution of the simulation in our model allows us to draw conclusions on the individuals in the modelled society.

D. J. Poza (✉)
INSISOC. E.I. Industriales, Universidad de Valladolid,
Pso del Cauce s/n, 47011 Valladolid, Spain
e-mail: poza@insisoc.org

R. del Olmo
Department of Civil Engineering, Higher Polytechnic School, University of Burgos,
C/Villadiego, S/N, 09001 Burgos Spain,
e-mail: rdelolmo@ubu.es

C. Hernández et al. (eds.), *Managing Complexity*, Lecture Notes in Management and Industrial Engineering, DOI 10.1007/978-3-319-04705-8_36,

ABM has some advantages that make it suitable for social research [2, 6, 7, 16]. We can summarize them as follows: it permits a more natural and transparent description of the modelled systems; it allows a realistic incorporation of explicit relationships from a geographic environment; it permits modelling local interactions, which implies considering the relationship between the micro-definition of the individuals and the macroscopic behaviour of the system; it captures the emerging behaviour, i.e. the appearance of new patters as a consequence of interactions at an elemental level [20]; and the possibility of including other dimensions (e.g. economical, social, geographical, technological, etc.) in a single model.

Social simulation is accepted in many scientific environments, including Archaeology, due to its versatility to simulate individuals' birth and death, their activities and their ways of interaction. As Barceló [3] puts it: "*This methodology is considered fundamental in Prehistoric Archaeology, whose aim, human action in a remote past is by definition non-observable. Artificial societies allow us to represent hypothetical models of what could have happened, which facilitates the analysis and the comparison of those models*".

This article is organized in two parts: First, we will analyse the relevance of social simulation in Archaeology, which will enable us to justify the importance of ABM when modelling and simulating prehistoric societies. Secondly, we will describe an application of ABM in the study of Ethnicity in a prehistoric society.

2 Archaeology and Agent-Based Social Simulation

Through the application of social simulation in Archaeology we can simulate the behaviour of individuals in past societies. "*Archaeologists and historians have started to transform social theories into computer programs by simulating social beings—obviously not humans—that 'live' in virtual environments that result from the generalization of social theory or historic data. By means of the implementation of social events as computational agents and their mutual influences as interactions we want to discover that collective action can be described and explained as something non-accidental and non-chaotic*" [3].

In the following paragraphs we will cite some examples and applications of ABM in this scientific domain that can be found in literature specialized in social and historic simulation.

In their project EOS, Doran and Palmer [14, 15] studied the emergence of social order in the Palaeolithic Age. Doran [13] continued this work to explain some features associated with the increase of social complexity. Other works related to hunter-gathering societies were conducted by Barceló [4, 5] to study how local interactions facilitate cultural diffusion and ethnic differentiation, and Del Castillo and Barceló [12]. De La Cruz et al. [10] developed a general model of social behaviour of a fishing-hunting-gathering society with strict social norms and no political institutions.

Dean et al. [11] use the model Sugarscape by Epstein y Axtell [17] as the basis for establishing the population dynamics in the Anasazi society (Arizona) between centuries IX and XIV. Lake [24] and Kohler et al. [23] studied spatial processes and resource management in ancient societies. Other similar works are: Gumerman et al. [19], Kohler et al. [22] and Johnson et al. [21].

To complete this brief review, we will cite the works by Christiansen and Altaweel [8] and Wilkinson et al. [25], related to the evolution and decadence of the settlements in Mesopotamia during the Bronze Age.

The case of study by Barceló et al. [5] that we describe in the following section evidences ABM's capabilities to shed light on Archaeology research. Traditionally, ancient societies have been tackled through universal theoretical models. However, the use of Agent-based Modelling in the study of ancient societies has recently increased. This tool allows researchers to create artificial societies based on archaeological, ethnological and historical data and observe how they interact in a virtual environment. By means of computer simulations, one can analyse the parameters that affect the outcome of the simulations, which may lead to identify yet unknown social relationships.

3 A Case of Study: the Emergence of Ethnicity and Cultural Differentiation in Patagonian Hunter-Gatherer Societies

Ethnicity is the historical emergence of groups of people who share cultural characteristics in such a way that members within a group regard themselves and are regarded by others as truly distinctive [9]. Ethnic differentiation is a consequence of how social agents have interacted over a period of time. Particularly, in the case of Patagonia's hunter-gatherer societies—target of this study—geographic separation among groups is believed to be crucial in the emergence of cultural groups; the further the groups, the less interactivity between groups and therefore, the more cultural differences between groups.

3.1 Description of the Model

The model consists of a population of a fixed number (N) of agents that move randomly in a virtual space searching for economic resources and interacting with other agents. Each agent represents a household in the Patagonian hunter-gatherer society under study. All the agents are identical, except for two properties:

- Capacity ($C_{j\in N}$): This parameter represents each agent's ability to exploit natural resources: $C_{j\in N} \in [0,1]$
- Cultural identity [1]: an integer vector of k features (i.e. cultural dimensions) that can take any value {1, 2, …, r} within a set of r cultural traits. In other words,

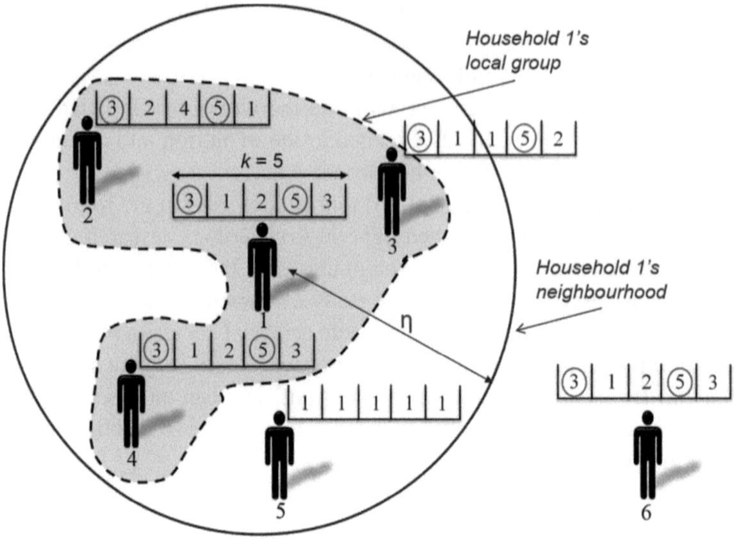

Fig. 1 An example of a local group from household 1's viewpoint for a given neighbourhood radius η. We have consideredk=5 cultural dimensions (which means that the cultural identity vectors have length 5) and r=5 cultural traits (i.e. each cultural dimension can take a value between 1 and 5). The parameter cultural proximity δ takes the value 0.4 which means that the agents within agent 1's neighbourhood need to share, at least, 2 (out of 5) cultural traits to be considered a member of her local group. This condition holds for agents 2, 3 and 4, but not for agent 5. Therefore, agents 2, 3 and 4 will cooperate with agent 1 in the current simulation period. Agent 6 belongs to the same ethnic group as agent 1 (as they share, at least, a proportion of δ cultural traits), but she is outside her local group and thus she will not cooperate in the current simulation period

each agent has k features each of which represent a cultural dimension (e.g. language, religious beliefs, etc.) and, each cultural dimension can take r different values (cultural traits).

The virtual space where the agents interact is a finite toroidal grid made up of regular patches. The agents interact with other agents from their neighbourhood as long as they belong to the same cultural group. The size of the neighbourhood is given by the parameter η.

In order to determine what agents belong to the same cultural group, the model computes the cultural proximity within the agents. Cultural proximity is measured by the relative number of shared cultural dimensions. That is to say, two households consider they belong to the same ethnic group if the number of cultural dimensions they share is above a certain threshold δ, which is another parameter of the model (cultural proximity). Therefore, an agent's local group (i.e. the set of agents with whom she will interact during the current simulation period) consists of other culturally-close agents within her neighbourhood (see Fig. 1).

The model considers increasing returns-to-cooperation (given by a parameter θ) which means that families get more resources when they cooperate than when they act individually (i.e. θ > 1 represents the fact that the amount of resources collected by two agents that cooperate is more than twice the amount of resources they would collect if they acted individually). Therefore, the output (i.e. the amount of resources) that an agent j gets depends of her own capacity c_j, the sum of the capacities of the agents within her local group $G_j(t)$ and the value of the parameter return-to-cooperation θ according to equation 1:

$$O_j(t) = c_j \left(\sum_{k \in G_j(t)} c_k \right)^{\theta - 1} \text{ with } \theta \geq 1 \tag{1}$$

Each agent j has a surplus $S_j(t+1)$ that depends on three factors (Eq. 2):

- The output obtained at the current period $O_j(t)$ given by Eq. (1).
- The remaining surplus from previous periods $S_j(t)$ multiplied by a depreciation factor $(1-\rho)$
- The amount of resources each agent consumes every period S_0:

$$S_j(t+1) = O_j + S_j(t)(1-\rho) - S_o \tag{2}$$

Culture diffusion and ethnic differentiation occurs through local interactions within the population. On the one hand, culture diffuses within the agents that interact in the same local group by copying—with certain probability $\mathbf{p_{diff}}$—the most frequent trait in her cultural group. This process favours the emergence of culturally-close groups. On the other hand, culture may evolve by means of a local mutation process, in which an agent can mutate one of her cultural traits with certain probability $\mathbf{p_{mut}}$. When this occurs, the new cultural traits are copied by all the agents in her local group.

The agents may die at some point during the simulation due to two reasons:

- Death by old age: the agents have certain life expectancy: their maximum age follows a Poisson distribution with mean a particular life expectancy (another parameter of the model). The agents' age increases at every simulation step. The agents die when they reach their maximum age.
- Death of starvation: At every simulation step, the agents consume an amount of surplus equal to S_0. They die when they run out of surplus ($S_j=0$).

Independently from the reason why an agent dies, she is replaced by another agent in such a way the number of agents remains constant (and equal to N). The new agent will inherit the cultural dimensions of an agent in the population chosen through a roulette wheel (i.e. the probability that an agent is replicated is proportional to her surplus).

The simulation takes place as follows: at a time period t the agents move randomly to one of their neighbour patches. Then, they determine what their local group is (i.e. agents within a distance η that are culturally-close, that is to say, agents that share at least a proportion of δ cultural traits). The cooperation of the agents that belong to the same local group may result in an increasing return-to-cooperation, according to Eq. 1. Then, the agent's surplus at the current simulation period is computed through equation 2. Afterwards, with probability p_{diff}, she copies a trait (the mode of the traits of her local group); and, with probability p_{mut}, she mutates one of her traits and spreads it to her local group. The agents are replaced either if they do not have enough surplus or if they reach their maximum age. Finally, the number and the size of the ethnic groups (i.e. all the agents in the population that share at least a proportion of δ cultural traits) at this period of the simulation are computed.

3.2 Results

This relatively simple model is the first approach to a more complex model that will permit Archaeology researchers to test different social theories by means of computational simulation. The simulations performed revealed the influence of the parameters increasing returns-to-cooperation (θ) and cultural proximity (δ) on the emergence of ethnic groups[1]. In fact, it was found that both parameters have a crucial effect on the outcome of the simulations (i.e. the number and the size of the ethnic groups).

For low values of δ, the agents only need to share a few cultural traits to be considered members of the same ethnic group. This gives rise to one simple ethnic group in which cultural mutation and local diffusion processes are not enough to break the group's ethnic identity. However, when the value of δ is increased, cultural mutation and local diffusion split the population into several ethnic groups.

On the other hand, when the value of the parameter θ is low, the size of the emergent ethnic groups is small. Low values of θ mean low returns-to-cooperation. In fact, for $\theta \approx 1$, cooperating with the local group does not provide more surplus than acting individually. Consequently, the replication process facilitates the reproduction of any emerged cultural differentiation and therefore we observe a higher number of ethnic groups. For larger values of θ, cooperating within a local group yields more surplus and therefore the agents who cooperate get more replications in the future generations, which splits the population into a smaller number of ethnic groups.

One of the conclusions we can draw from this brief study is that ABM is a really relevant burgeoning methodology in social simulation in general and in Archaeology in particular. A proof of that is the model by Barceló et al. [5] that we have described in this paper. It is an outstanding example of how ABM and Archaeology can be jointly used in the study of the emergence of ethnicity and cultural differentiation in ancient societies.

This model is currently being extended in order to simulate a more realistic behaviour. In fact, in the current version of the model, the agents always cooperate with other agents (as long as they are culturally similar). However, being culturally similar might not be a sufficient condition to cooperate. In order to make the agents

---

[1] All the experiments were conducted with the following parameters: $N=50$; $\eta=2$; $k=8$; $r=8$; pdiff=0.8, pmut=0.05; S0=0.4 and $\rho=0.5$.

decide whether to cooperate or not, the patches will be endowed with random amounts of resource and different levels of accessibility to those resources. This has an important implication: the agents will act by themselves unless they cannot get enough resources to survive (because there are too little resources, they are hard to collect, or the agents' capacities are low). If this is the case, they will ask other culturally-close agents for help. However, this demand for help might be rejected by other agents if their capacity does not even let them collect the resources they need to survive.

Acknowledgement The authors acknowledge support from the Spanish Ministry of Science and Innovation Project CSD2010-00034 (SimulPast CONSOLIDER-INGENIO 2010) and by the Regional Government of Castile and Leon (Spain) Project VA056A12-2.

References

1. Axelrod R (1997) The dissemination of culture. J Confl Resolut 41(2) 203–226
2. Axtell RL (2000) Why agents? On the varied motivations for agent computing in the social sciences. In: Macal CM, Sallach D (eds) Proceedings of the workshop on agent simulation: applications, models, and tools. Argonne National Laboratory, Argonne, pp 3–24
3. Barceló JA (2010a) Sociedades artificiales para el análisis de procesos sociales en la prehistoria. CPAG 20:123–148
4. Barceló JA, Cuesta F, Del Castillo F (2010b) Simulating social, economic and political decisions in a hunter-gatherer group. The case of prehistoric Patagonia. Computer Applications in Archaeology Conference. http://www.leidenuniv.nl/caa/caa_proceedings.htm. Leiden. Accessed Feb 2013
5. Barceló JA, Cuesta F, Del Castillo F et al (2010c) Simulating prehistoric ethnicity. The case of Patagonian hunter-gatherers. In: Contreras F, Melero M (eds) Fussion of cultures. Computer applications and quantitative methods in archaeology.
6. Bonabeau E (2002) Agent-based modeling: methods and techniques for simulating human systems. Proc Natl Acad Sci USA 99(2):7280–7287
7. Bousquet F, Le Page C (2004) Multi-agent simulations and ecosystem management: a review. Ecol Modell 176(3–4):313–332
8. Christiansen J, Altaweel M (2006) Simulation of natural and social process interactions: an example from Bronze Age Mesopotamia. Soc Sci Comput Rev 24(2):209–226
9. Cohen AP (2000) Signifying identities: anthropological perspectives on boundaries and contested values. Routledge, London
10. De La Cruz DE, Noriega P et al (2010) Normas en sociedades cazadoras-pesacadoras-recolectoras. Argumentos para el uso de la simulación social basada en agentes. CPAG 20:149–161
11. Dean JS, Gumerman GS, Epstein JE et al (2000) Understanding Anasazi culture change through agent-based modelling. In: Kohler TA, Gumerman G (eds) Dynamics in human and primate societies. Oxford University Pres, New York, pp 179–205
12. Del Castillo F, Barceló JA (2012) Why hunter and gatherers did not die more often? Simulating prehistoric decision making. Proceedings of the Computer Applications and Quantitative Methods in Archaeology Congress. Amsterdam University Press
13. Doran JE (1999) Prospects for agent-based modelling in archaeology. Archeologia e Calcolatori 10:33–44.

14. Doran JE, Palmer M (1995a) The EOS project: integrating two models of palaeolithic social change. In Gilbert N, Conte R (eds) Artificial societies: the computer simulation of social life. UCL Press, London, pp 103–125
15. Doran JE, Palmer M (1995b) The EOS project: modelling prehistoric sociocultural trajectories. Proceedings of First International Symposium on Computing and Archaeology, pp 183–198
16. Epstein JM (1999) Agent-based computational models and generative social science. Complexity 4(5):41–60
17. Epstein JM, Axtell R (1996) Growing artificial societies: social sciences from the ground up. Brooking Institution Press, Washington
18. Gilbert N (2007) Agent based model. Sage, London
19. Gumerman G, Swedlund AC, Dean JS, Epstein JS (2003) The evolution of social behaviour in the prehistoric American Southwest. Artif Life 9(4):435–444
20. Holland JH (1998) Emergence. From chaos to order. Addison-Wesley, Reading
21. Johnson CD, Kohler TA, Cowan JA (2005) Modeling historical ecology, thinking about contemporaly systems. Am Anthropol 107:96–108
22. Kohler TA, Gumerman G, Reynolds RG (2005) Simulating ancient societies. Sci Am 293:77–84
23. Kohler TA, Kresl J, Van West CR, Carr E, Wilshusen R (2000) Be there then: a modelling approach to settlement determinants and spatial efficiency among late ancestral pueblo populations of the Mesa Verde region, U.S. southwest. In: Kohler TA, Gumerman GJ (eds) Dynamics in human and primate societies. Oxford University Press, New York, pp 145–178
24. Lake MW (2000) MAGICAL computer simulation of mesolithic foraging. In: Kohler TA, Gumerman GJ (eds) Dynamics in human and primate societies. Oxford University Press, New York, pp 107–143
25. Wilkinson TJ, Christiansen J, Ur J et al (2007) Urbanization within a dynamic environment: modelling bronze age communities in upper Mesopotamia. Am Anthropol 109(1):52–68

Complex Networks Applied to Communication Management in Emergency Crisis

Cristina Ruiz Martín, Mario Ramírez Ferrero, José Luis González Álvarez and David J. Poza

Abstract The Nuclear Emergency (NE) influence domain spreads over a large number of systems of different nature. Nuclear Emergency Plans (NEPs) are usually designed using a top-down approach, establishing a hierarchy working as a command chain. Although this provides advantages in terms of control, it might not be flexible enough to face unpredicted emergent behaviours due to complex interactions among the systems involved. Using a real case study, this work discusses the results of the communication and command chain network analysis to facilitate the assessment of key indicators to improve the NEP: resilience, adaptability and responsiveness.

Keywords Nuclear emergency · Nuclear emergency plan · Complex networks · Communication

1 Introduction

NEPs have to be designed for the compliance of a framework of laws, regulations and directives. Traditional approaches to warrantee this compliance use a top-down perspective to establish a hierarchical structure similar to a military

C. Ruiz Martín (✉)
Grupo INSISOC, Dpto. de Organización de Empresas y CIM, Escuela de Ingenierías Industriales, Universidad de Valladolid, Pso del Cauce S/N, 47011, Valladolid, Spain
e-mail: cruiz@insisoc.org

D. J. Poza
INSISOC. E.I. Industriales, Universidad de Valladolid, Pso del Cauce S/N, 47011 Valladolid, Spain
e-mail: poza@insisoc.org

M. Ramírez Ferrero
Grupo INSISOC, Universidad de Burgos, Burgos, Spain
e-mail: mario@insisoc.org

J. L. González Álvarez
Unidad de Protección Civil y Emergencias, Delegación del Gobierno en Castilla y León, Valladolid, Spain
e-mail: jlsmithbarrett@gmail.com

C. Hernández et al. (eds.), *Managing Complexity*, Lecture Notes in Management and Industrial Engineering, DOI 10.1007/978-3-319-04705-8_37,

command chain. This approach is focused on providing a quick, organised and effective response. It has advantages in terms of control and lack of behaviour uncertainty and information ambiguity. However, it makes more difficult to analyse the systems involved in the emergency because it does not allow management mechanisms to handle *emergence* due to complex interrelations among these systems. Unexpected or unmanaged *emergent behaviours* can modify the designed work of the system.

This NEP case study [1, 2] deals with the NE assuming the existence of an error-free command chain, the lack of undesired behaviours such as identity theft and that the countermeasures planed will be enough to control and mitigate the emergency. Nevertheless, after the accident at TEPCO's Fukushima Nuclear Power Plant (NPP), the International Atomic Energy Agency (IAEA) stated the needed of improvements to strengthen management systems, response arrangements and transparency and effectiveness of communication mechanisms [6].

In other works, information systems for emergency response have been proposed [13], multi-criteria decision making mechanisms have been published for radiological disasters [5, 9] or other scenarios such as e-government [12], methods for evacuation management support has been defined [7], studies about task time prediction have been carried out [8], but they do not tackle the problem under an organizational and communication management point of view. Knowledge management has to be considered during a NE because of its socio-technical dimension; understanding information and communication management is a prerequisite to solve the crisis [3].

Following the work at [10] and [11], the aim of this paper will be to discuss the results of the NEP Communication Network (CN) and Command Chain Network (CCN) analysis with several purposes: to propose resilience improvements, and to identify imprecisions, complementarities and synergies to develop a better comprehensive plan, taking into account that the role of people involved in higher levels of the hierarchy is more a coordinating than a directing one.

To present these objectives the following structure will be applied: Sect. 2 will describe the modelling and analysis methodology used; Sect. 3 will describe the NEP; Sect. 4 will discuss the results of the CN & CCN analysis on the basis of complex networks methodology; eventually, Sect. 5 will exhibit the conclusions and future lines of this work.

2 Modelling and Analysis Methodology

The methodology currently used in the roadmap to conceptualise and build the NEP model is based on Agent-Based Modelling (ABM) techniques, and it is summarised in Fig. 1 [4].

The first step was to build a theoretical model with the information provided by one of the responsible for the emergency plan [1, 2]. The people involved in the plan (agents), the communications among them (distinguishing the different media), the messages and the actions to be taken and the resources to be committed were identified.

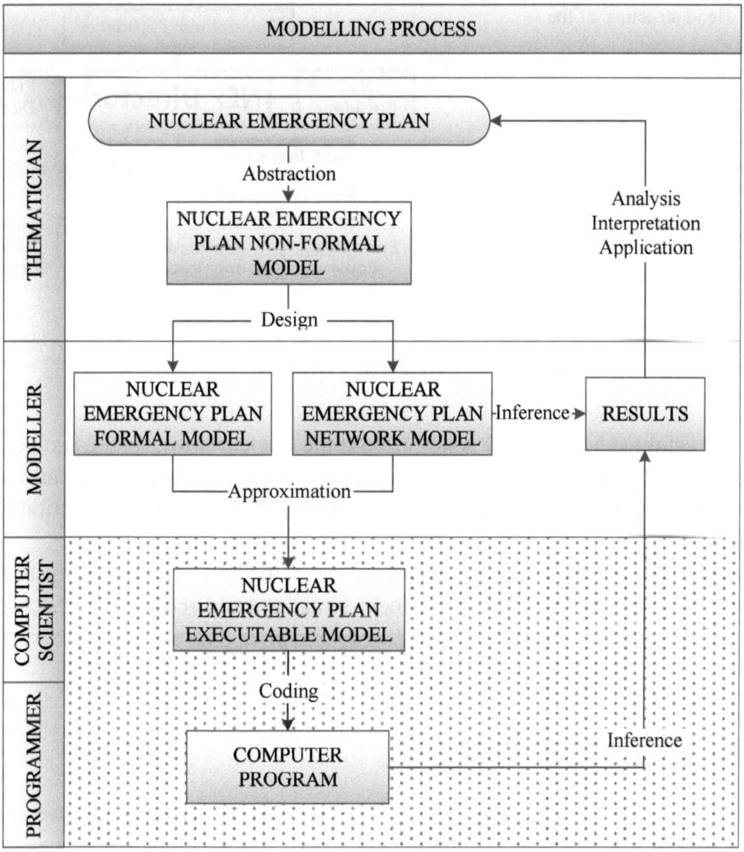

Fig. 1 Modelling process

Then, as a starting and reference point, the model was formalised assuming that only those communications and actions specified in the NEP exist. Communications not specified in the NEP could be an important source for unpredictable emergent behaviour, so they will be taken into account in future scenarios.

Once the formal model was developed, before building an executable model and carrying out dynamic analysis taking into account the interrelations among agents by actions (future works will tackle this issues), an initial study of the characteristics and properties of CN & CCN on the basis of complex networks was carried out. It provided metrics to extract important information at this first stage. The result discussion is explained in Sect. 4.

It is worth mention that, for this work, according to Fig. 1 scheme, the modeller, computer scientist and programmer roles have been handled by the same person. The inference process in the modeller role consists of obtaining the different parameters from the network.

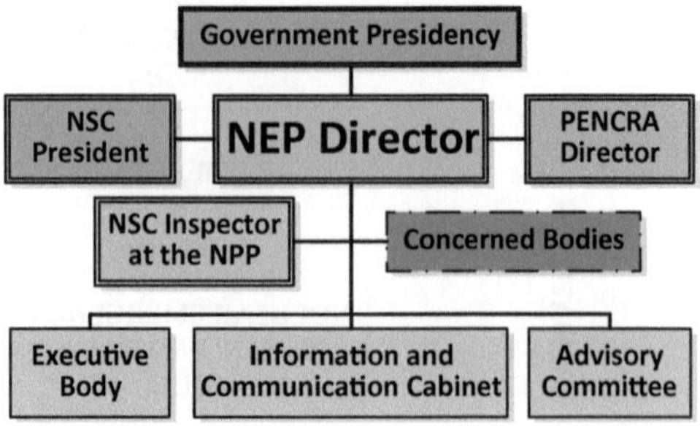

Fig. 2 Head structure of the NEP

3 Nuclear Emergency Plan Description

The NEP establishes 3 control zones around the nuclear power plant. Its location is such that these zones extend over 77 municipalities in 3 provinces of 3 autonomous regions of the country, each one with its own regional/local administration and organisms.

The NEP defines the organisational structure of the bodies involved, the resources to be committed, communication channels, situations and actions to be taken before (preventive), during (control and mitigation) and after (recovery) the emergency.

3.1 Organisational Structure

Depending on the implication of each organism in the emergency, two types of bodies are considered: competent and concerned. While competent bodies have direct action on the NE, concerned ones only act in some cases but may be required to inform or take some actions, so the NEP Director has to inform them about the emergency evolution.

Figure 2 represents the head structure of the NEP. The chair is delegated on the NEP Director, which is the overall responsible for the decisions taken, although higher National Government ranks can relieve him and assume control if necessary. Every single order must come from the NEP Director and be communicated following the hierarchy. No member accepts orders that do not come directly from any of his superiors or from the Public Security and Order Group, since they are also considered as authorities.

Nuclear Safety Commission (NSC) President and PENCRA Director are at the same level as NEP Director. They have to be informed about every action taken to solve the crisis and its evolution. NSC Inspector at the NPP communicates the measures taken to solve the incident at the nuclear power plant to NEP Director directly, but he is a member of the NSC. The Advisory Committee has to advise NEP Director in decision making. Information and Communication Cabinet is charge with the media. The Executive Body has to materialise the actions commanded by the NEP Director to solve de crisis, and it is constituted by several groups, (Radiological, Health, Logistical Support, Public Security and Order and Technical Assistance and

Coordination) each one with its own predefined structure. The Team Leaders of these groups are included in the head structure of the NEP. Additional Team Leaders from other institutions can be invited to join the Executive Body at the discretion of the NEP Director.

3.2 Resources and Communications

Depending on their nature, three kinds of resources are considered in the NEP: specific resources (only used in NEs, such as radiation measurement and surveillance equipment); not-specific resources (also used for other emergencies); and general use resources (used for a general purpose but also needed during a NE). These resources can also be classified distinguishing between those already available since the emergency beginning (radiological protection equipment, etc.) and those the NEP Director has to request and that will be available after a period of time (CBRN equipment, buses, etc.).

Two types of communications can be considered: those due to the NEP, mainly related to the command chain and information requests and reports, and those not planned but still appearing among or inside the systems. Focused on the first ones, it can be distinguished between the whole communication network, analysed at [10], and just the Command Chain Network (CCN) analysed at [11].

3.3 States and Actuations to Overcome the Emergency

Different situations to be declared at the discretion of the *NEP Director* are identified at the NEP. Within them, sets of actions are described relating to emergency declaration or conclusion, deployment and repositioning of teams, protection, information request, reporting, resolution of conflicts, and request of resources. It also details the operational behaviour to be followed during the whole emergency. The NE will always be governed centrally; teams may operate independently, but subordinated to the central government.

4 Discussion and Analysis of the Nuclear Emergency Plan

To build the NEP model used in this research, the people involved in the nuclear emergency (agents), their hierarchy, the messages and commands, the senders and receivers and the actions to be developed by the agents have been identified.

At this step we detected that one of the groups of the Executive Body (Logistical Support Group) has not documented its structure and each member's responsibility. Inside the group, this fact can cause that the adaptation time for new hires will be longer. Between the remaining groups, it carries out that other stakeholders don't know the way they work and they would not be able to detect possible inefficiencies.

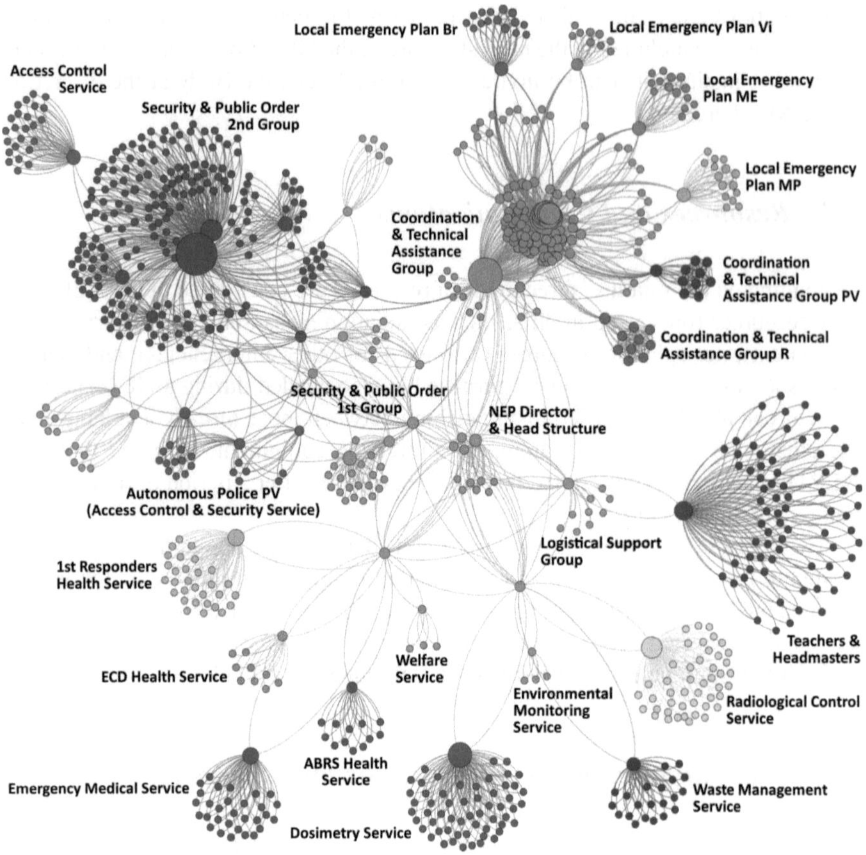

Fig. 3 NEP's communication network once the emergency has been declared, the executive body is assembled but no further action has been taken

The assumptions made to build the NEP were very restrictive. It was developed assuming that there would not be contradictory or incompatible commands, everyone involved in the nuclear emergency will abide the commands as soon as possible without undesirable behaviours. Due to the great number of agents involved in the emergency, the implementation of a management communication system would reduce the probability of receiving contradictory or incompatible orders and the probability of communication channels collapse. Furthermore, it would facilitate the information managing. The functional requirements to this system are: (1) a data base implementation (with different query and write permissions) with the orders and actions assigned to the teams. (2) An auto-confirmation system for the information reception or action developed that writes this information on the data base.

Once the model was built, a Communication and a Command Chain Network analysis has been done on the basis of complex networks. Figure 3 show the CN once the emergency has been declared, the executive body is assembled but no further action has been taken [10, 11]. These analyses have provided metrics to evaluate and improve the NEP.

It has been detected that the CN is not resilient enough. There are some services (Health Group & Logistical Support Group) whose CCN is handled by a single communication channel, so a failure in this channel can cause the isolation of some agents, services or groups. To improve the NEP resilience is necessary to implement communication redundancies to avoid problems caused by a single communication channel breakdown.

The centrality and prestige analysis has shown that in general, the most important nodes are the ones expected from the NEP design. There is an exception. The PageRank (a measure of how important is a node taken into account its connections—number of agents a node can command and number of nodes he can be commanded by) in the CCN in the deployed scenario (once all teams have been positioned) points out the most important node is the Local Emergency Plan Head of one municipality as he is the only one with a PageRank value higher to 75 % of the maximum, but the design indicates the most important node should be the NEP Director. The explanation for this result is that this municipality is comparatively big and the Local Emergency Plan Head has a relatively high number of agents following his commands. On the other hand, he can be commanded by all the workers at Coordination and Technical Assistance Group and by two teams from Security and Public Order Service located at his municipality.

There are several workers at Coordination and Technical Assistance Group who can order all the Local Emergency Plan Heads. This has revealed the possible existence of authentication problems of the personnel when an order is edict.

Acknowledgments Authors have been supported by Spanish Ministry of Science and Innovation (CONSOLIDER CSD2010-00034), Regional Government of *Castilla y León* (VA056A12-2), BPMSat and University of Valladolid. Special thanks to *Delegación del Gobierno en Castilla y León* and INSISOC's colleagues.

References

1. B.O.E. (Spanish Government Gazette) (Jan 1985) Law 2/1985, 21 January, on Civil Protection
2. B.O.E (Spanish Government Gazette) (Sept 2009) Royal Decree 1428/2009, September 11th, on the modification of the nuclear emergency basic plan approved in the Royal Decree 1546/2004
3. Dogan H, Henshaw M, Ragsdell G (2011) The risk of information management without knowledge management: a case study. J Inform Knowl Manage 10(4):393–408
4. Galán JM, Izquierdo LR, Izquierdo SS, Santos JI, del Olmo R, López-Paredes A, Edmonds B (2009) Errors and artefacts in agent-based modelling. J Artif Soc Soc Sim 12(1):1
5. Geldermanna J, Bertscha V, Treitza M, Frenchb S, Papamichailb KN, Hämäläinenc RP (2009) Multi-criteria decision support and evaluation of strategies for nuclear remediation management. Omega Int J Manage Sci 37:238–251
6. Langlois L (2012) Accepted November 26th 2012, unpublished. IAEA action plan on nuclear safety. Energy Strategy Review
7. Lv Y, Huangb GH, Guob L, Li YP, Daid C, Wangc XW, Sunb W (2013) A scenario-based modeling approach for emergency evacuation management and risk analysis under multiple uncertainties. J Hazard Mater 246–247:234–244

8. Park J, Jung W (2007) A study on the development of a task complexity measure for emergency operating procedures of nuclear power plants. Reliab Eng Syst Safe 92:1102–1116
9. Parlak AI, Lambert JH, Guterbock TM, Clements JL (2012) Population behavioral scenarios influencing radiological disaster preparedness and planning. Accident Anal Prev 48:353–362
10. Ramirez M, Ruiz C, Gonzalez JL, Lopez A (2013) Modelling of a nuclear emergency plan: a systems of systems engineering approach. 8th International Conference on System of Systems Engineering (SoSE). http://www.insisoc.org/SoSE2013/. (in press)
11. Ruiz C, Gonzalez JL, Ramirez M, Lopez A (2013) Modelling of a nuclear emergency plan: command chain network analysis. Intelligent systems and decision making for risk analysis and crisis response: proceedings of the 4th international conference on risk analysis and crisis response, Istanbul, Turkey, 27–29 August 2013
12. Shan S, Wang L, Li L, Chen Y (2012) An emergency response decision support system framework for application in e-government. Inform Technol Manage 13:411–427
13. Turoff M, Chumer M, Van de Walle B, Yao X (2003) The design of a dynamic emergency response management information system (DERMIS). J Inform Technol Theory Appl 5(4):1–35

Localization Based on Business Interactions Through a Simulated Annealing Algorithm

Rosa Mª Sánchez-Saiz, José Manuel Galán and José Ignacio Santos

Abstract This work is aimed at presenting a simulated annealing algorithm as a decision support tool for the localization problem of stores in metropolitan areas. The approach is based on the empirical estimation of attraction and repulsive forces that emerge as a consequence of the spatial interaction among businesses. Quantification of these externalities is carried out by means of networks modelling techniques. The methodology is illustrated with a case study in the city of Turin (Italy).

Keywords Simulated annealing · Localization problem · Complex networks · Externalities · Decision support tools

1 Introduction

Economic and industrial organization literature has analysed and postulated theories about the relationships that exist among the different types of business and how they affect each other in their location decisions. Many of these models lack of a formal formulation to quantify these interactions. Recently, some authors have proposed methodologies to begin to understand quantitatively the nature of these phenomena [3 ,4, 7, 8].

Location patterns of firms do not seem the result of individual independent decisions or purely random events, but they are, at least partly, consequence of repulsive, attractive or mutual indifference forces among them [5].

Repulsive forces are usually found in business focused on input or customer markets, and on those that do not work for profit, such as foundations, associations, schools, universities and health centres. In these cases, firms make production

R. Mª. Sánchez-Saiz (✉) · J. M. Galán · J. I. Santos
Departamento de Ingeniería Civil.EPS, Área de Organización de Empresas,
UBU. C. Villadicgo S/N, 09001 Burgos, Spain
e-mail: rsanchez@ubu.es

J. M. Galán
e-mail: jmgalan@ubu.ess

J. I. Santos
e-mail: jisantos@ubu.es

C. Hernández et al. (eds.), *Managing Complexity*, Lecture Notes in Management and Industrial Engineering, DOI 10.1007/978-3-319-04705-8_38,

and pricing decisions in response to how customers are spatially located, which conditions spatial patterns. Some of the classical models of spatial competition, as Hotelling [6], show that geographic concurrence can emerge as consequence of competing stores trying to capture sales in central areas.

However, there are other activities characterized by agglomerative patterns, for example:

1. Companies that produce differentiated products that may vary in several aspects, requiring that buyers inspect them in person. This applies to car dealers, nightlife districts, advertising, clothing stores, etc. In these cases competitors often locate close together creating clusters.
2. Companies in dynamic environments, in terms of products or production cycles, which require the use of expensive specialized machinery. In these cases concentration may make possible reducing production costs by means of sharing input factors or the appearance of suppliers specialized in particular operations that improve the efficiency of the whole cluster.

In the most general cases there is a combination of repulsion and attraction forces among competitors.

In this paper we briefly analyse the foundations of some methodologies that capture empirically aggregation and dispersion phenomena, and the forces of attraction and repulsion. Those methodologies go beyond the relations between activities of the same type and generalize the idea to interactions among different types of businesses, e.g. searching for possible complementarities. We illustrate in a particular case study the use of this formalization together with a fitness function and a simulated annealing algorithm as a decision making tool to find candidate locations that explicitly takes into account interaction externalities.

2 Method

2.1 Network-Based Model of Attraction and Repulsion Forces

A recent attempt to study quantitatively the relationships of interaction between businesses corresponds to Duranton and Overman [3]. Although they focus on the spatial distribution of industries, and not on the distribution of commercial stores in urban areas like our case, their approach is interesting because it makes use of the statistical properties of the empirical data to infer agglomeration measures. Inspired by Duranton and Overman work, Jensen et al. [4, 7, 8] proposed an original network-based approach to quantify empirically the intensity of the attraction and repulsion forces between commercial stores.

Suppose an urban area with N_T stores corresponding to a set of sites $T = \{T_i; i = 1,\ldots,N_T\}$ where they are placed. The subset $A = \{A_i; i = 1,\ldots,N_A\}$ is formed by the N_A sites occupied by stores of type A. For each site $S \in A$, we define the variables $N_T(S,r)$ and $N_A(S,r)$ as the number of respectively total and A-stores, including S, whose distance from the site S is less than or equal to a radius r. The spatial interaction between A-stores is quantified by an intra-attraction coefficient $M_{AA}(r)$ as follows:

$$M_{AA}(r) = \frac{N_T - 1}{N_A(N_A - 1)} \sum_{i=1}^{N_A} \frac{N_A(A_i,r)}{N_T(A_i,r)} \tag{1}$$

The coefficient $M_{AA}(r)$ is an average over the N_A A-stores of the result of dividing the local aggregation ratio $\frac{N_A(A_i,r)}{N_T(A_i,r)}$ and the total aggregation ratio $\frac{N_T - 1}{N_A - 1}$. When $M_{AA}(r) > 1$ we say that the type of stores A shows spatial aggregation, i.e. the local aggregation average of A-stores is greater than the expected value if they were randomly distributed in the set of sites T. On the other hand, when $M_{AA}(r) < 1$ we say that the type of stores A shows spatial dispersion.

We can easily generalize the last measure to all sites $S \in T$ and any pair of types of stores A and B. Now, the spatial interaction between A-stores and B-stores is quantified by an inter-attraction coefficient $M_{AB}(r)$ as follows:

$$M_{AB}(r) = \frac{N_T - N_A}{N_A N_B} \sum_{i=1}^{N_A} \frac{N_B(A_i,r)}{N_T(A_i,r) - N_A(A_i,r)} \tag{2}$$

Where $N_T(A_i,r)$, $N_A(A_i,r)$ and $N_B(A_i,r)$ are the number of respectively total, A-stores and B-stores whose distance from the site A_i is less than or equal to the radius r. In this case, when $M_{AB}(r) > 1$ we say that the type of stores A attracts B-stores, i.e. the local aggregation average of B-stores around A-stores is greater than the expected value assuming that B-stores were randomly distributed; and consequently we say that the type of stores A repulses B-stores when $M_{AB}(r) < 1$.

The intra and inter attraction coefficients can be used to model the spatial interaction between commercial stores as a network [4, 8]. It is not difficult to see the matrix of the attraction coefficients as an adjacency matrix of a weighted directed network [10]. The nodes of this interaction network are the set of types of stores, and the weights of the links between any pair of nodes A and B is defined as:

$$\propto_{AB} = \log(M_{AB}) \tag{3}$$

The logarithmic transformation of the attraction coefficients makes easy the interpretation of the link weights: a positive value means a relationship of attraction and negative one a relationship of repulsion.

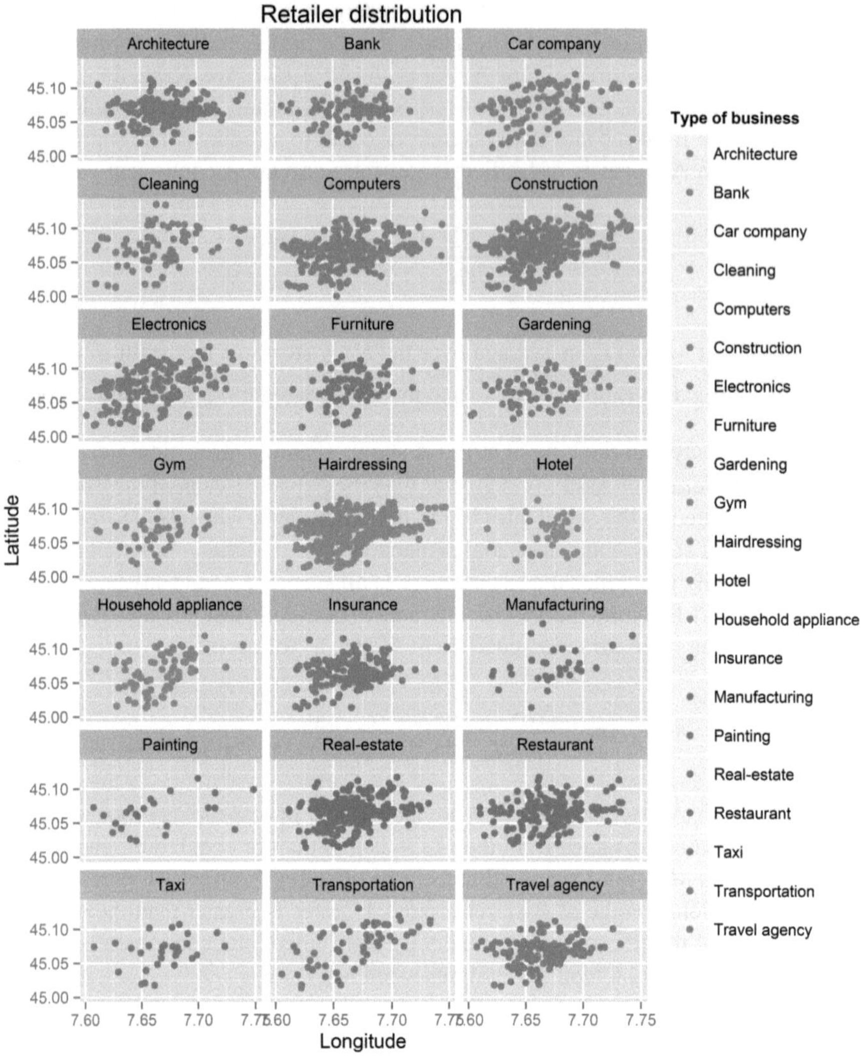

Fig. 1 Different types of business considered in our analysis and their spatial location

3 Localization Decision Making Using a Simulated Annealing Algorithm

3.1 Case Study

To illustrate our approach we have considered interactions among 21 types of business (see Fig. 1) located in the area of Turin, an Italian city in the north of the

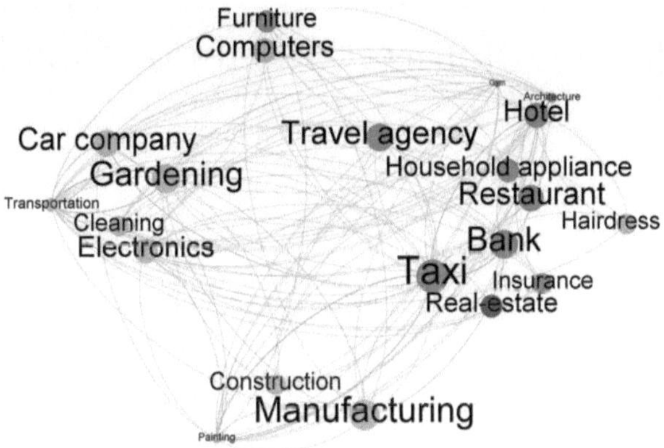

Fig. 2 Interactions among the different types of business considered. Colour of nodes represents different classes found after applying a community detection algorithm. Node sizes show the number of stores of each type included in the analysis, green links represent positive associations while pink links represent negative associations. The figure only contains the main significant interactions, those in the range [−0.2, 0.2] are not included

country. Our database contains 3866 georeferenced stores ranging from 541 in the case of the most numerous type, to 23 in the case of the less.

We have computed the attraction matrix among the different types of stores following the explained methodology. The results of the main types of interactions are summarized in Fig. 2 as a graph, where each element α_{AB} of the matrix is represented as a link showing the interaction between A and B type of business.

3.2 Search of Candidate Locations Using a Simulated Annealing Algorithm

Once business interaction has been quantified by means of the attraction matrix, Jensen [7] proposes a mathematical index to measure the fitness of a candidate place for a new store of a given type of business. This index is based on the assumption that current store localizations correctly capture positive and negative interaction externalities through the empirically estimated matrix. Jensen defines the index as follows: Let be x, y the coordinates of a geographic point candidate to hold a type of activity A, then the quality of this point is given by:

$$Q_A(x,y) = \sum_B \propto_{AB} \left(neA_{AB}(x,y) - \overline{neA_{AB}} \right) \tag{4}$$

Where $neA_{AB}(x,y)$ represents the number of stores in the neighbourhood of (x,y), and $\overline{neA_{AB}}$ represents the average number of B stores around an A store (again given a neighbourhood defined by a radius).

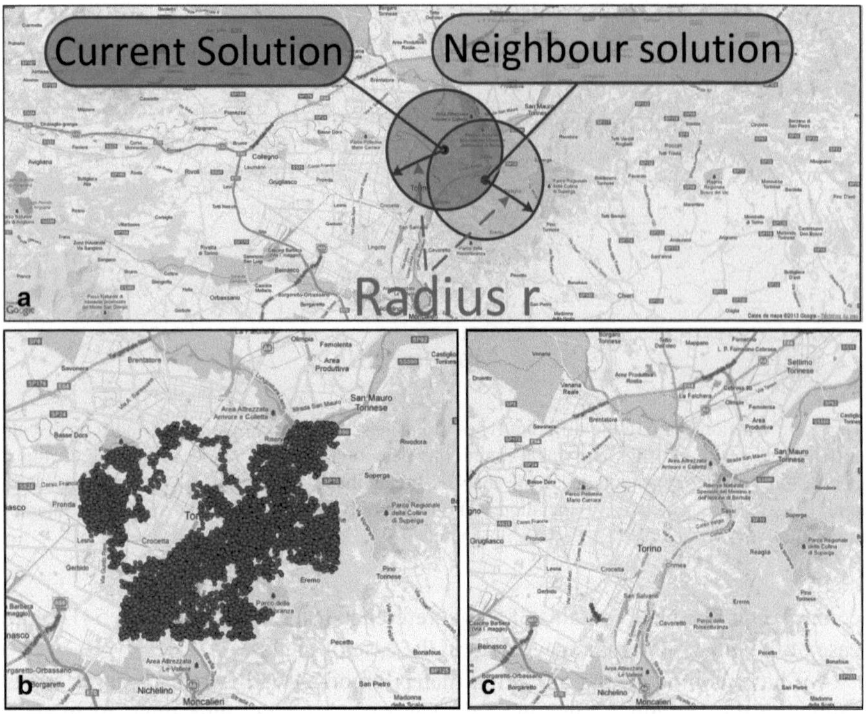

Fig. 3 Panel **a** sketches the mechanism to create a neighbour solution. Panel **b** represents solutions visited searching for a new bank location and considering F fitness function. Parameters used are $T_0=200$, $T_f=0.2$ following a Cauchy cooling schedule [2] and 10 solutions for each temperature. Panel **c** shows the best solutions found

We modify this Q index proposing a fitness function to locate an A store $F_A(x,y)$. This function is minimum in those locations in which the new place interacts with a neighbourhood in the same empirical average relation found in the city, and it increases if the relation is different. While Q index optimizes considering that if an interaction between two businesses is positive, the more in the neighbourhood the better, F implicitly assumes that average empirical relation is an optimum, so it penalizes any type of deviations. Jensen reports correlation between Q and the success of retail stores in the case study of Lyon [7], but as far as we know there is not any study that compares correlation indexes of Q versus F.

$$F_A(x,y) = \sum_B \left(neA_{AB}(x,y) - \overline{neA_{AB}} \right)^2 \tag{5}$$

In order to find good candidates according to F index we use a simulated annealing algorithm [1, 9], a probabilistic metaheuristic for global optimization, considering F as fitness function. The method is based on searching in the neighbourhood of a previous solution considering an adaptive criterion to accept a new solution. Acceptance mechanism is controlled by the temperature variable which determines the probability of acceptance of a solution that is worse than the previous one. Temperature is initialized in high value, T_0, and is being decreased according to an specified cooling schedule α(.) until a final temperature T_f is reached. The algorithm always accepts as new solution a better solution, but it is also possible to accept a worse solution depending on the following equation:

$$P_{\text{acceptance}} = e^{\left(-\frac{\delta}{T}\right)} \tag{6}$$

Where δ is the difference between the F index of the new and old solutions, and T is the current value of the temperature. This strategy initially allows escaping from local optima when temperature is high, and converging through an optimum in the final iterations. A neighbour solution is generated selecting randomly a new point (x',y') inside the radius r of the current solution (x,y) (see Fig. 3).

References

1. Cerný V (1985) Thermodynamical approach to the traveling salesman problem: an efficient simulation algorithm. J Optim Theory Appl 45:41–51. doi:10.1007/BF00940812
2. Dave PH, Dave HB (2008) Design and analysis of algorithms. Pearson Education/Dorling Kindersley (India), Delhi
3. Duranton G, Overman HG (2005) Testing for localization using micro-geographic data. Rev Econ Stud 72(4):1077–1106. doi: 10.1111/0034-6527.00362
4. Gómez S, Jensen P, Arenas A (2009) Analysis of community structure in networks of correlated data. Phys Rev E 80(1). http://journals.aps.org/pre/abstract/10.1103/PhysRevE.80.016114
5. Hoover EM, Giarratani F (1984) An Introduction to regional economics. McGraw-Hill
6. Hotelling H (1929) Stability in competition. Econ J 39(153):41–57
7. Jensen P (2006) Network-based predictions of retail store commercial categories and optimal locations. Phys Rev E—Stat, Nonlinear, Soft Matter Phys 74(3). http://journals.aps.org/pre/abstract/10.1103/PhysRevE.74.035101
8. Jensen P (2009) Analyzing the localization of retail stores with complex systems tools. Lect Note Comput Sci 5772:10–20. doi:10.1007/978-3-642-03915-7_2
9. Kirkpatrick S, Gelatt CD, Vecchi MP (1983) Optimization by simulated annealing. Science 220(4598):671–680. doi:10.1126/science.220.4598.671
10. Newman MEJ (2003) The structure and function of complex networks. SIAM Rev 45(2):167–256. doi:10.1234/12345678

Methodology for the Strategic Capacity Planning in Universities

Rocio de la Torre, Amaia Lusa and Manuel Mateo

Abstract This paper introduces the Strategic Capacity Planning problem in universities and proposes a methodology for solving the problem. The methodology includes the characterization of the problem, the design of a mathematical model, the data collection and pre-analysis, the model solving and the introduction of uncertainties. Besides the use for staff capacity planning, the methodology can be used to assess the impact that different strategies may have on the personnel costs and structure.

Keywords Strategic capacity planning · Quantitative methods · Staff planning · KIO/KIF · KIBS

1 Introduction

The number of Spanish universities has increased considerably in the last decades [4, 7]. This growth has been accompanied by changes in the European higher education (European Credit Transfer and Accumulation System—ECTS), an increasing concern about the quality of university tasks (teaching, research, technology knowledge transfer, etc.), and financial problems in public universities. Moreover, the latter has been accentuated by the global economic crisis, and particularly, by the Spanish economic crisis.

Apart from the external changes, there have been several substantial changes in the structure of the university during the last years. Also, the academic staff has

R. de la Torre (✉) · M. Mateo
Department of Management/ETSEIB, Universitat Politècnica de Catalunya,
Av. Diagonal 647, planta 7, 08028, Barcelona, Spain
e-mail: maria.rocio.de.torre@upc.edu

A. Lusa
Department of Management/ IOC/ETSEIB, Universitat Politècnica de Catalunya,
Barcelona, Spain
e-mail: amaia.lusa@upc.edu

M. Mateo
e-mail: manel.mateo@upc.edu

C. Hernández et al. (eds.), *Managing Complexity*, Lecture Notes in Management and Industrial Engineering, DOI 10.1007/978-3-319-04705-8_39,

been notably modified [4]. One of the most relevant changes has been the importance given to the research and the knowledge and technology transfer (KTT) tasks, apart from the teaching tasks. This, obviously, has a great influence on the definition of the ideal structure of the university (composition of the workforce and, in particular, of the academic staff).

The strategic management was first developed in universities as a result from the experience gained in companies. According to Llinàs-Audet et al. [7], while the strategic management was adopted by some European universities during the eighties of the last century, it was not introduced in Spanish universities until the nineties.

Several studies show that the number of strategic practices in universities is increasing [1–3, 6, 8, 12]; these, of course, are highly influenced and constrained by the legal and academic regulations of each university.

Llinàs-Audet et al. [7] present an analysis about the current state of strategic planning in the Spanish universities. Moreover, the authors discuss on the effectiveness of the management tools implemented to date. They state that "there are not definitive standard formalized procedures to guide higher education institutions in this process". In line, Corominas and Sacristán [4] note that "in the literature predominates outline and repeated proposals that frequently are not a result of a rigorous analysis of reality or are unreasoned". Rowley et al. [10] affirm that often strategic planning in universities is rather short term, or alternatively, a schedule for the resolution of specific problems.

Thus, strategic capacity planning in Spanish universities is a hot topic and very timely, as it will require changes in current funding policies, human resources policies [11] and academic policies. Some authors, such as Hunt et al. [5], argue that the use of strategic planning allows universities a better use of their resources, and therefore they achieve greater institutional success (most internationally, creating a better and innovative academic environment, etc.).

Surprisingly, to the best of our knowledge, a formalized procedure for the strategic capacity planning in universities has not been previously proposed.

This paper presents a procedure for solving the strategic capacity planning problem in universities, and a mathematical model (which is part of the procedure) for one of the variants of the problem.

The organization of the rest of the paper is as follows: Section 2 contains a brief description of a variant of the problem. Section 3 presents a methodology for strategic capacity planning in universities. Section 4 includes a mathematical program for the problem described in Sect. 2. Finally, conclusions and references are detailed in Sects. 5 and 6 respectively.

2 Strategic Capacity Planning in a University

The problem consists in determining the capacity (i.e., staff, both in quantity and type) for a long time horizon. The required capacity (or a probabilistic distribution of it) is assumed to be known and it refers only to the teaching capacity. Of course, the cheapest way to cover the required capacity would be with part time staff (they work in another company or organization and do some lectures at the university, at a quite low cost). However, even if this was legal, it would be detrimental to the quality (teaching is not their main job) and to the overall activity of the university (including, of course, research and KTT tasks). To avoid this, an ideal academic staff structure should be defined.

The specific characteristics of the problem are detailed below:

- The system should have a certain excess of capacity.
- The problem is considered to be dynamic, i.e. the capacity can vary along the horizon.
- The resources in strategic capacity planning are human resources and, in this case, only academic staff (and not administration staff) is considered.
- Each worker belongs to a unit (e.g., department) and to a category. The capacity and the types of task that he/she is able to do may depend on the category.
- There are two types of categories: temporary and permanent categories.
- Regarding temporary categories, it is mandatory for a member of the staff to progress to a top category once a certain period of time is completed.
- Regarding permanent categories, it is considered that a member of the staff is able to progress to another category once the merits required in the top category are reached. A person can access to a category from a lower one or from outside the organization.
- The considered decisions that influence the variation of the capacity include the creation of posts (in certain categories and units) and the non-renewal of contracts.
- There is a maximum budget for personnel costs.
- The evaluation criteria of the solution include the personnel cost and the composition of the academic staff (according to a preferable structure).
- Some data can be uncertain.

3 Methodology of Strategic Capacity Planning in Universities

The proposed methodology is developed to help the organization to:

- Define the long-term workforce planning, considering uncertainties.
- Evaluate the impact of strategic decisions taken prior to the capacity planning decisions, as for example the implementation of specific personnel policies (e.g., stabilization or promotion plans, collective dismissal, etc.), the creation or elimination of degrees, changes in the units (e.g., departments) configuration, among others. The phases that make up the methodology are represented in Fig. 1 and briefly described after the figure.

Phase I: Problem's Characterization A scheme of the most relevant characteristics of the strategic capacity planning in Knowledge Intensive Organizations

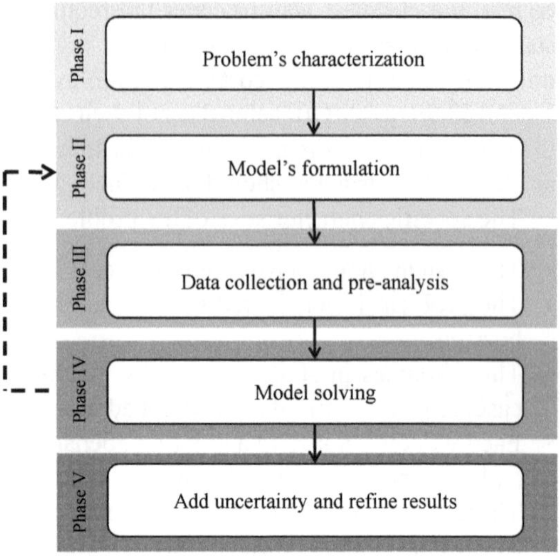

Fig. 1 A methodology scheme for the strategic capacity planning problem in universities. (Source: authors)

(KIOs) was presented in Martinez et al. [9]. According to the different characteristics (organizational structure, workforce, capacity decisions, demand, service level, costs associated to the capacity decisions, financing, uncertainty, planning horizon, and evaluation criteria), the problem gives rise to different variants. The first step is to identify the variant, describing the specific characteristics of the problem.

Phase II: Model's Formulation Once the problem has been characterized, next phase consists in designing a mathematical optimization model. Each characteristic of the problem carries an associated set of variables and constraints. To illustrate this phase, Sect. 4 provides the model formulation for the particular case of the university described in Sect. 2.

Phase III: Data Collection and Pre-Analysis The objective of this phase is to define the sources of information for each type of data, and process the information as required by the model.

Phase IV: Model Solving In this phase the aim is to solve the mathematical model. The implementation and test of the model are included in the phase.

Phase V: Add Uncertainty and Refine Results This phase is applied if uncertainty is considered to be relevant (for example, in the required capacity, which may have different scenarios; in the probability that a member of the academic staff is able to progress to another category; or in the available budget). Stochastic models and simulation may be used (or combined) depending on the case.

4 Model's Formulation

The mathematical model for the problem described in Sect. 2 is presented (Table 1 and 2).

Table 1 Data base description

Parameter	Description
K	Set of categories
K'	Subset of temporary categories $(\forall k \in K \mid k \subseteq ([1,k_1] \cup [k_1+1,k_2]))$
K''	Subset of permanent categories $(\forall k \in K \mid k > k_2+1)$
c_{kt}	Cost in (mu/worker) associated to the category k in period t. $[\forall t;\ \forall k \in K]$
v_t	Cost in (mu/hour) associated to part time lecturers in period t. $[\forall t;\ \forall u]$
C_{ut}	Required capacity for the unit u, in period t. $[\forall t;\ \forall u;\ \forall k \in K]$
h_{kt}	Capacity associated to each worker in the category k in period t. $[\forall t;\ \forall k \in K]$
L_{ukt}	Expected personnel cuts (for instance, due to retirement) in the unit u, category k, in period t. $[\forall t;\ \forall u;\ \forall k \in K]$
Γ_k^+	Set of categories to which it is possible to access from the category k. $[\forall k \in K]$
Γ_k^-	Set of categories from which it is possible to access to the category k. $[\forall k \in K]$
r_{uskt}	Proportion of workers in unit u, that can promote, as maximum, from the category s to the category k, in period t. $[\forall t]$
UP_{kt}, LP_{kt}	Preferable bounds for the proportion of academic staff that belongs to the category k in the period t. This condition is not rigid, but non-compliance is penalized
α_{ut}	Excess of capacity that should have, at least, the unit u in the period t $[\forall t]$. Note that, even if it is not usual, this parameter could be negative (if a shortage in the capacity is allowed, this would mean a worsening in the service level)
λ_{kt}	Penalty associated to the discrepancy between the preferable and the actual composition of academic staff of the category k, in the period t. $[\forall t]$
μ_t	Penalty associated to the maximum discrepancy between the preferable and the actual composition of the academic staff, in the period t. $[\forall t]$
ω	Penalty associated to the maximum discrepancy between the preferable and the actual work force
B_t	Planned budged of the salaries of the academic staff for the period t. $[\forall t]$

Table 2 Variables of the model

Variable	Description
$\mathrm{w}_{\mathrm{ukt}}$	Integer variable that indicates the number of workers of the unit u, category k and period t. $[\forall t;\ \forall u;\ \forall k \in K]$
A_{ut}	Integer variable that indicates the number of hours given by part time workers in the unit u in period t. $[\forall t;\ \forall u]$
$\mathrm{Q}_{\mathrm{uklt}}$	Integer variable that indicates the number of workers who access to the category l from the category k, in the unit u, in the period t. $[\forall t;\ \forall u;\ \forall k \in K]$
w_{ukt}^+	Integer variable that indicates the number of new workers in the unit u and category k, in the period t from the labor market. $[\forall t;\ \forall u;\ \forall k \in K]$
w_{ukt}^-	Integer variable that indicates the number of off workers (excluding the previously forecasted) in the unit u and the category k, in the period t. $[\forall t;\ \forall u;\ \forall k \in K]$
$\delta_{kt}^+, \delta_{kt}^-$	Positive and the negative discrepancy, respectively, between the preferable and the actual composition of the academic staff in the category k in the period t. $[\forall t;\ \forall k \in K]$
δ_t	Maximum discrepancy (positive or negative), between the preferable and the actual composition of the academic staffin all categories in period t (i.e. $\delta_t = max_k(\delta_{kt}^+, \delta_{kt}^-)$). $[\forall t]$).
Δ	Maximum discrepancy between the preferable and the real composition of the academic staff (i.e. $\Delta = max_t(\delta_t)$)

Model

$$[MIN]\ z = \sum\nolimits_{\forall u,t}\left[A_{ut}\cdot v_t + \sum\nolimits_{\forall k}\left(w_{ukt}\cdot c_{kt}\right)\right] + \sum\nolimits_{\forall k,t}\left[\lambda_{kt}\cdot\left(\delta_{kt}^{+}+\delta_{kt}^{-}\right)\right] + \sum\nolimits_{\forall t}\mu_t\cdot\delta_t + \omega\cdot\Delta \quad (1)$$

$$\sum\nolimits_{\forall k} w_{ukt}\cdot h_{kt} + A_{ut} \geq (1+\alpha_{ut})\cdot C_{ut} \quad \forall u,t \quad (2)$$

$$w_{ukt} \leq r_{uskt}\cdot w_{ust} - L_{ukt} \quad \forall u,t;\quad s\in\Gamma^{-}, k\in K'\setminus k=\{1,k_1\} \quad (3)$$

$$w_{ukt} \leq r_{uskt}\cdot w_{ust} - L_{ukt} + w_{ukt}^{+} \quad \forall u,t;\quad s\in\Gamma^{-}, k\in K';\quad k=\{1,k_1\} \quad (4)$$

$$w_{ukt} = w_{ukt-1} - L_{ukt} + \sum\nolimits_{s\in\Gamma_k^{-}} Q_{uskt} - \sum\nolimits_{l\in\Gamma_k^{+}} Q_{uklt} + w_{ukt}^{+} - w_{ukt}^{-} \quad \forall u,t;\quad \forall k\in K'' \quad (5)$$

$$Q_{uskt} \leq r_{uskt}\cdot w_{uskt} \quad \forall u,t;\quad \forall s\in K'' \mid \Gamma_s^{+}\neq\{\varnothing\};\quad \forall k\in\Gamma_s^{+} \quad (6)$$

$$\sum\nolimits_{\forall u} w_{ukt} \geq \left(LP_{kt}\cdot\sum\nolimits_{\forall u,k} w_{ukt}\right) - \delta_{kt}^{-} \quad \forall t;\quad \forall k\in K \quad (7)$$

$$\sum\nolimits_{\forall u} w_{ukt} \leq \left(UP_{kt}\cdot\sum\nolimits_{\forall u,k} \omega_{ukt}\right) + \delta_{kt}^{+} \quad \forall t; \forall k\in K \quad (8)$$

$$\delta_t \geq \delta_{kt}^{+} \quad \forall t;\ \forall k\in K \quad (9)$$

$$\delta_t \geq \delta_{kt}^{-} \quad \forall t;\quad \forall k\in K \quad (10)$$

$$\Delta \geq \delta_t \quad \forall t \quad (11)$$

$$\sum\nolimits_{\forall u}\left[A_{ut}\cdot v_t + \sum\nolimits_{\forall k}\left(w_{ukt}\cdot c_{kt}\right)\right] \leq B_t \quad \forall t \quad (12)$$

$$w_{ukt},\ A_{ut},\ Q_{uklt},\ w_{ukt}^{+},\ w_{ukt}^{-} \geq 0 \quad (13)$$

$$\delta_{kt}^{-},\ \delta_{kt}^{+},\ \delta_t,\ \Delta \geq 0 \quad (14)$$

Equation (1) presents the objective function. The aim is to minimize the costs associated to the salaries as well as the discrepancies between the ideal and the actual composition of the academic staff. Equation (2) limits the minimum available capacity; (3) and (4) balances the number of temporary staff members of each unit, category and time of the university; the permanent staff members of each unit, category and time of the university are balanced in (5); (6) imposes an upper bound to the number of workers to be promoted. Constraints regarding the ideal composition of the academic staff are included in equations (7) and (8). Constraints (9) to (11) correspond to the maximum discrepancies between all categories and periods, to avoid, insofar as possible, that the discrepancy be concentrated in few categories or periods (assuming that it is preferable a regular distribution of the discrepancy). Constraint (12) limits the personnel costs per period. Finally, (13) and (14) impose that variables are non-negative.

5 Conclusions

This paper presents a methodology to deal with the problem of the Strategic Capacity Planning in universities and illustrate one of the steps by presenting a mixed integer lineal programming model for a variant of the problem. The methodology can be used not only for planning capacity, but also for assessing the impact of different university strategies (personnel policies, studies offered by the university, etc.). This can be done by changing some parameters of the model. For instance, in case of assessing the impact of personnel policies, it is necessary to modify the proportion of workers that can promote (r_{uskt}), the preferable bounds for the composition staff (UP_{kt}, LP_{kt}), and the planned budget of the salaries (B_t).

Acknowledgements Supported by the Spanish Ministry of Economy and Competitivity (project DPI2010-15614).

References

1. Agasisti T, Arnaboldi M, Azzone G (2008) Strategic management accounting in universities: the Italian experience. High Educ 55(1):1–15
2. Clark BR (1998) Creating entrepreneurial universities: organizational pathways of transformation. IAU Press by Pergamon, Oxford
3. Clark BR (2003) Sustaining change in universities: continuities in case studies and concepts. Tert Educ Manage 9:99–116
4. Corominas A, Sacristán V (2010) The strategic crossroads of Spanish public university. Revist de Educación 355:57–81
5. Hunt CM, Oosting KW, Stevens R, Loudon D, Migliore RH (1997) Strategic planning for private higher education. Haworth, Binghamton
6. Jarzabkowski P (2003) Strategic practices: an activity theory perspective on continuity and change. J Manage Stud 40:1

7. Llinàs-Audet X, Girotto M, Solé F (2010) University strategic management and the efficacy of the managerial tools: the case of the Spanish universities. Revista de Educación 355:33–54
8. Lounsbury M (2001) Institutional sources of practice variation: staffing college and university recycling programs. Adm Sci Q 46(1):29–56
9. Martinez M, Lusa A, Mas M, De La Torre M, Mateo M (2012) Strategic capacity planning in KIOs: a classification scheme. In 6th International Conference on Industrial Engineering and Industrial Management, Vigo, Spain.
10. Rowley D, Lujan H, Dolence M (1997) Strategic change in colleges and universities: planning to survive and prosper. Jossey-Bass, San Francisco
11. Santiago P, Brunner JJ, Haug G, Malo S, Pietrogiacomo P (2009) OECD reviews of tertiary education: Spain. OECD, Paris
12. Shattock M (2003). Managing successful universities. Society for Research in Higher Education & Open University Press, London

MIX
Papier aus verantwortungsvollen Quellen
Paper from responsible sources
FSC® C105338

If you have any concerns about our products,
you can contact us on
ProductSafety@springernature.com

In case Publisher is established outside the EU,
the EU authorized representative is:
Springer Nature Customer Service Center GmbH
Europaplatz 3, 69115 Heidelberg, Germany

Printed by Libri Plureos GmbH
in Hamburg, Germany